DEPLOYING MOBILE WiMAX

DEPLOYING MOBILE WiMAX

**Max Riegel, Dirk Kroeselberg, Aik Chindapol
and Domagoj Premec** (contributor)

of Nokia Siemens Networks

A John Wiley and Sons, Ltd, Publication

This edition first published 2009
© 2009 John Wiley & Sons Ltd.,

Registered office
John Wiley & Sons Ltd, The Atrium, Southern Gate, Chichester, West Sussex, PO19 8SQ, United Kingdom

For details of our global editorial offices, for customer services and for information about how to apply for permission
to reuse the copyright material in this book please see our website at www.wiley.com.

Library of Congress Cataloging-in-Publication Data

Riegel, Max.
  Deploying mobile WiMAX / Max Riegel, Aik Chindapol, Dirk Kroeselberg, Domagoj Premec.
    p. cm.
  Includes bibliographical references and index.
  ISBN 978-0-470-69476-3 (cloth)
  1. IEEE 802.16 (Standard) 2. Ad hoc networks (Computer networks) 3. Wireless
Internet. I. Chindapol, Aik. II. Kroeselberg, Dirk. III. Title.

  TK5105.77.R54 2009
  004.67–dc22

                                                                2009030045

A catalogue record for this book is available from the British Library.

ISBN 978-0-470-69476-3 (H/B)

Set in 10/12pt Times Roman by Thomson Digital, Noida, India
Printed and Bound in Great Britain by CPI Antony Rowe, Chippenham, Wiltshire

Contents

About the Authors

Max Riegel is head of WiMAX and IEEE standardization at Nokia Siemens Networks. He is engaged in the WiMAX Forum as co-chair of the Networking Working Group and participated in the 16ng Working Group of the IETF as technical advisor and contributor. He has more than 25 years' of experience in the telecommunications industry and more than 15 years' of professional experience in the technical and operational issues of the Internet. After more than 10 years in management positions for telecommunication product development within several companies, he joined IETF standardization in 1998. Being engaged in IEEE 802 standardization since 2000 and in the WiMAX Forum since 2004, he is an expert in broadband wireless Internet access with deep involvement in Mobile WiMAX radio and network standardization from its very beginning. He has a Dipl.-Ing. degree in Electrical Engineering from the Technical University Munich, Germany.

Dirk Kroeselberg received a diploma degree in Mathematics and Computer Science from the University of Giessen, Germany, in 1997. He joined Siemens in 1998, working on a broad range of communication security topics for Siemens Corporate Technology and Siemens Mobile Phones. He has been with Nokia Siemens Networks since April 2007, being responsible for WiMAX Forum network standardization. He is co-chairing the work on network architecture evolution within the WiMAX Forum Network Working Group (NWG) and leads the standardization efforts within NWG for emergency services support and WiMAX-SIM that are now part of the Release 1.5 WiMAX network specifications.

Aik Chindapol received his Ph.D. in Electrical Engineering from the University of Washington in 2000. He joined Siemens Mobile Networks, Siemens Corporate Research and Nokia Siemens Networks in 2001, 2003 and 2007, respectively. Prior to WiMAX, he was previously engaged in research and standardization of GSM/GPRS/EDGE, Wi-Fi and Bluetooth. During 2006–2009, he led teams from Siemens and Nokia Siemens Networks in the successful standardization of IEEE 802.16, Mobile WiMAX Radio Release 1.5, and the 2009 revision of the IEEE 802.16 standard. His current research interests include network coding, error control coding, radio resource management and performance optimization of mobile broadband systems.

Contributor to Chapter 5 on Mobility:

 **Domagoj Premec** is a research engineer in the field of IP technologies for Nokia Siemens Networks and is based in Zagreb, Croatia. His current work encompasses IP mobility management in the context of wireless packet networks as well as IPv6 transition. He was actively involved in the standardization of WiMAX Forum Network Architecture Release 1 and Release 1.5 and he also participates in the IETF working groups related to mobility management. He received a MSc degree in Information Science and Linguistics from the University of Zagreb, Croatia.

Preface

The Internet and the World Wide Web have changed the way we retrieve information. Formerly, books were written to provide as much information as possible to readers, and it was the main intention of authors to provide very complete descriptions of their topics. A book stood as the primary carrier of all the knowledge. Therefore, it became the privilege of only very few people with access to huge libraries to leverage multiple sources and gain knowledge of a particular topic. Nowadays, the Internet allows everyone with the will to learn to retrieve almost any information available from around the world. It is no longer difficult to search for a reference among millions of pages of literature or to browse through the latest technical specifications. The scarcity of information has long gone and been replaced somehow by abundance and confusion. Ironically, the problem has shifted from not having any information to having too much of it.

When planning the contents of this book, we took into account that books have changed their role in handling information. They are no longer the primary source of information; they have found their prevalent role in transforming plain information into usable knowledge by providing a proper structure and overview of information available on the Internet. In the end, books serve as a friendly interface between enthusiastic readers and the machine-like Internet. The selection of the most relevant aspects, the arrangement and the presentation of facts between the front and back cover create a framework for knowledge, to which the reader can easily accumulate further information from other sources. This book was written with this paradigm shift in mind.

The imagination of potential readers plays an important role in the whole project. It sets the assumption about prior knowledge and the intentions and expectations of the readers. Apart from undergraduate and graduate students who may be exploring Mobile WiMAX for the first time, experienced telecommunication engineers having to promptly understand the basic functionality and performance of the technology and form the intended readership for this book. Making use of the technology was more of our focus than providing a comprehensive base for further enhancements of the technology itself. Nevertheless, the book may form a good foundation to the topic for developers starting implementation work.

This book is aligned to the specifications of the WiMAX Forum and IEEE 802.16 so it should be easy to step back and forth between the book and the original sources. Our desire is to provide an overview of the underlying principles of the mobile networking technology and the rationale behind the approaches in order to make it easier for readers to understand the Mobile WiMAX specifications and look for decisive details. All of us were engaged for years in the initial development of the technology both inside our company and in the standardization organizations creating the specifications. We spent many hundreds of hours in discussions of alternative solution proposals and we were often part of the decision process to determine the best approach. Many of the points raised during the discussions and especially those affecting the particular decisions may never appear in the text of the final specification. Indeed, they are essential for the reader's understanding, in order to create knowledge from the information provided. This spirit of the meetings and countless conference calls guided our description of the technical facts in this book.

'Nobody is perfect', so if we have missed an important aspect, or should have gone deeper into particular deployment scenarios, or have omitted anything in the final review, we would greatly appreciate your feedback. Not only did the Internet change our daily lives and the way we retrieve information, but also, fortunately, it provides us with the means to establish a communication link between ourselves and the readers of this book. Messages can be sent to us at authors@deployingmobilewimax.com or by visiting www.deployingmobilewimax.com.

Acknowledgements

The authors would like to acknowledge a number of colleagues from Siemens, Nokia and Nokia Siemens Networks who put their time and effort in contributing to numerous technical discussions, pointing out non-trivial tips and tricks and eventually reviewing the contents of this book. In particular, we would like to express our deep gratitude to Dr. Achim Brandt for his long and fruitful cooperation in WiMAX standardization and extremely helpful comments on the mobility aspects covering both network and radio perspectives. We are also extremely grateful to Domagoj Premec for his contributions not only to our mobility chapter but to the technology as well. We would also like to acknowledge Hannes Tschofenig, who not only put a tremendous amount of effort into reviewing and improving large parts of the security and service provisioning sections, but also served as a most valuable discussion partner throughout the process of writing.

In addition, the authors are indebted to the detailed and thoughtful comments received from Ray Jong-A-Kiem, Andrea Bacioccola, Zexian Li, Daniele Tortora, Giovanni Maggi, Scott Probasco, Martin Bokämper, Devaki Chandramouli, Richard Wisenoecker, James Winterbottom and Oleg Marinchenco.

We also would like to express our sincere appreciation to the team at John Wiley & Sons in keeping us focused and stayed on track throughout the production process. Special thanks go to Katharine Unwin and Sarah Tilley for their outstanding encouragement and professionalism.

We are extremely grateful to our families for their patience and endless support.

The authors and contributor welcome any comments, questions or suggestions to the contents of this book. Feedback can be sent to authors@deployingmobilewimax.com or by visiting www.deployingmobilewimax.com

List of Acronyms

2G	Second Generation (mobile systems)
3G	Third Generation (mobile systems)
3GPP	Third-Generation Partnership Program
3GSM	Third-Generation GSM (services)
4G	Fourth Generation (mobile systems)
A&A	Authentication and Authorization
AAA	Authentication, Authorization and Accounting
A-DPF	Anchor Data Path Function
AES	Advanced Encryption Standard
AF	Application Function
A-GPS	Assisted GPS
AK	Authentication Key
AKA	Authentication and Key Agreement
AMC	Advanced Modulation and Coding
AMR	Adaptive Multi-Rate
AR	Access Router
ARQ	Automatic Retransmission Request
ASN	Access Service Network
ASN-GW	ASN Gateway
ASP	Application Service Provider
ATM	Asynchronous Transfer Mode
AuC	Authentication Center
AVP	Attribute Value Pair
BB	Broadband
BE	Best Effort
B-ISDN	Broadband Integrated Services Digital Network
BNG	Broadband Network Gateway
BS	Base Station
BTC	Block Turbo Code
BU	Binding Update
BW	Bandwidth
CA	Certificate Authority
CBC	Cipher Block Chaining
CC	Convolutional Code

CCM	Counter with CBC-MAC
CCoA	Collocated Care-of Address
CDD	Cyclic Delay Diversity
CDMA	Code Division Multiple Access
CHAP	Challenge Handshake Authentication Protocol
CID	Connection Identifier
CINR	Carrier-to-Interference and Noise Ratio
CMAC	Cipher-based Message Authentication Code
CMIP	Client Mobile IP
CN	Correspondent Node
CoA	Care-of Address/Change of Authorization
CPE	Customer Premises Equipment
CPS	Common Part Sublayer
CQICH	Channel Quality Information Channel
CR	Core Router
CRC	Cyclic Redundancy Check
CRL	Certificate Revocation List
CS	Convergence Sublayer
CSMA/CA	Carrier-Sensed Multiple Access and Collision Avoidance
CSN	Connectivity Service Network
CTC	Convolutional Turbo Code
CTR	Counter Mode Encryption
CUI	Chargeable User Identity
C-VID	Customer VLAN Identifier
CWG	Certification Working Group
DAB	Digital Audio Broadcasting
DCCA	Diameter Credit Control Application
DCD	Downlink Channel Descriptor
DECT	Digital Enhanced Cordless Telecommunications
DER	Distinguished Encoding Rules
DES	Data Encryption Standard
DHCP	Dynamic Host Configuration Protocol
DL	Downlink
DNS	Domain Name Server
DPF	Data Path Function
DSA	Dynamic Service Addition
DSC	Dynamic Service Change
DSD	Dynamic Service Deletion
DSL	Digital Subscriber Line
EAP	Extensible Authentication Protocol
EAPOL	EAP Over LAN
ECB	Electronic Code Book
EDGE	Enhanced Data rates for GSM Evolution
EIK	EAP Integrity Key
EIR	Equipment Identity Register
EMSK	Extended Master Session Key

EMU	EAP Method Update
EPC	Evolved Packet Core
e-rtPS	extended rtPS
ERT-VR	Extended RT-VT
ETH	Ethernet
ETH-CS	Ethernet Convergence Sublayer
ETS	Emergency Telecommunications Services
ETSI	European Telecommunications Standards Institute
E-UTRAN	Enhanced UMTS Terrestrial Radio Access Network
EVC	Ethernet Virtual Connection
FA	Foreign Agent
FA-CoA	Foreign Agent Care-of Address
FAP	Femto Access Point
FBSS	Fast Base Station Switching
FCC	Federal Communications Commission
FCH	Frame Control Header
FDD	Frequency Division Duplexing
F-FDD	Full-duplex FDD
FFT	Fast Fourier Transform
FQDN	Fully Qualified Domain Name
FUSC	Fully Used Subchannelization
GARP	Generic Attribute Registration Protocol
GERAN	GSM EDGE Radio Access Network
GETS	Government Emergency Telecommunications Services
GMSH	Grant Management Subheader
GPCS	General Packet CS
GPRS	General Packet Radio Service
GPS	Global Positioning System
GPSK	Generalized Pre-Shared Key
GRE	Generic Routing Encapsulation
GRWG	Global Roaming Work Group
GSM	Groupe Speciale Mobile
GSMA	GSM Association
HA	Home Agent
H-ARQ	Hybrid ARQ
HCS	Header Check Sequence
HELD	HTTP-Enabled Location Delivery
H-FDD	Half-duplex FDD
HLR	Home Location Register
HLS	Home Location Server
HMAC	Hashed Message Authentication Code
HNP	Home Network Prefix
HO	Handover
HoA	Home Address
HSPA	High-Speed Packet Access
HSS	Home Subscriber Server

HTTP	Hypertext Transport Protocol
IANA	Internet Assigned Numbers Authority
IBS	Integrated Base Station
ICMP	Internet Control Message Protocol
IE	Information Element
IEEE	Institute of Electrical and Electronics Engineers
IETF	Internet Engineering Task Force
IGD	Internet Gateway Device
IKE	Internet Key Exchange
IMEI	International Mobile Equipment Identity
IMPI	IP Multimedia Private Identity
IMPU	IP Multimedia Public Identity
IMS	IP Multimedia Subsystem
IMSI	International Mobile Subscriber Identity
IP	Internet Protocol
IP-CAN	IP Connectivity Access Network
IP-CS	IP Convergence Sublayer
IPoETH	IP over Ethernet
IP-TV	IP Television
IS-95	Interim Standard 95
ISC	IMS Service Control
ISDN	Integrated Services Digital Network
ISF	Initial Service Flow
ISP	Internet Service Provider
ITU	International Telecommunication Union
ITU-T	ITU Telecommunication Standardization Sector
L2	Layer 2
LA	Location Agent
LAN	Local Area Network
LBS	Location-Based Services
LC	Location Controller
LDPC	Low-Density Parity Check (code)
LMA	Local Mobility Anchor
LMSC	LAN/MAN Standards Committee
LOST	Location-to-Service Translation protocol
LPF	Local Policy Function
LR	Location Requester
LS	Location Server
LTE	Long-Term Evolution
MAC	Media Access Control
MAG	Mobility Access Gateway
MBS	Multicast/Broadcast Service
MCC	Mobile Country Code
MDHO	Macro Diversity Handover
MEF	Metro Ethernet Forum
MGW	Media Gateway

MIMO	Multiple Input, Multiple Output
MIP	Mobile IP
MLP	Mobile Location Protocol
MM	Mobility Management
MME	Mobility Management Entity
MMS	Multimedia Message Service
MN	Mobile Node
MNC	Mobile Network Code
MO	Management Object
MS	Mobile Station
MSB	Most Significant Bit
MS-ID	Mobile Station Identifier
MSK	Master Session Key
NAI	Network Access Identifier
NAP	Network Access Provider
NAS	Network Access Server
NAT	Network Address Translation
NDS	Network Domain Security
NENA	National Emergency Number Association
NICS	Network Implementation Conformance Statement
NRM	Network Reference Model
nrtPS	non-rtPS
NRT-VR	Non-RT-VR
NSP	Network Service Provider
NWG	Network Working Group
OAM	Operation And Maintenance
OCS	Online Charging Server
OCSP	Online Certificate Status Protocol
OEM	Original Equipment Manufacturer
OFCS	Offline Charging System
OFDM	Orthogonal Frequency Division Multiplexing
OFDMA	Orthogonal Frequency Division Multiple Access
OMA	Open Mobile Alliance
OMA-DM	OMA Device Management
OS	Operating System
OTA	Over The Air
PAP	Password Authentication Protocol
PAPR	Peak-to-Average Power Ratio
PBU	Proxy Binding Update
PCC	Policy and Charging Control
PCEF	Policy Control Enforcement Function
PCRF	Policy and Charging Rules Function
P-CSCF	Proxy Call State Control Function
PDA	Personal Digital Assistant
PDF	Policy Distribution Function
PDFID	Packet Data Flow Identity

PDN	Packet Data Network
PDU	Protocol Data Unit
PF	Policy Function
PG ID	Paging Group ID
PHS	Payload Header Suppression
PHY	Physical Layer
PIDF-LO	Presence Information Data Format Location Object
PKI	Public Key Infrastructure
PKM	Privacy Key Management
PLMN	Public Land Mobile Network
PMIP	Proxy Mobile IP
PMIPv4	Proxy Mobile IPv4
PMIPv6	Proxy Mobile IPv6
PMN	Proxy Mobile Node
PMP	Personal Media Player
POTS	Plain Old Telephone System
PPA	Prepaid Agent
PPC	Prepaid Client
PPP	Point-to-Point Protocol
PPPoE	PPP over Ethernet
PPS	Prepaid Server
PSAP	Public Safety Answering Point
PSC	Power Saving Class
PSK	Pre-Shared Keying
PSTN	Public Switched Telephony Network
PUSS	Partially Used Subchannelization
QAM	Quadrature Amplitude Modulation
QoS	Quality of Service
QPSK	Quadrature Phase Shift Keying
RADIUS	Remote Access Dial-In User Service
ROHC	Robust Header Compression
ROI	Return On Investment
RRM	Radio Resource Management
RRP	Registration Response
RRQ	Registration Request
RRSI	Received Signal Strength Indication
RSA	Rivest, Shamir and Adleman
RTG	Receive/transmit Transition Gap
rtPS	real-time Polling Service
RT-VR	Real-Time Variable Rate
SA	Security Suppression
SAE	System Architecture Evolution
S-CSCF	Serving Call State Control Function
SCTP	Stream Control Transmission Protocol
SDFID	Service Data Flow Identity
SDU	Service Data Unit

SET	SUPL Enabled Terminal
SF	Service Flow
SFA	Service Flow Authorization (entity)
SFID	Service Flow Identifier
SFM	Service Flow Management (entity)
SGSN	Serving GPRS Support Node
SHA	Secure Hash Algorithm
SIF	SUPL Initiation Function
SIM	Subscriber Identity Module
SIP	Session Initiation Protocol
SLA	Service Level Agreement; SUPL Location Agent
SLP	SUPL Location Platform
SMS	Short-Message Service
SNI	Service Network Interface
SNMP	Simple Network Management Protocol
SNR	Signal-to-Noise Ratio
SOAP	Simple Object Access Protocol
SPFID	Service Profile Identity
SPI	Security Parameters Index
SPR	Subscriber Profile Repository
SPWG	Service Provider Working Group
STBC	Space–Time Block Coding
SUPL	Secure User Plane Location
SVC	Switch Virtual Connection
S-VID	Service Provider VLAN Identifier
TCP	Transmission Control Protocol
TDD	Time Division Duplexing
TDM	Time Division Multiplex
TE	Terminal Equipment
TEK	Traffic Encryption Key
TELEX	Teleprinter Exchange
TEV	Test Equipment Vendor
TFTP	Trivial File Transfer Protocol
TISPAN	Telecoms & Internet converged Service & Protocols for Advanced Networks
TLS	Transparent LAN Service; Transport Layer Security
TLV	Tag/Length/Value
TMSI	Temporary Mobile Subscriber Identity
TTG	Transmit/receive Transition Group
TTLS	Tunneled TLS
TWG	Technical Working Group
UCD	Uplink Channel Descriptor
UDP	User Datagram Protocol
UDR	User Data Record
UE	User Equipment
UGS	Unsolicited Grant Service

UICC	Universal Integrated Circuit Card
UL	Uplink
ULP	Userplane Location Protocol
UMTS	Universal Mobile Telecommunication System
UNI	User Network Interface
URI	Uniform Resource Identifier
URL	Uniform Resource Locator
USB	Universal Serial Bus
USI	Universal Service Interface
USIM	Universal Subscriber Identity Module
UTRAN	UMTS Terrestrial Radio Access Network
VLAN	Virtual Local Area Network
VLS	Visited Location Server
VoIP	Voice over IP
VPLS	Virtual Private LAN Service
VPN	Virtual Private Network
VSA	Vendor-Specific Attribute
VSP	VoIP Service Provider
WCDMA	Wideband Code Division Multiple Access
WIB	WiMAX Initial Bootstrap
WiBro	Wireless Broadband
Wi-Fi	WLAN based on IEEE 802.11
WiMAX	Worldwide interoperability for Microwave Access
WLAN	Wireless Local Area Network
WLP	WiMAX Location Protocol
WMF	WiMAX Forum
WPS	Wireless Priority Service
WRI	WiMAX Roaming Interface
WRX	WiMAX Roaming exchange
XML	Extensible Markup Language

1

Introduction

This chapter will introduce the aim and positioning of Mobile WiMAX in relation to other telecommunication technologies and will provide a presentation of the specification structure and document identifier applied for Mobile WiMAX. The chapter concludes with an overview of the intention and content of each of the chapters of this book.

1.1 WiMAX in the Telecommunication Markets

Mobile WiMAX has left its phase of 'superhype' and converted to real network operation and successful telecommunication business. The roots of Mobile WiMAX go back to springtime 2002, when startup companies appeared in the IEEE 802 standardization arena and introduced the idea of multi-megabit broadband Internet access over a cellular infrastructure.

The promise was that a new mobile radio access technology would outperform the highly developed 3G radio technologies. It took some time and quite a number of disputes both inside and outside the IEEE 802 standardization committees until two projects were established, one known as the IEEE 802.20 Mobile Broadband Wireless Access (MBWA) Working Group and the other finally leading to the Mobile WiMAX radio standard by creating the IEEE 802.16e amendment to the IEEE 802.16 metropolitan area network interface specification. About a year after the standardization work was initiated, adoption of the IEEE 802.16e specification by the Korean WiBro initiative caused enormous public awareness resulting in huge hype around Mobile WiMAX. Many believed that Mobile WiMAX would become the preferred fourth generation (4G) radio technology superseding all the CDMA-based 3G technologies of the UMTS.

Despite the similarities of the radio interface, Mobile WiMAX follows a different paradigm of telecommunications (Figure 1.1).

From the beginning, the Mobile WiMAX technology aimed at a different deployment model, similar to the 'Digital Subscriber Line' model introduced on the wire-line networking area by cable and xDSL technologies.

1.1.1 Integrated Services Digital Networks

Traditionally all telecommunication networks were specially designed for particular services, and there were quite a number of them. The PSTN is the biggest example. It is a huge network

Deploying Mobile WiMAX   Max Riegel, Dirk Kroeselberg, Aik Chindapol and Domagoj Premec
© 2009 John Wiley & Sons, Ltd

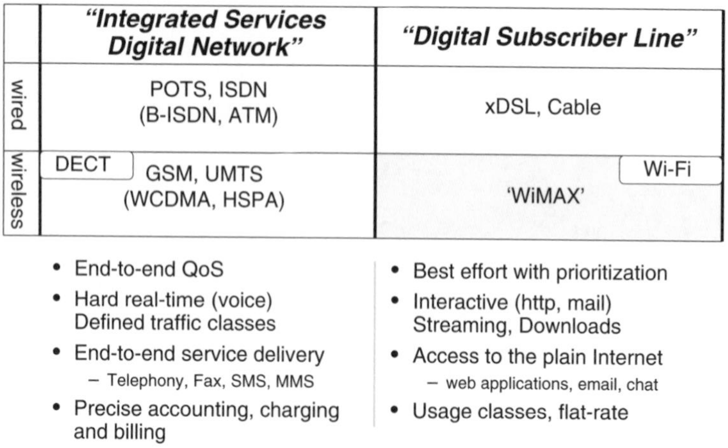

	"Integrated Services Digital Network"	"Digital Subscriber Line"
wired	POTS, ISDN (B-ISDN, ATM)	xDSL, Cable
wireless	DECT ⎤ GSM, UMTS (WCDMA, HSPA)	'WiMAX' Wi-Fi

- End-to-end QoS
- Hard real-time (voice) Defined traffic classes
- End-to-end service delivery
  − Telephony, Fax, SMS, MMS
- Precise accounting, charging and billing

- Best effort with prioritization
- Interactive (http, mail) Streaming, Downloads
- Access to the plain Internet
  − web applications, email, chat
- Usage classes, flat-rate

Figure 1.1   Telecommunication market segmentation

which is highly optimized to the telephony service by providing circuit-switched connections between customers. Other dedicated networks were constructed, e.g. for the TELEX service or for packet data services. All the networks were designed to ensure perfect delivery of their particular services, and these services built the economic base for operation of the network. Service-specific sophisticated accounting, charging and billing models were implemented to generate revenues for telecommunications operators based on the usage and value of the services provided to customers. The more the services were 'consumed' by the customers, the higher the income generated for the operators. The model did not change when telecommunications converted from analog transmission to digital transmission. Digitization of analog signals like voice or facsimile enabled the replacement of the analog networks by a digital counterpart. Moreover, with everything becoming digital, a single digital network became capable of providing all kinds of services over a common digital transmission system. ISDN became the ubiquitous wire-line telecommunication system in the world, even when for economic reasons the last mile of the network, aka the subscriber line, remained analog in most parts of the world.

To untie the terminal from the subscriber line, allowing usage of telecommunication services at any time at any place, was a huge desire of both the telecommunication operators and the customers. Cordless telephony by replacing the telephone cord by a local short-range radio transmission system was an intermediate step toward full mobility of terminals. DECT is the most advanced technology for cordless systems and resembles many of the concepts of ISDN. Full mobility of terminals and services was finally reached by GSM and its sibling IS-95 leading to an explosion of the mobile telecommunication market. UMTS and its ancestor GSM form today the technical base of most of the traditional cellular telecommunication services represented mainly by telephony and short-message service (SMS).

1.1.2 The Digital Subscriber Line to the Internet

Apart from traditional paradigms, the Internet introduced in the 1990s a disruptive telecommunication model. Primarily being an interconnection system of data networks, the Internet introduced a connectionless global telecommunication network based on the concept of best-effort delivery. Best effort means that the end-to-end network does not guarantee delivery of information. If transmission fails, e.g. due to congestion in transmission nodes, the end systems have to detect it and take countermeasures to correct these transmission failures.

Such a communication system became feasible by the increasing computing power in the terminals. By shifting the reliability to the edge of the network, the place to create and to provide communication services was also shifted to the edge, moving it essentially out of the domain of the network operator. Every user of the Internet is able to become a service provider as well, and that feature was and is the foundation of the World Wide Web, which put an attractive graphical user interface with ubiquitous linking and embedding facilities on top of the Internet. The World Wide Web was the application which brought the Internet into everyone's daily life and since then the Internet has been used for all kinds of communication and information including telephony and television.

Access to the Internet can be established through the traditional ISDNs, but the connection and service delivery model does not fit well with the nature of the Internet. Much more appropriate for access to the Internet is the DSL model, which provides a broadband leased-line service to the Internet. Data can be sent and received at any time without taking care of the connection and the user gets the feeling of instant access to all resources on the Internet. According to the cost structure of the underlying network infrastructure, a monthly flat fee, sometimes combined with some transfer volume restrictions, replaced the strict usage-based charging of the traditional telecommunication networks.

DSL-type networks are widely deployed based on cable or xDSL technologies with a transmission speed of multiple megabits per second (Mbit/s) and high competition among network providers to provide ever higher speed connectivity to their customers. Having experienced the comfort of mobile telephony, customers are looking also for mobility for their broadband Internet access. Wi-Fi, aka WLAN, based on IEEE 802.11, became a huge success by allowing people to just sit down and connect to the Internet wirelessly without being tied by an Ethernet cable to the modem of the DSL.

1.1.2.1 Mobile WiMAX, the Wireless Subscriber Line to the Internet

Providing broadband Internet connectivity wirelessly over a much wider range than a couple of meters around a Wi-Fi access point is the driving idea behind Mobile WiMAX. It is intended to become the wireless counterpart to the cable and xDSL networks and to enable a new area of telecommunication by opening up the Internet for mobile terminals as well as new kinds of mobile applications.

The technology is designed to cope with the biggest challenge for operation of a mobile DSL: the huge growth of data volumes as experienced so far in the wired networks and now also appearing in the mobile networks. Starting mainly from wireless DSL deployments, the support of mobility functions like network detection and selection, handover, roaming and paging allows for the evolution toward nomadic and mobile service offerings for notebook computers and handheld devices.

1.2 Mobile WiMAX Specifications

IEEE 802.16 as well as the WiMAX Forum [2] create the technical specifications of Mobile WiMAX. IEEE 802.16 is a working group of the IEEE 802 LMSC, which is in charge of the

PHY and MAC specifications of the radio interface. The remaining parts of the access network functionalities are specified by the standardization activities in the WiMAX Forum, which refers not only to IEEE 802.16 but also to appropriate specifications of other standardization organizations, e.g. the IMS system of 3GPP. In addition to the technical specifications, the WiMAX Forum also develops the certification process for Mobile WiMAX equipment to ensure interoperability between different implementations.

Apart from its technical work, the WiMAX Forum is active in the promotion, marketing and regulatory areas to support the worldwide acceptance of the Mobile WiMAX technology.

1.2.1 Specification Areas

Mobile WiMAX technical specifications cover three main areas: radio, network and roaming. For each of the areas one or more interfaces are defined, which provide reference points for interoperability. Figure 1.2, a schematic figure of an access network, shows the location and the relation of the interfaces to each other.

The three areas of interoperability in Mobile WiMAX consist of:

- **Radio interface**: The radio interface denoted by R1 in Figure 1.2 defines the interface between the mobile terminal or subscriber equipment and the base station (BS) of the access network. The interface consists of three parts: the PHY and MAC according to the IEEE 802.16 specification as well as network layer functions, which are defined as part of the Mobile WiMAX network specification.

  Certification is provided for all the three parts defining the air interface R1, the PHY layer, the MAC layer and the network functions for establishing the IP configuration and carrying user payload over the air.
- **Network interfaces**: Mobile WiMAX defines distinct logical network entities for the access serving network, called ASN, as well as for the connectivity serving network, called CSN.

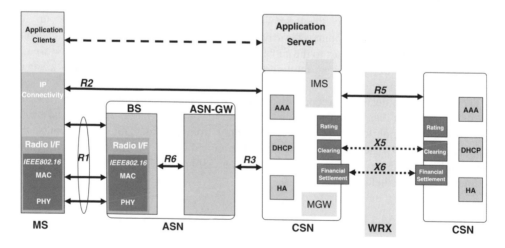

Figure 1.2 WiMAX specification areas

The ASN consists of a number of base stations (BSs) connected to at least one ASN gateway (ASN-GW), which also anchors the interface to the CSN. A standardized interface between CSNs exists for roaming purposes. The interfaces denoted R2, R3, R5 and R6 in Figure 1.2, as well as R4 and R8 (not depicted in the figure), denote reference points for interoperability in the network.

Network interoperability tests are provided for all network interfaces to ensure the proper operation of network equipment from different manufacturers.

- **Roaming interfaces**: In addition to the radio and network interfaces, Mobile WiMAX supports standardized roaming interfaces to facilitate worldwide roaming support among the WiMAX operators. Roaming is supported by a WiMAX roaming exchange (WRX) network, which mediates standardized procedures and messages between WiMAX operators to enable connectivity provisioning for subscribers in foreign networks. The roaming architecture is defined by the interfaces X2–X6 as well as R5. Figure 1.2 only depicts X5, X6 and R5, which pass through the WRX.

Interoperability testing is provided across all specified interfaces to ensure proper operation of the Mobile WiMAX roaming exchange.

1.2.2 Development Process for WiMAX Interoperability

WiMAX interoperability relies on certification and interoperability testing, which is the final outcome of the standardization work in the WiMAX Forum in addition to the radio interface specification stemming from IEEE 802.16. In an extension of the commonly used three-stage standardization process, the WiMAX Forum is creating the necessary documentation for interoperability and certification in six stages, depicted in Figure 1.3.

The stages define distinct steps for the development of the WiMAX Forum specifications:

- **Stage 1: Requirements**: First, the functional requirements for the WiMAX terminals and the access network are specified. As usual, new functional requirements are add-ons to the existing specifications; the specifications are published in releases, with each new release comprising a number of functional enhancements to the previous release.
- **Stage 2: Architecture**: The development of new functions starts with the design of the overall architecture of the solution and the definition of basic message flows.
- **Stage 3: Protocols and procedures**: Based on the Stage 2 results, protocols and detailed procedures of the solution are specified in Stage 3. Implementation of functions in products can start with the completion of this stage, which provides all the necessary information.

  While Stage 2 and Stage 3 specifications provide a choice of implementation options for particular functions, as in the case of the IEEE 802.16 radio interface specification, a profile document lists the selected options for certification and interoperability testing.
- **Stage 4: Conformance statements**: For the development of the interoperability tests a conformance statements document details and lists all the functions which are subject to the tests. The implementation conformance statements are captured in a derived document, which allows manufacturers to specify which of the listed functions are supported by their implementations.

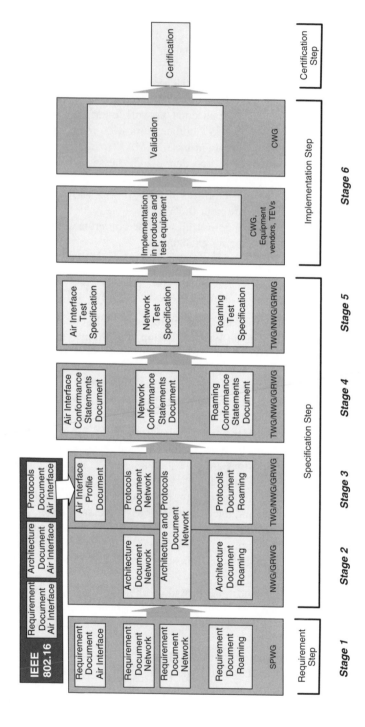

Figure 1.3 WiMAX Forum six-stage development process

- **Stage 5: Test procedures**: This stage of the development process defines all the test purposes and procedures for each of the functions listed in the conformance statements.
- **Stage 6: Certification process**: This stage consists of the detailed processes of the certification and interoperability testing based on the results of Stage 5 as well as the templates for capturing and documenting the outcome of the certification and interoperability testing.

The six-stage development process of the WiMAX Forum generates all the paperwork necessary to develop and verify interoperable implementations of the Mobile WiMAX technology.

1.2.3 Documentation Structure of the WiMAX Forum

The documentation structure of the WiMAX Forum is closely aligned to the six-stage development process. It defines a separate series for each of the stages of the specification areas (Figure 1.4).

All the technical specifications are captured in the series starting with the capital 'T'. The first digit of the series number denotes the technical area – air interface, network, roaming – while the second digit reflects the particular stage to which the documents belong. The series number starts with a capital 'C' for the certification documents. Figure 1.5 shows the complete set of series numbers currently defined by the WiMAX Forum.

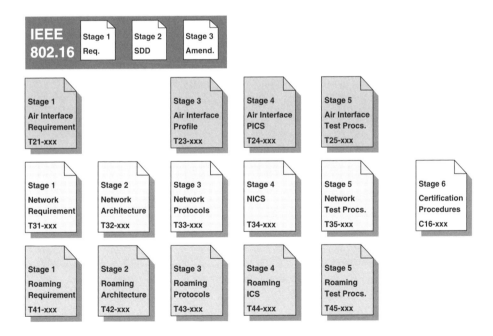

Figure 1.4   Documentation structure

A1x WiMAX Forum Processes Administrative Aspects	C1x Certification Aspects	T1x Overall Deployment, Application and OAM
A11: Policies and Procedures A12: Release Plans	C11: Instructions C12: Policies C16: Procedures	T11: End-to-end Requirements T12: WMF Credentials (X.509) T13: Application level issues T14: Spectrum policies issues
T2x Air Interface	**T3x Network**	**T4x Roaming**
T20: Air Interface Guidelines T21: Air Interface Requirements T22: Air Interface Stage 2 T23: Air Interface Specifications T24: Conformance Statements T25: Test Procedures T26: Reserved T27: Spectrum Issues and Coexistence	T30: Network Guidelines T31: Network Requirements T32: Architecture T33: Protocol Specifications T34: Conformance Statements T35: Test Procedures T36: Reserved T37: Interworking Specifications	T40: Roaming Guidelines T41: Roaming Requirements T42: Architecture T43: Protocol Specification T44: Conformance Statements T45: Test Procedures T46: Reserved T47: Interworking Specifications T48: Agreements T49: Information

Figure 1.5 WiMAX Forum documentation: series numbers

The series numbers form the base of the document identifiers used by the WiMAX Forum. Within a series, each document is identified by a serial number, the release number to which it belongs, as well as a version number indicating maintenance updates of the document. Document identifiers of approved WiMAX Forum documents are preceded by the string 'WMF-', work in progress contains 'DRAFT-' instead of 'WMF-'.

Figure 1.6 explains the anatomy of document identification in the WiMAX Forum.

The document identifier of approved specifications starts with the letters 'WMF' in front of the series number and the serial number of the specification within its series, followed by the release number it was created for, and concludes with a version number, indicating the maintenance cycles it stems from. Initial editions of specifications carry the version number 'v01'.

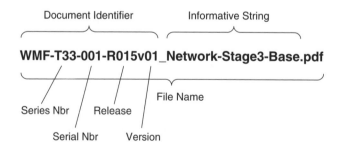

Figure 1.6 Anatomy of the WiMAX Forum document identification

An informative string indicating the title of the specification is appended to the document identifier to form the file name of the document. The informative string is added to enable easier browsing and searching in file archives for particular specifications. The document identifier would be sufficient, as it is unique and concise, but human beings often appreciate additional information which is more meaningful to them.

The complete specification of the WiMAX Forum documentation structure and identification, as well as the release structure and process, is available from [112] and [113].

1.3 About This Book

The structure of this book is guided by the aim to provide an overview of the technology and a foundation for stepping deeper into specifications and deployment planning. It follows a top-down approach by providing first the big picture and general principles and then delving deeper into technical issues throughout the course of the book.

The book mostly covers what is contained in Release 1.5 of the WiMAX Forum. It consists of eight chapters starting with the positioning of Mobile WiMAX in the telecommunication landscape and an introduction to the overall structure of the specifications, and concludes with a look at future enhancements of the network and radio architecture as visible in the release planning of the WiMAX Forum and accompanying specifications in IEEE 802.16.

Chapter 2 provides an introduction to the Mobile WiMAX network architecture as a result of the design of a mobile network fully aligned to the Internet network and business model for offering broadband IP services. In addition to the presentation of the WiMAX roaming architecture and the wireless DSL deployment option, the chapter contains a comparison between Mobile WiMAX and the enhanced system architecture evolution of the 3GPP.

The following chapter addresses the security and subscription-related aspects of Mobile WiMAX starting with an explanation of the authentication procedures for devices, as well as for subscriptions, in addition to an overview of the EAP methods and the certificates and public key infrastructure which are used for it. The chapter comprises as well considerations about the overall security design in WiMAX and the identities which are part of or involved in the security procedures. The chapter concludes with a description of the AAA protocols and the AAA routing issues.

Services and the management of services from a Mobile WiMAX network are the topics of Chapter 4. After an introduction to the AAA-related aspects of accounting and charging, the QoS concept and QoS management with and without the involvement of a dedicated PCC function are explained. After laying the foundation of service provisioning, the chapter presents the supporting location information, IMS and emergency services in more detail.

Chapter 5 provides an introduction to the mobility support in Mobile WiMAX, which is composed of ASN-anchored mobility functions and CSN-anchored Mobile IP support. The chapter elaborates all the PMIP and CMIP versions used in the architecture, and also explains the network architecture option without Mobile IP support and functions in the ASN to restrict mobility support for particular subscribers to particular regions, as needed to fulfill regulatory requirements for fixed wireless access from a mobile network.

The basics of the WiMAX radio interface are covered in Chapter 6. It addresses the functions in the PHY layer as well as in the MAC layer, and provides insights into MAC layer functionality for network entry and initialization, for connection management and QoS support, and for mobility support comprising handover functions as well as the extensions dealing with the sleep mode and the idle mode. Throughout the whole chapter, detailed references are made to the Release 1.0 profile of the IEEE 802.16 radio specification.

The extensions of the radio interface in the system profile Release 1.5 are the topics of Chapter 7. It presents the additional features which were added to build the radio interface specification for Release 1.5. Finally, Chapter 8 provides future perspectives of Mobile WiMAX.

2

Network Architecture

This chapter will provide a comprehensive introduction to the rationale and design considerations of the Mobile WiMAX network architecture, including a presentation of the network reference architecture, control plane functions as well as data path and transport plane design, and the identifiers used by the technical specifications. Furthermore, the chapter covers the WiMAX roaming architecture and support of Ethernet services and concludes with an architectural comparison between Mobile WiMAX and the 3GPP LTE/SAE technology.

2.1 Providing Access to the Internet

In the past two decades, the Internet has introduced a major shift in telecommunication paradigms, first in fixed networks and now in the mobile telecommunication networks as well.

2.1.1 Traditional Operator Networks

Traditional operator networks are following architectures adapted to the operational and business needs of operators, which deploy the full value chain of a telecommunication network comprising the radio access infrastructure, the control infrastructure in the core as well as the application servers, producing the offered telecommunication services.

The design of the traditional operator network is widely influenced by the integration of all stages of telecommunication offerings within a single operation, as shown in Figure 2.1.

The network is structured in a radio access, a control and a services part. Value is generated by providing services to subscribers over a mobile radio access infrastructure deeply controlled by the service mediation functions in the core network to enable complex business models with fine-grained customized offerings and billing. The radio access network part is mainly considered as a prerequisite to enable profits from services. As the cost of the radio access increases with the bandwidth, operators focus on highly valuable services requiring only small transfer volumes over the air. The Short-Message Service (SMS) is an typical example.

Deploying Mobile WiMAX   Max Riegel, Dirk Kroeselberg, Aik Chindapol and Domagoj Premec
© 2009 John Wiley & Sons, Ltd

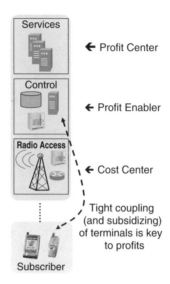

Figure 2.1   Traditional operator value chain

2.1.2 Internet Access Networks

The Internet introduced a different kind of operator model by splitting service provisioning from network access. From its end-to-end paradigm, the IP used for best effort, i.e. uncontrolled transport of information throughout the ubiquitous network, the control of service access by controlling the transport of information became unfeasible. Without end-to-end control of the delivery path between the service creation entity and the consumer, it was not feasible to charge subscribers for particular services.

The Internet brought a split between the content or service provider and the network access provider. With service provisioning becoming its own business not tied to the access line, network access evolved into an independent business. It even split into several dedicated business areas enabling operators to focus on their most profitable assets by leveraging services of other operators for parts not belonging to their intended core business. A quite common differentiation of the network access operation is the split between the network access provider (NAP) and the network service provider (NSP), as shown in Figure 2.2.

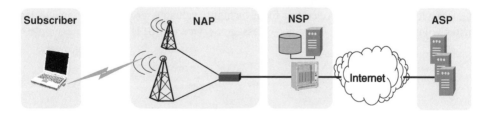

Figure 2.2   Internet access architecture

The NAP operates the access network infrastructure, i.e. the wired and wireless network equipment between the user or subscriber and the interface to the core network with all its supporting functions like network management systems and wholesale accounting. It usually acts as proxy for the authentication of users entering the network by forwarding the credentials offered by the users to the business operations selling network access to users. Only when the service provider acknowledges the credentials then the NAP will establish access connection and charge the service provider for the connectivity based on a wholesale contract.

The NSP deals with all aspects of the customer relation, e.g. with marketing and customer care and billing, as well as with the network equipment for access control and configuration to the Internet, policing, accounting, and roaming and mobility support across access networks.

2.1.2.1 Broadband Operator Value Pattern

Splitting the telecommunication network operation into multiple, independently operated roles provided the foundation for ever-increasing bandwidth in the access networks.

Formerly the telecommunication operation was mainly defined by the services provided over the network access and the intention to strive for high-valued services with low-bandwidth requirements in the pre-Internet area. The independence of the services from the access network created competition among the access providers to offer the most attractive Internet access.

With services based on best-effort connectivity and the robustness and reliance of the Transport Control Protocol (TCP) of the Internet, the main differentiator among network access offerings became the bandwidth of the bit-pipe to the Internet. More advanced offerings may also optimize for low round-trip delay, but the importance of short round-trip delays for the responsiveness of Internet services is not so commonly known and therefore rarely used in consumer marketing. Figure 2.3 shows the paradigm shift of the Internet with each radio access, ISP operation as well as application provisioning becoming a source of profit.

2.1.2.2 Network Operator Relations

Operators focusing on and dealing only with a piece of the whole telecommunication infrastructure demand more openness and flexibility in interacting with other kinds of operator businesses. Figure 2.4 shows what kinds of interactions happen in a network scenario, where the whole chain from service creation to delivery to the subscriber is split into the role models of ASP, NSP and NAP.

A common arrangement requirement is that a NSP has contracts with multiple NAPs offering last-mile connectivity between the ISP core and the customer. Vice-versa, a NAP may deserve connectivity to multiple NSPs to allow it to make best use of the infrastructure and reach a high level of usage of it. The better the access infrastructure used, the higher the utilization and cost-effectiveness of transport will be over the last mile. In particular, when setting up new equipment the growth of utilization directly influences the return on financial investment. High growth rates in throughput require frequent upgrades of the access infrastructure. Sharing the access infrastructure among multiple NSPs is an efficient way to increase the profitability of the investments by reaching the ROI (Return On Investment) faster due to higher traffic volume.

Additional relations exist between NSPs when offering network access to their subscribers in foreign regions. Usually such network access in foreign regions is provided by roaming agreements, when no direct relationship between the local NAP and the home NSP of the subscriber

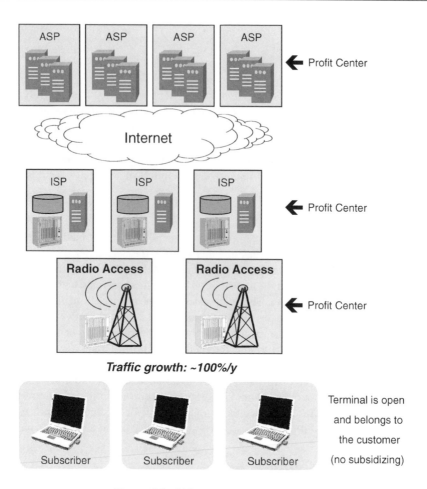

Figure 2.3 Value pattern of the Internet

exists. Roaming denotes the use of another NSP for offering network services to own subscribers. To enable roaming among NSPs, either direct technical and business relations have to be in place, or the operators make use of brokers, who manage the relations among a larger group of NSPs.

Possibly there is also the demand for a closer relation between NSPs and particular ASPs, which benefits from better control of the access line, e.g. by providing a better class of service for a real-time service like telephony or priority for streaming media like IP-TV. Such service providers may deploy direct connections that bypass the Internet to enable better than best-effort QoS for providing their applications to customers. All the operator relations mentioned here are depicted in Figure 2.4.

2.1.3 Mobile WiMAX Networking

Mobile WiMAX networks are aligned to the principles explained above for providing Internet access. The end-to-end network is built from radio access networks operated by the NAPs, by

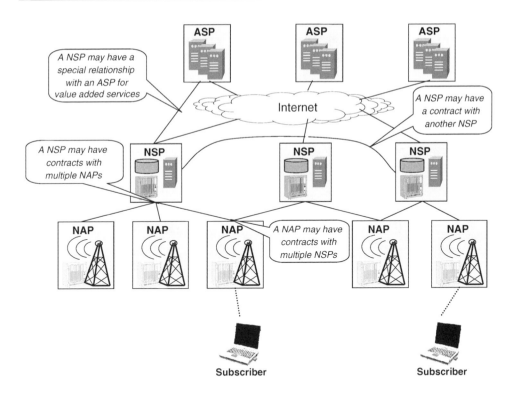

Figure 2.4 Operator relations in the Internet model

the core and connectivity serving portions operated by the NSP and by the Internet or dedicated ASPs providing the communication and application services to subscribers. Real networking as shown in Figure 2.5 comprises other networking functions like the backbone networks for interconnection of the networks of NAPs, NSPs and ASPs, as well as support functions like the WiMAX roaming exchange (WRX) for the exchange of subscriber identities and credentials to enable roaming access across all WiMAX radio access networks.

Even though Mobile WiMAX has 'mobile' in its name, the network not only is designed for mobile terminals, but also serves portable, nomadic and even fixed deployments. Instead of establishing multiple dedicated radio access networks, one for each particular deployment scenario, Mobile WiMAX provides a common radio access technology serving all kinds of usage scenarios and telecommunication businesses from the same infrastructure. Network entry and handover functions are not only used to provide mobility to handheld terminals, but also leveraged for load balancing purposes in the access infrastructure to cope with the transient high-capacity demand of fixed and nomadic users, or to provision users on-the-fly with newly subscribed services.

Mobile WiMAX is a high-performance mobile broadband access technology, which fits well with modern broadband network architectures. It does not comprise a complete telecommunication architecture including a detailed specification of the services to be delivered to end customers, like GSM, but consists of the specification of a radio access network based on the IEEE 802.16 radio interface produced for deployment in common broadband IP architectures.

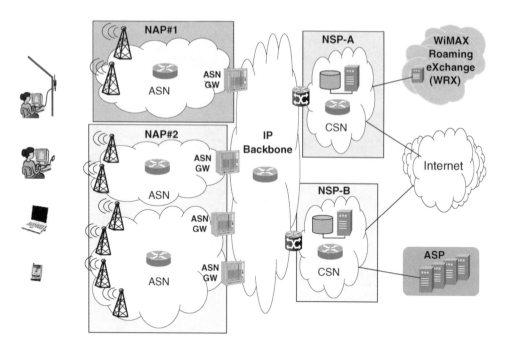

Figure 2.5   Mobile WiMAX end-to-end networking

Therefore, it specifies the radio interface behavior and network functions in the access network as well as the interface to the NSP, including the necessary supporting functions inside the NSP network.

2.2 Mobile WiMAX Network Reference Architecture

To allow manufacturers of Mobile WiMAX networks to gain competitive advantage in designing the best performing and most economical implementation of the network equipment, nevertheless providing interoperable products, the specification is based on a logical representation of a network architecture.

Aligned to the functional differentiation of service provider roles in the Internet model, the Mobile WiMAX access network architecture introduces logical network entities, which reflect the networking functions usually performed by the NAP and by the NSP. The logical network models are called ASN (Access Serving Network) and CSN (Connectivity Serving Network). They cover only the networking parts of the operation of a NAP and NSP, but do not comprise the business supporting systems nor the operation supporting systems, which are also necessary for successful NAP or NSP businesses. While the ASN fully covers the networking functions of a WiMAX NAP, the specification of the CSN is limited to the supporting functions which are necessary to allow a NSP to connect to an ASN:

- **CSN**: The specification of the CSN addresses the NSP part of the authentication, authorization and accounting process, the IP address management for the connected MSs, and the policy and QoS management based on SLAs between the subscribers and NSP. It further addresses mobility management across multiple ASNs, support of roaming subscribers by other CSNs and the connectivity of subscribers to the Internet and directly connected ASPs.

- **ASN**: The ASN comprises the IEEE 802.16 radio interface and the related supporting network functions, including the procedures for network detection and selection, network entry and admission control, and handover support to neighboring base stations. ASN also covers the functions for radio resource management, QoS and policy enforcement, and the session and mobility management of the link to the attached MSs as well as the operation of the foreign agent (FA) in the ASN and forwarding to the selected CSN. Further functions are the accounting client, control of the idle mode and paging, as well as the client for proxy MIP.

The specification of the Mobile WiMAX network does not dictate how the functions in the CSN or ASN are to be implemented, but details the interfaces between the functional units. The well-defined interfaces between the functional units establish the reference points of the architecture.

2.2.1 Reference Points in the Mobile WiMAX Network

Interoperability is achieved by the definition of reference points in the architecture, which represent interfaces between networks or between network elements inside a network.

2.2.1.1 Reference Points between Business Entities

Business needs define the main reference points in the Mobile WiMAX architecture. Reference points mark the borders of operational entities in the distributed Internet access model, as shown in Figure 2.6:

- **R1** represents the interface between the terminal or subscriber equipment and the radio access network. Denoting the air interface as the R1 reference point covers IEEE 802.16 PHY and MAC in addition to the network layer protocols for configuration and transport of user payload across the air interface.
- **R3** defines the interface between the ASN of the Mobile WiMAX access provider and the CSN of the NSP. R3 contains control protocols for the authentication, authorization and accounting (AAA) of the subscriber, including IP configuration and policy control functions, as well as for the dynamic establishment and relocation of the user data path to realize network sharing and wide area mobility. The dynamic establishment and relocation of the user data path is an optional function of the R3 reference point, which may not be necessary for smaller networks and fixed and nomadic usage models.
- **R5** specifies an interface between CSNs of different NSPs to enable network access to roaming subscribers. In the roaming case, foreign networks are providing service to subscribers who do not have a direct relationship with the NSP of the foreign network. When entering the network, the visited NSP forwards the credentials of the foreign subscriber over R5 to her/his home NSP and receives from the home NSP over R5 the confirmation and the configuration for providing network access. Depending on the contractual framework between the involved NSPs, either the roaming subscriber may get Internet connectivity directly from the visited NSP, or services from the home NSP are provided over the optional data path of the R5 interface.

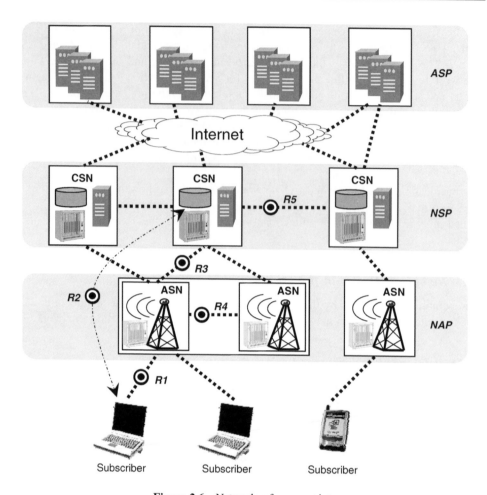

Figure 2.6 Network reference points

- **R2** defines a direct interface between the subscriber and her/his home NSP, which is necessary, for example, for the secure exchange of the credentials of the subscriber in the authentication process to the home NSP. The R2 interface carries control information only.

The reference points R1, R2, R3 and R5 mentioned above represent business relations between the parties involved in providing and consuming WiMAX network services.

2.2.1.2 Reference Points between Access Network Entities

In addition to the four reference points for business-related interfaces, a further three reference points labeled R4, R6 and R8 specify interfaces inside the ASN to enable interoperable implementations of equipment for the Mobile WiMAX radio access network.

Usually the radio access network is built up of a large number of base stations, each serving an area whose size depends on the design and surrounding of the base station and the used

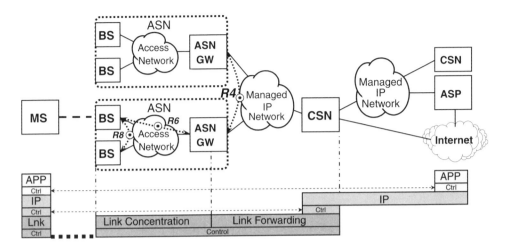

Figure 2.7 Functional model of the access network

spectrum, and also on the demand of transfer capacity in the area. Base stations are the gateways between the radio links to the subscribers and the wired part of the access network for aggregating and forwarding the traffic and for concentrating the links into higher order network interfaces. To hide the specifics of the radio interface to the NSP but allow local control and optimization of the traffic forwarding inside the ASN, the WiMAX architecture deploys a link access concentrator denoted the ASN gateway (ASN-GW) at the edge of the radio access network to the NSP. The ASN-GW acts as the local anchor of mobile subscriber stations in the ASN to allow handover between base stations of an ASN without any involvement or even notice by the CSN. The ASN-GW also terminates the R3 interface to the interconnected CSNs.

Mobile WiMAX specifies the split of ASN functions into functions located in the BS and those located in the ASN-GW to enable standardized interfaces inside the ASN (Figure 2.7):

- **BS**: It comprises the functions of the PHY layer and the MAC layer of the IEEE 802.16 radio interface including QoS enforcement in the scheduler for transmission over the air and the related radio resource management. Furthermore, the encryption and decryption of the user payload, the paging agent for performing paging and idle mode actions, as well as the data path functions and the handover functions for intra-ASN mobility management, are located in the BS.

  The BS represents one instance of the IEEE 802.16 MAC typically on top of one instance of the IEEE 802.16 PHY, i.e. one cell.[1] Real BS equipment usually serves multiple sectors from one unit, i.e. multiple cells, and represents the equivalent of multiple BSs in the Mobile WiMAX network model.

- **ASN-GW**: It is the location for the authenticator, key generation and key distribution, the paging controller and the service flow management and classification of the user payload.

---

[1] In cases like a multi-carrier BS, one instance of the MAC may be combined with multiple PHYs.

In addition to the data path functions and the handover functions for intra-ASN as well as inter-ASN, the ASN-GW comprises the access router, foreign agent and PMIP client for the MIP-based R3 interface, the DHCP proxy and a relay function for RRM messages to other BSs.

When access networks are too large in terms of number of concurrent users or the aggregated data rate to be handled by a single ASN-GW, the access network can be equipped with multiple ASN-GWs, or can be split into multiple ASNs, when a more distinct operational segregation is preferred. Direct interconnection between the ASN-GWs allows local mobility management across multiple ASNs without relocating the forwarding anchor for the R3 interface. Nevertheless, relocation of the anchor to the new ASN-GW is still possible and may be later performed for optimization of the traffic path.

Figure 2.7 exposes the reference points inside the radio access network and indicates that they define logical interfaces between network entities, which do not necessarily require a dedicated physical network interface for each reference point:

- **R4** defines an interface between ASN-GWs, which allows extension of the coverage area or the capacity of a WiMAX radio access network by direct interconnection between multiple ASN-GWs, each serving its own ASN or a part of a larger ASN. The R4 reference point comprises control signaling as well as a data path for forwarding user traffic.
- **R6** denotes the interface between the BS and the ASN-GW. It carries user traffic as well as control information between the ASN-GW and the BS.
- **R8** is the interface between BSs inside an ASN for direct exchange of control information to allow more effective management of the radio access resources inside an ASN.

As indicated at the bottom of Figure 2.7, the ASN is just providing the access links of subscriber stations or MSs to the CSNs. Each MS is connected by a dedicated link to its IP anchor in the CSN, where the access router resides and the IP address assignment is controlled. The ASN itself is unaffected by the IP addressing. In the ASN, traffic is forwarded depending on the identity of the MS and the address of the IP anchor in the CSN. The figure also demonstrates the end-to-end principle of IP networking, where applications reside in the hosts at the edge of the network and all application-related information and signaling are transparently carried across the network.

2.2.1.3 Reference Point Details

The structure of the reference points in the WiMAX architecture follows real network interfaces by concurrently performing multiple message transfers over the same interface. Other network architectures define reference points for single functions and protocols. In Mobile WiMAX, however, each of the reference points represents the complete set of functions and protocols that are carried between the connected network entities.

As depicted in Figure 2.8, reference points comprise multiple functions and may deploy a common protocol or multiple different protocols to realize the set of functions of a reference point. An example of a common protocol for multiple functions is RADIUS for the AAA process, while Mobile IP (MIP) is an example of a protocol for a single function. In addition to the control functions, the transfer of user payload may be part of a reference point, and if present it forms the data path part.

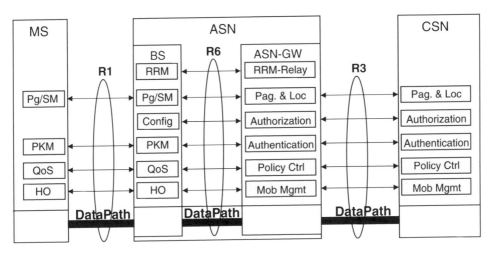

Figure 2.8   WiMAX reference point structure

The Stage 2 and Stage 3 specifications of the Mobile WiMAX network are not structured reference point by reference point, but deal with network function by network function across all involved reference points to provide a better insight into the end-to-end functional behavior of the access network. A functional structure of the specification more easily handles the distinction between mandatory and optional features. Features marked as optional are optional on all reference points, and keeping all reference points together for a particular feature makes it easier for developers to implement or skip such optional features.

Except for the radio interface with R1 and R2, none of the other reference points of Mobile WiMAX is mandatory. It is up to the implementers and operators to decide which of the reference points to implement in the network. Usually most of the available reference points will be exposed in the equipment and networks to make best use of the standardization of Mobile WiMAX.

The WiMAX Forum test specifications support certification and interoperability testing for each reference point individually to enable certification of higher integrated equipment which does not expose every reference point.

2.2.2 The Mobile WiMAX Network Reference Model

To avoid ambiguities and exaggerated complexity in the implementation and deployment choices of the Mobile WiMAX network architecture, the specifications of the Mobile WiMAX network refer to a schematic form of the network architecture. The reference model of the network as shown in Figure 2.9 exposes only the specified network entities and the reference points between the entities, but does not depict any implementation choices inside the network entities or the transport architecture between them.

The network reference model (NRM) differentiates between the control plane and data path. While all reference points carry control information, only R1, R3, R4, R5 and R6 transfer user payload.

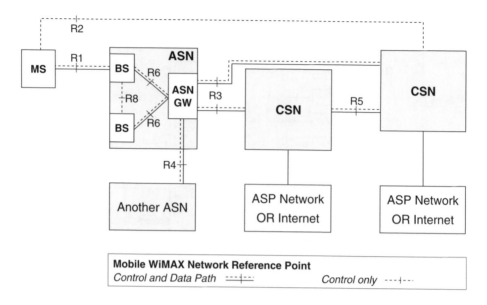

Figure 2.9   Mobile WiMAX NRM

2.2.3 Control Plane Functions

The control plane comprises all the functions to establish, maintain and tear down mobile connectivity in a cellular network based on the IEEE 802.16 standard [1]:

- **Network entry discovery and selection/reselection** defines the procedures of a MS to scan and detect the available WiMAX access networks, as well as to choose the preferred NAP for connecting to the home NSP.
- **IP addressing** comprises the assignment of IPv4 as well as IPv6 addresses to MSs via DHCP, MIP or stateless autoconfiguration in the case of IPv6.
- **WiMAX key hierarchy and distribution** specifies the generation, derivation and the distribution of all the keying material needed in the Mobile WiMAX access network.
- **Authentication, authorization and accounting** cover the procedures for the network access authentication and authorization according to the user profile of the subscription, and the accounting procedures for measuring and signaling the usage of the access network by a particular subscriber.
- **Network entry and exit** define the procedures for establishing initial connectivity to a Mobile WiMAX network after network entry detection and selection, and the procedures for terminating the network connectivity orderly.
- **QoS and SFID management** define the procedures for creation, modification and deletion of the initial and other service flows, as well as the management of service flow identifiers (SFIDs) and the static QoS policy provisioning.

- **ASN anchored mobility management** provides the handover support of radio links across BSs inside a single ASN or across ASNs without changing the anchor ASN.
- **CSN** *anchored mobility management* defines mobility management based on Mobile IP across ASNs with the mobility anchor located in the CSN.
- **Radio resource management** is a function inside an ASN to increase the deployment efficiency of the available radio resources.
- **Paging and idle mode MS operation** covers the control procedures for location update, paging, and entering and leaving the idle mode according to the specifications in IEEE 802.16.
- **IPv6 and simple IP support** provide operational specifics for IPv6 support and the interconnection of ASN and CSN when Mobile IP is not deployed for the R3 interface, respectively.

To facilitate interoperation between networks and between equipment based on different releases of the WiMAX specifications, negotiation functions are in place on R4, R6, R8 as well as on R3 and R5 to determine the most appropriate mode of operation.

2.2.4 Data Path

User payload is carried by the data path, which is also usually denoted as the user plane in mobile architectures. A technique called tunneling is applied across the whole radio access network up to the CSN to realize packet forwarding independently of the addresses used in the user payload. Each of the MSs resides on a dedicated virtual circuit up to the CSN. The tunneling across R3, R4, R6 and R1 establishes the circuit behavior between MS and CSN.

Different protocols are used for tunneling across the reference points. The convergence sublayer (CS) on top of the IEEE 802.16 MAC layer provides the encapsulation of user payload across the air interface. In the wired part of the radio access network GRE is applied for the R4 and R6 interfaces, and IP-in-IP or GRE encapsulation is used on the R3 interface, when dynamic establishment and relocation of the tunnel by Mobile IP is required.

When mobility support is not required on R3, static tunneling is configured and the choice of the applied tunneling protocol can be left to the peer-to-peer agreements of the involved operators. 'XXX' in the protocol stacks of Figure 2.10 indicates that there are multiple choices for the protocol depending on whether mobility support is needed or not.

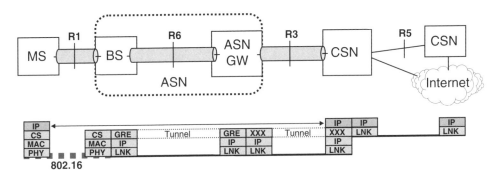

Figure 2.10   Data path tunneling

A data path may be part of the R5 reference point when user payload has to be forwarded between CSNs. Native IP forwarding between networks is used as subscriber stations are anchored in the CSN, i.e. IP addresses out of the address range of the CSN are assigned to the MSs.

2.2.5 Transport Plane

Mobile architectures separate user traffic from control traffic for the establishment, management and termination of user connections. All traffic originating or terminating in the MS, including configuration information for the MS, is carried over tunnels inside the ASN, to keep user traffic separate from the information exchanges needed for operation of the mobile access network.

WiMAX networks deploy a common transport infrastructure for both the control and user traffic. By shielding the user traffic inside tunnels, users are not able to address any of the network nodes in the access network, making malicious attacks on network nodes out of IP-based terminals nearly impossible.

In alignment to the terminology of 'control plane' and 'data path', the transport infrastructure to interconnect the network entities in the mobile access network is commonly called the 'transport plane'. As shown in Figure 2.11, Mobile WiMAX deploys an IP-based transport plane, which allows the establishment of a Mobile WiMAX access network based on any links that are able to carry IP packets. Unlike other access network technologies, Mobile WiMAX does not require special links in the access network, like TDM (e.g. for GSM), ATM (e.g. for UMTS) or Ethernet (e.g. for DSL).

2.2.6 Mobile WiMAX Identifiers

The control plane of Mobile WiMAX requires means to denote and identify the involved objects. Instead of telephony-related identifiers like the ITU-T E.164 addresses (international

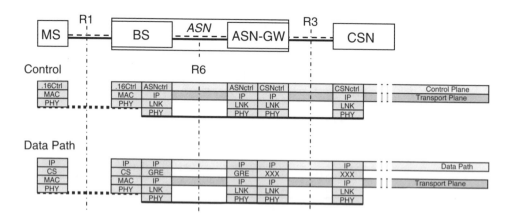

Figure 2.11   Traffic planes in mobile networks

telephone numbers), Mobile WiMAX uses identifiers that are well known in data communication networks or widely used in the Internet:

- The **subscriber** of the WiMAX service is identified by a network access identifier (NAI) as
  specified by [5]. The identifier usually follows the form 'user@home-nsp-domain'. It may be
  decorated with some additional fields for defining operator relations in roaming scenarios.
- The **subscriber station** used for gaining access to WiMAX services is identified by the MS
  ID as specified by [1]. The MS ID is a 48-bit unique identifier commonly assigned by the
  manufacturer from its MAC address range. It is used whenever an action addresses the
  terminal or connections to the terminal, e.g. in 802.16 management messages for network
  entry or handover, as well as in many other control messages in the ASN.
- The **network access provider** offering the radio access for WiMAX services is identified by a
  globally unique 24-bit identifier called the operator ID, as defined in [1]. It can be obtained as
  an IEEE 802.16 Operator ID by the IEEE Registration Authority [117], which allows several
  forms of assignment. It may be either assigned as a globally unique identifier by the IEEE
  authorities or derived from the ITU E.212 MCC–MNC assignments. In cases where no public
  service is offered, it can be arbitrarily taken from a pool of reserved IDs for private usage.
- The **base stations** providing the point of attachment for the radio links to the subscriber
  stations are identified by 48-bit unique identifiers, which are built according to the
  specification for the BS ID in [1]. The upper 24 bits contain the operator ID of the NAP,
  while the lower 24 bits are assigned by the NAP to differentiate between its BSs. With regard
  to real deployments, a BS represents one cell of radio coverage, not the physical unit
  commonly called base station.
- **Network service providers** are identified by unique 24-bit operator IDs, obtained the same
  way as those for the NAPs. As both NSP IDs and NAP IDs are taken from the same number
  space, the numbers are unique in both domains. Furthermore, in cases where ASN and CSN
  belong to the same operator, the operator can choose to use the same ID as NAP ID and NSP ID.
- The **IP address of the subscriber station** is assigned by the NSP either from its globally
  unique address range or from a private IP address range. In NAP sharing scenarios, different
  NSPs may use the same private address range for assigning IP addresses to their subscriber
  stations. The Mobile WiMAX ASN is able to handle overlapping IP address spaces, i.e. two
  different MSs connected to different NSPs assigned the same IP address.
- For paging, BSs are grouped in **paging groups**, and each group is assigned a 16-bit paging
  group ID (PG ID) by its NAP. When a subscriber station is paged within a paging group,
  paging occurs across all BSs belonging to that group.
- The **paging controller, which manages the paging process** and the idle state of the
  subscriber station, is identified by a 48-bit unique identifier, called the paging controller ID.
  Details on the use of the paging controller ID in network re-entry and location update
  messages can be found in [1].
- The **authenticator** controls the identity of the subscriber entering the network and generates
  the keys required for encryption of the messages going over the air. It resides in the ASN-GW
  and is uniquely identified by the authenticator ID, which is formed from a routable IPv4 or
  IPv6 address or from a 48-bit MAC address.
- The **anchor of the data path** of a subscriber station inside an ASN is not necessarily
  collocated in the ASN-GW, which hosts the authenticator of the subscriber station. The data
  path anchor, to which the CSN sends downlink packets, may reside in another ASN-GW and

may be relocated during handover processes. The function is identified by the anchor data path function ID, which is formed from a routable IPv4 or IPv6 address or from a 48-bit MAC address.

Apart from the identifiers mentioned above, a couple more are defined for management of the service flows and over-the-air connections. Details are available in section 4 of [115]. Also, in the remainder of this book, identities and identifiers for a number of specific network technologies, like security or QoS, are considered in more detail.

2.3 Mobile WiMAX Roaming Architecture

Roaming describes a technical and business relationship between two NSPs by which subscribers of one provider can connect and gain access to network services of the other provider. The relations of the NSPs are described by home NSP for the provider owning the subscribers and visited NSP for the provider granting access to other providers' subscribers.

2.3.1 Roaming Functions

Roaming consists of four main functional areas:

1. Providing access to the networking services.
2. Collection, aggregation and exchange of usage information.
3. Billing and financial settlement procedures between providers.
4. Interconnections between home and visited networks.

Each of the four areas requires dedicated functions inside the service provider as well as interactions between the service providers, as shown in Figure 2.12.
   The Mobile WiMAX network architecture supports roaming by introducing the reference point R5 between two CSNs for the exchange of AAA information between providers and optionally also the transfer of the user payload through the visited NSP up to the home NSP.
   Beyond the interfacing of the networking functions of two CSNs, roaming involves a few more functions, which the WiMAX Forum described in its WiMAX Roaming Interface (WRI) specifications [120, 121].

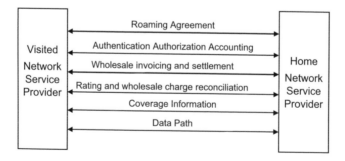

Figure 2.12   Roaming interactions

Roaming requires first that there is a roaming agreement in place between the involved operators. The roaming agreement is a contract among the partners supporting roaming, which details the terms and conditions for the technical part as well as the business side. The WiMAX Forum provides a proposal for the legal part [131] as well as a template for completion of all the technical details [132], which are required for the technical setup and day-to-day operation.

The networking base of roaming is the AAA message exchanges between the visited and the home NSP. Forwarding sensitive business information like AAA messages is bound to the existence of a roaming agreement between the peers.

Roaming also requires that usage of the network resources is recorded, rated, aggregated and agreed between the roaming partners. Finally, the generation of financial invoices and the exchange of payments for the use of others' network resources must be in place.

2.3.2 Roaming Models

There are multiple models for roaming arrangements. Roaming may be unilateral when a NSP provides network services to customers of another NSP, or it may be bilateral when the customers of two NSPs make use of the others' NSP network infrastructure. More often the relations are multilateral when a group of NSPs builds a consortium allowing customers of each of the partners in the consortium to use the networks of all the other partners.

2.3.2.1 WiMAX Roaming Exchange

Roaming may build on peer-to-peer connections between the involved service providers, or may involve an intermediary, known as WiMAX roaming exchange (WRX), to manage the business relations in a larger roaming scenario. With the help of a roaming exchange a service provider does not need to establish roaming relationships with all its partners, but only one contractual relationship with a roaming exchange which has contracts in place with all the other service providers. A roaming exchange may provide all or a subset of the following services:

- Roaming contract management.
- Authentication support.
- Wholesale rating.
- Financial clearing and settlement.
- Invoicing.
- Fraud management.
- Payload traffic routing.

Multiple roaming exchanges may compete or cooperate with each other, and service providers may leverage all or only a part of the services which roaming exchanges offer. More about roaming models and the organizations involved can be found in [119].

The WRX establishes a new entity between the visited NSP and the home NSP in the WiMAX network reference model (Figure 2.13), which is interconnected to the CSNs by the R5 reference point. It intercepts and routes the AAA messages by its AAA proxy, derives from the AAA messages the information required for rating and clearing, and behaves toward the connected CSNs exactly according to the specification of R5, probably even providing the data path for the forwarding of user payload.

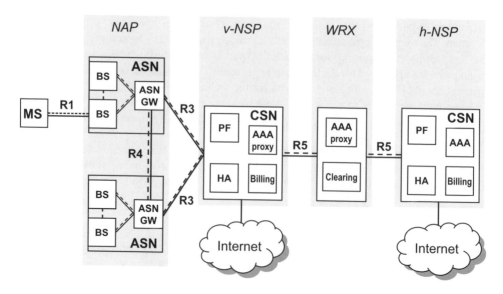

Figure 2.13   Roaming reference model

The WRX acts like an AAA proxy entity for the R5 reference point and does not require any extension to the WiMAX network specification. However, the insertion of a WRX in the path between visited NSP and home NSP introduces some additional operational requirements and restrictions, such as a more complicated process in the MS for the discovery and selection of the most appropriate NAP and visited NSP for gaining access to the home NSP when multiple choices are available.

2.3.2.2 Network Discovery and Selection

When entering the network, a mobile subscriber station executes network discovery and selection to determine the most appropriate access network for connecting to its NSP. Network sharing and roaming allow for more complex network structures and impose additional challenges for establishing access to the services.

Figure 2.14 depicts a simple example of a network configuration, which may be quite common and allows to explain the essential pieces of the network discovery and selection procedures in Mobile WiMAX. Two network access providers, denoted NAP_1 and NAP_2, respectively, provide radio access in the area where the MS resides. Neither of the access providers has direct contractual relations with the service provider of the MS, but both are connected to NSPs which have roaming agreements in place with the home NSP, hNSP.

NAP_1/vNSP_1 is an operator with both an ASN as well as a CSN for providing radio access and connectivity to IP services and to the Internet. It is a roaming partner of hNSP and provides connectivity to customers of hNSP.

NAP_2 does not operate its own core infrastructure but provides network access services as wholesale offerings to NSPs. It has business relationships with the visited NSPs, vNSP_A and vNSP_B, both being in roaming partnerships with hNSP.

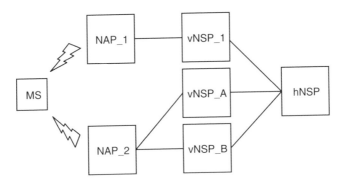

Figure 2.14 Network sharing and roaming scenario

The challenge for the MS is to discover the network structure, then select and establish the most favorable path to its home NSP. The solution consists of four procedures:

1. **NAP discovery** is the process where the MS discovers all available NAPs in its area by scanning and decoding the periodic radio control messages sent out by the base stations which comprise the BS-ID. The upper 24 bits of the BS-ID contain the operator ID of the NAP. The MSB of the lower 24 bits of the BS-ID indicates whether there are one or more distinct NSPs connected to the NAP.

   NAP_1 in the example above would indicate in the MSB of the lower 24 bits that the only connected NSP belongs to the same operator, while NAP_2 would indicate that the NAP serves multiple NSPs and further steps are needed to detect the identities of the connected NSPs.

2. **NSP discovery** provides the list of NSPs connected to the NAPs. Each NAP periodically broadcasts its list of connected NSPs, or provides the list on demand to the MS. The list consists of the 24-bit operator IDs of NSPs and potentially also the complete, human-readable names of the NSPs, to provide meaningful information to the user of the MS for manual selection, if such manual selection is supported.

   Usually MSs are preprovisioned with a preference list of NAPs, with which the home NSP has contractual relationships, as well as a preference list of the NSPs, which are in roaming partnerships with the home NSP. Both lists are used by the MS to define the preferred NAP and NSP for establishing network access.

3. **NSP enumeration and selection** is the process of structuring the information retrieved by NAP discovery as well as NSP discovery in aligning to the preprovisioned preferences and drawing a conclusion about the preferred NAP and NSP by manual or automatic selection. In the case of manual selection, the information obtained by the discovery process in the current location is supplemented by preprovisioned information about operator relationships and presented to the user for decision making.

   Manual selection is also used when the user intends to connect to a particular NSP, e.g. to exercise an initial provisioning procedure, or to use the network on a 'pay per use' basis.

4. **ASN attachment** is the final step in performing network entry to the selected NAP and to signal to the NAP which NSP the MS should be connected to. Decoration of the NAI is used in roaming scenarios to signal to the NAP both the home NSP and the selected visited NSP.

When 'user@home.nsp' is the form of the NAI for direct access to the home NSP, then the decorated form of the NAI in roaming cases looks like 'home.nsp!user@visited. nsp'. The NAP derives information about the home NSP and about the selected visited NSP from the decorated NAI for AAA routing during network entry.

The visited NSP is not involved in the network discovery and selection process itself. Information is either preprovisioned by the home NSP or derived from the scanning process in the coverage area. However, the visited NSP forwards the AAA messages either directly or with the involvement of a WRX to the home NSP for AAA purposes, when it detects from the NAI that the user belongs to a roaming partner.

A thorough introduction to the operational aspects of roaming in Mobile WiMAX networks is provided by [118].

2.3.3 WiMAX Roaming Interface

The WiMAX Roaming Interface (WRI) includes functions for wholesale rating, clearing, financial settlement and fraud management as well as the AAA and traffic forwarding functions of the R5 reference point. It is important to note that the WRI is based on wholesale billing and does not deal with retail billing. Retail billing is applied by the home NSP to create invoices for each individual customer and is based on the retail pricing plan, which home NSPs provide to their customers. Wholesale billing accumulates the usage records of all customers of a particular NSP into one bill based on the wholesale pricing plans negotiated between the visited NSP and home NSP.

The following topics are shown in Figure 2.15, and are addressed as part of the WRI set of documents:

- The **AAA proxy service** logical function performs the correlation and aggregation of session records, validates the records according to the established roaming agreements and transfers the aggregated session reports over the X2 reference point to the wholesale rating logical function. The P1 reference point, by which the AAA proxy service retrieves the accounting records, is a private interface of the NSP and is not standardized. The specification of the AAA proxy service logical function and the data format and transfer protocol of the X2 reference point is contained in [122] and [123].
- The **wholesale rating** logical function calculates and assigns monetary values to the session records delivered at the X2 reference point. Calculation of the monetary values is based on the wholesale pricing plan negotiated between the roaming partners. The rated sessions are forwarded over the X3 reference point to the clearing logical function. The specification of the wholesale rating logical function and details of the X3 reference point are contained in [124] and [125].
- The **clearing** logical function process extracts and validates accounting data received over the X3 reference point from the own wholesale rating logical function as well as over X5 from the clearing process located in the roaming peer. The process comprises the correction of errors and agreeing on the data in the accounting data received from the roaming peer and sending the results over the X4 reference point to the financial settlement logical function.

  The clearing process and the details of the X5 reference point are described in [126, 127]. The X4 reference point is covered only in [126].

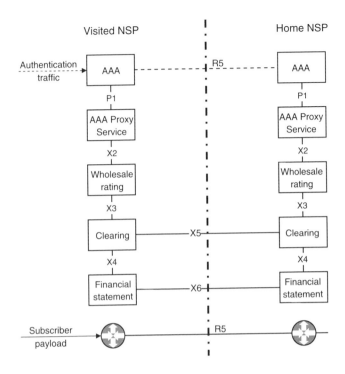

Figure 2.15 WiMAX roaming interface architecture

- The **financial settlement** logical function addresses the settlement of financial responsibilities. It receives from the clearing process validated accounting data and generates from the data a monthly report which is exchanged with the roaming peer and leads to the transfer of funds if the report is validated by the peer. It also comprises the exchange of foreign currencies and the tracking of outstanding financial obligations. The financial settlement logical function and the X6 reference point are specified by [128] and [129].
- The **fraud management** logical function aims to detect abnormal usage patterns in the data records going into the aggregation and wholesale rating process. It leads to reports about potential fraudulent activities and to the first opportunity to perform fraud prevention.
  The fraud management logical function is not covered in the first release of the WRI.
- **Interconnection** [130] is that part of the WRI specification which deals with the connection options for the transfer of AAA messages, the financial settlement and clearing information, and the user payload carried over the data path. It covers connections based on VPN technology and dedicated connections between NSPs and connections via a WRX.

Each of the logical functions can either be incorporated into the network of the home and visited operators or be performed by the WRE on behalf of the operators.

2.4 Ethernet Services Support

Ethernet services (often called carrier Ethernet services) are widely deployed telecommunication services for establishing and connecting the private networks of corporations or public authorities. Moreover, they allow service providers to offer telecommunication services on top of the infrastructure of another network operator. The services are realized by transparent Layer 2 connectivity based on Ethernet, which is well known for its widespread availability, its high scalability in terms of bandwidth as well as network size, its excellent support of all kinds of network layer protocols and its leading edge cost position.

While basic Ethernet is standardized by IEEE 802, the Metro Ethernet Forum (MEF) established a set of specifications describing the model and characteristics of the services, defining the architectural framework and the network interfaces, and providing a framework for the operation and management of Ethernet services. In addition, the ITU-T has adopted Ethernet in its specification as a transport technology for global public telecommunication networks and is continuing to extend the scope of the specifications.

Mobile WiMAX supports fixed, nomadic, portable and mobile broadband access to IP services over a single cellular network. As an extension to its basic deployment, Mobile WiMAX is able to support Ethernet services from the same radio access infrastructure. Offering Ethernet services in addition to IP services allows operators to participate in the growing carrier Ethernet market in regions or cases where an appropriate wired infrastructure is not available. In particular, Ethernet services over Mobile WiMAX may help DSL operators to extend their Ethernet-based access networks to connect customers without wired phone lines or customers requiring portable or mobile access.

2.4.1 Ethernet Services

The MEF established specifications for the definition of Ethernet services and related service attributes to facilitate comprehensive SLAs between service providers and customers. Based on the concept of a point-to-point circuit carrying Ethernet, denoted the Ethernet virtual connection (EVC), the MEF defines in [133] three kinds of basic Ethernet services, namely the Ethernet line service, the Ethernet LAN service and the Ethernet tree service:

- The **Ethernet line service** is a point-to-point connection carrying Ethernet frames between two customer interfaces of the metro Ethernet network (Figure 2.16). It is often used for substituting legacy time division multiplex (TDM) private lines by less expensive Ethernet private lines.

  The Ethernet private line service is becoming an Ethernet virtual private line service as multiple private line services are multiplexed onto a single Ethernet interface at the provider edge. For example, an ISP may use the virtual private line service to provide Internet services over a single Ethernet interface to multiple customers. Such a configuration fits well with the bandwidth hierarchy of Ethernet interfaces.

- The **Ethernet LAN service** provides multipoint-to-multipoint connectivity for Ethernet frames across a number of customer interfaces (Figure 2.17). Essentially, it is like an extension of the customer's own LAN infrastructure.

  It provides the base for the transparent LAN service (TLS), which enables full transparency for any Ethernet control protocols allowing customers to define new virtual LANs (VLANs) across their metro Ethernet networks without involving the Ethernet service provider. The E-LAN service may also be used because of its inherent replication function for multicast and broadcast frames to deliver multicast services from a service provider's head-end to multiple customers.

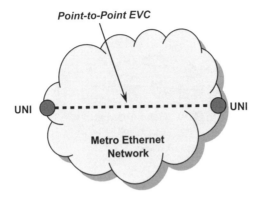

Figure 2.16   Ethernet line (E-Line) service

The E-LAN service is well suited when all connected end stations belong to the same enterprise and reside in the same security domain. Direct connectivity across the TLS from any end station to any other end station is feasible within enterprise LANs, but public access networks usually restrict direct connectivity between end stations.

- An **Ethernet tree service** type distinguishes between leaf UNIs and root UNIs, as depicted in Figure 2.18. Leaf UNIs are restricted to the exchange of data only with root UNIs, but never directly with another leaf UNI. Root UNIs can exchange data with any leaf UNI and with any other root UNI. The E-Tree service is built upon a rooted multipoint Ethernet virtual connection.

The E-Tree service could be useful for providing public broadband access or for delivering multicast services efficiently to multiple UNIs. However, there is one deficiency of the E-Tree service in its current definition for establishing public access networks. IPv6 operation requires direct host-to-host connectivity for neighbor discovery messages to allow secure neighbor discovery, as explained in [138]. Therefore the Mobile WiMAX network specification defines a 'public access mode' of the E-LAN service, which is equivalent to an E-Tree service with secure neighbor discovery enabled.

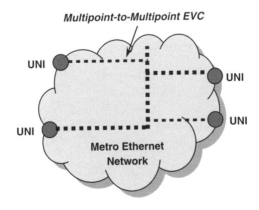

Figure 2.17   Ethernet LAN service

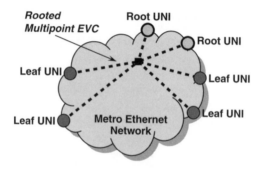

Figure 2.18   Ethernet tree service

2.4.2 Ethernet Standards

The basic standards for Ethernet services stem from the Higher Layer LAN Protocols Working Group P802.1 of the IEEE 802 LAN/MAN Standards Committee. The IEEE 802.1D standard on media access control (MAC) bridges defines the basic forwarding and filtering behavior of today's switched Ethernet, which enables end stations to communicate transparently with each other across multiple LANs. IEEE 802.1Q adds to basic bridging the concept of segregating LAN traffic into VLANs, which allows the definition of multiple isolated LANs on the same bridged LAN infrastructure.

- **IEEE 802.1Q-2005** (Virtual Bridged Local Area Networks) [134] adopts from 802.1D the generic bridge architecture, the internal sublayer service, the major features of the filtering and forwarding process, the Rapid Spanning Tree Protocol and the Generic Attribute Registration Protocol (GARP) and adds the definition of VLAN services, the required extensions to the filtering database and forwarding process, the definition of an extended frame format able to carry VLAN identifiers as well as priority information, the GARP VLAN Registration Protocol and the Multiple Spanning Tree Protocol.

  The standard supports up to 4094 VLANs on the same bridged LAN infrastructure, which is usually sufficient for corporate networks but limits its applicability for larger provider networks.

  VLANs are commonly used inside larger organizations to establish on the common infrastructure isolated LANs for particular applications and organizational units across widespread campuses and sites.

  Two amendments to IEEE 802.1Q expand the applicability of VLANs for large operator networks and even across multiple operator networks.

- **IEEE 802.1ad-2005** (Amendment 4: Provider Bridges) [135] amends the IEEE 802.1Q-2005 standard to enable service providers to offer the capabilities of IEEE 802.1Q virtual bridged LANs to a number of customers with no need for alignment across the customers and only minimal interaction between the operations, administration and maintenance (OAM) of the service provider and the OAM of the customers.

  The standard introduces another layer of traffic segregation by appending the customer VLAN identifiers (C-VID) to a service provider VLAN identifier (S-VID), allowing the service provider to encapsulate the particularities of the configuration of the customer VLANs in a VLAN ID assigned and managed by the service provider. It addresses also the extension to the frame format for inclusion of the customer VLAN tag as well as the service provider VLAN tag, and reserves distinct MAC address spaces for customer L2 control protocols and service provider L2 control protocols, respectively.

VLAN stacking (commonly called Q-in-Q) is the most widely used method in today's carrier Ethernet service deployments inside provider networks, but it lacks the scalability for crossing multiple operator domains and backbone applications due to the 12-bit limit ($\sim$4000) of the S-VID.

- **IEEE 802.1ah** (Amendment 7: Provider Backbone Bridges) [136] is an amendment to IEEE 802.1Q-2005 and IEEE 802.1ad-2005 allowing provider networks to scale up to 2^{24} ($\sim$16 million) service VLANs by introducing provider backbone bridges, which are compatible and interoperable with the provider bridges defined by IEEE 802.1ad. At the edge of a provider bridged network, the provider backbone bridge encapsulates customer MAC frames containing the C-VID and a S-VID by assigning a 24-bit service instance identifier inside backbone MAC frames containing the MAC addresses of the source and destination backbone bridges. At the other end of the backbone, the provider backbone bridge extracts the customer MAC frame from the backbone MAC frame and forwards it, according to the service instance identifier, to the addressed provider bridged network. Due to the encapsulation of customer MAC frames within backbone MAC frames, the protocol is commonly called MAC-in-MAC.

IEEE 802.1ah is not the only viable solution for extending Ethernet services over large-scale networks. Other protocols like the Virtual Private LAN Service specified by RFC 4761 and RFC 4762 may be used as well for the interconnection of customer and provider networks.

Figure 2.19 shows the frame formats deployed by these Ethernet standards. The standards build on each other and set up a hierarchical structure for carrying Ethernet over ever-larger network infrastructures.

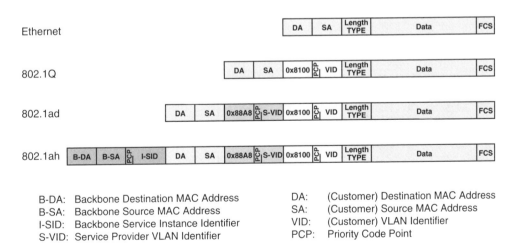

Figure 2.19 Ethernet frame formats

2.4.3 Ethernet Services Support in Mobile WiMAX

Primarily designed for access to the Internet and for delivery of IP-based services, the Mobile WiMAX reference architecture is also suitable for Ethernet services. The distinction between the operator role of a NAP and a NSP, respectively, as well as the definitions of the ASN and the CSN as logical network representations, are applicable for Ethernet services (Figure 2.20). The main difference to the IP services network architecture is the modification of the data path between the MS and the CSN for the transport of Ethernet frames. The protocols chosen for the data path for IP Services can carry Ethernet as well.

The transport of Ethernet frames over the IEEE 802.16 radio interface between the MS and the BS requires the deployment of the Ethernet specific part of the packet convergence sublayer (Ethernet-CS) instead of the IP-CS. The Ethernet-CS supports the transport of Ethernet frames within IEEE 802.16 MAC frames and extends the classification capabilities of the CS to Ethernet-specific header fields like source and destination MAC addresses, Ethernet priority and VLAN-ID, in addition to all the IP header fields supported by the IP-CS.

GRE is used as the tunneling protocol between the BS and the ASN-GW, which supports the encapsulation of Ethernet frames as well as the encapsulation of IP packets. GRE is also a tunneling protocol option for the MIP data path between ASN-GW and CSN. Ethernet frames are transferred in the payload of GRE when Mobile IP is applied for dynamic tunnel management between the ASN-GW and the CSN. WiMAX makes use of an optional extension to the Proxy Mobile IPv4 (PMIPv4) protocol to signal the mobile host with its 48-bit MS-ID instead of an IPv4 address for enabling this unusual deployment of PMIPv4.

When dynamic tunneling is not required any Ethernet transport protocol can be used between ASN-GW and CSN, which is able to forward Ethernet frames based on the MS-ID instead of the destination MAC address, and which preserves the customer VLAN ID assignment. Forwarding based on MS-ID instead of destination MAC address is required because Ethernet frames contain destination MAC addresses other than the MS-ID, when the MS is not the end station but contains a bridge forwarding the frames to end stations behind the MS.

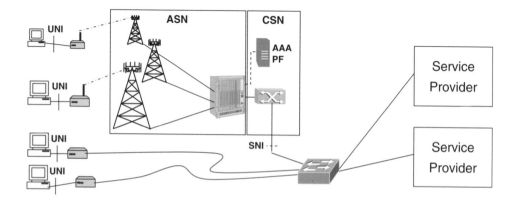

Figure 2.20 Wireless access to Ethernet services

2.4.3.1 The Role of the CSN for Ethernet Services

For Ethernet services, the CSN contains the bridging function for forwarding Ethernet frames between MSs and the Ethernet service provider. The CSN establishes for each of the connected MSs a dedicated (virtual) bridge port to ensure compliance with standard bridging behavior. In alignment to the behavior of IP services all Ethernet payload frames going over the radio interface are forwarded across the ASN to a bridge port in the CSN to enable the NSP to take full control of the traffic of the subscribers.

With the ASN just building Ethernet virtual connections between the MS and the CSN, the configuration of the bridge in the CSN determines which kind of Ethernet service is provided to the customer. An E-Line service is realized by forwarding the traffic of a MS to a single other end station either on the wireless side or on the wired side of the network. E-LAN service involves multiple other end stations, and the applied forwarding behavior in the bridge determines whether the enterprise LAN behavior or an E-Tree-like public access mode is realized.

The bridge does not have to cope with mobility issues in the access network, as it is located behind the mobility anchor in the CSN, which manages the relationship between the MS and its assigned bridge port. From the perspective of the bridge, the MS always appears on the same bridge port regardless of the physical location of the MS in the access network. Support of Ethernet services over a cellular infrastructure including mobility support does not impose any new requirement on the centralized bridging functionality in the CSN.

Even when mobility support itself is not demanded, the control functions of a mobile network provide benefits, such as the instantaneous provisioning of any kind of Ethernet service to any place in the access network, which are unknown in current fixed networks.

2.4.3.2 Reuse of the Mobile WiMAX Control Plane

The approach to realize Ethernet services by just extending the data path in Mobile WiMAX for forwarding Ethernet frames and by centralizing the bridging function in the CSN allows the nearly complete reuse of the control plane of IP services. Only minor modifications are necessary in the control plane in addition to the extension of the data path to support Ethernet services in Mobile WiMAX. Modifications to the control plane of IP services are defined for the following functions:

- For **AAA**, the authentication procedures are applicable to Ethernet services without modification, but authorization is extended by Ethernet-specific parameters for the definition of the service. The accounting model for IP services, which is based on the status of the IP session, is not applicable to Ethernet services. However, the accounting procedure itself can be reused. Accounting of Ethernet services is triggered by the status of the Ethernet connectivity, which is tied to the establishment of the link layer session and the service flows over the Ethernet CS.
- **Policy control and QoS** are extended by Ethernet-specific attributes to support the classification and the QoS configuration of service flows carrying Ethernet frames. A completely new function for Ethernet services that does not exist for IP services is the processing of VLAN tags in upstream and downstream directions according to configuration parameters provisioned as part of the authorization process.

- The **PMIPv4** specification is amended to make it applicable for the transport of Ethernet frames over PMIPv4-managed tunnels. A new attribute is deployed in the specification to allow signaling of the MS-ID in the Mobile IP.

All the other functions of the Mobile WiMAX control plane are used for Ethernet services without any modifications. Only authorization, accounting, QoS management and CSN-anchored mobility management require Ethernet-specific extensions, which will coexist with the related IP-specific functions inside the same access network.

2.4.3.3 Coexistence of Ethernet Services with IP Services

The coexistence of Ethernet services with IP services is highly important in allowing a Mobile WiMAX access network to support IP services as well as Ethernet services concurrently over the same infrastructure.

Support of Ethernet does not require a new network architecture or any new functional entities in the network. The support of Ethernet services is smoothly integrated into the Mobile WiMAX network architecture and Ethernet services can be operated in parallel to the IP services in the same network, as depicted in Figure 2.21.

The alignment of the Ethernet service provisioning model as closely as possible with the model for IP services not only reduced the necessary standardization efforts for developing the specifications, but also enables the development of WiMAX network equipment, which

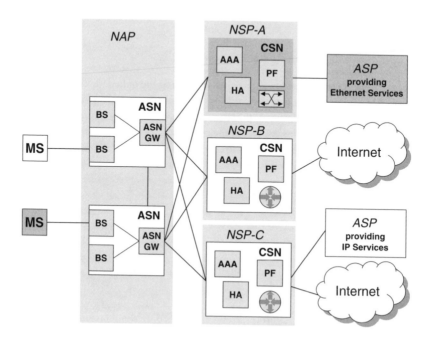

Figure 2.21   Coexistence of Ethernet services with IP services

serves IP as well as Ethernet from the same box. Such equipment is suitable for any kind of WiMAX access network and provides high commonality, which finally leads to better scale of economies. Moreover, it prepares operators for the extension of their services portfolio into broader telecommunication markets.

With the biggest part of the investment of a network operator going into the access infrastructure, it is important to enable the most flexible deployment of the access network. Participation in the growing market for Ethernet services without the prerequisite of huge investment may contribute considerably to a sustainable business case for a wireless broadband access provider. Mobile WiMAX supports an integrated approach to provide fixed, nomadic and mobile IP services and Ethernet services from the same network infrastructure.

2.4.4 Deployment Options

2.4.4.1 Metro Ethernet Services

Even when it is relatively easy to become a WiMAX service provider by implementing just a single CSN, dealing with the operation of a Mobile WiMAX network and the effort necessary to establish all the contractual frameworks to make use of the wholesale access connectivity of WiMAX NAPs may exceed the capabilities of some enterprises and smaller operators. Such enterprises, which are looking only for some point-to-point connectivity in areas where they do not have a wired infrastructure, may be very interested in the deployment of metro Ethernet services. Mobile WiMAX is able to provide transparent Ethernet connectivity over a wireless access network in exactly the same way as metro Ethernet network providers are offering wired Ethernet services today.

Metro Ethernet network providers may extend their business by implementing the CSN functions for Ethernet services in their networks and may become wireless Ethernet service providers without the heavy investment required to establish their own wireless access infrastructure, by leveraging the wholesale WiMAX connectivity offered by NAPs.

2.4.4.2 Wireless DSL Networks

The established DSL networks deploy Ethernet services based on Q-in-Q technologies in the access network to aggregate customer traffic, to segregate different services and to provide even transparent LAN connectivity with support for customer-assigned VLAN-IDs.

The Broadband Forum specification TR-101 [137] describes the Ethernet-based access network architecture supporting multiple configurations to connect customer equipment and customer networks to the services of a DSL operator. The specification describes the processing and forwarding of Ethernet frames between the T reference point to the customer and the V reference point to the broadband network gateway (BNG). To facilitate different modes within the same access network infrastructure, C-VIDs and S-VIDs are used in a DSL-specific manner:

- For **business TLS users**, a unique S-VID is assigned at the T interface and preserved throughout the network. For this kind of service the DSL access network strictly follows the recommendations in IEEE 802.1ad for the usage and assignment of the VLAN tags.

- For **business or residential users** with E-Line access to the DSL services, unique S-VIDs or a unique C-VID/S-VID combination are assigned and preserved throughout the access network depending on scalability requirements or usage of the C-VID by the customer.
- For **residential users** with shared access to the DSL services (E-LAN, public access mode) unique S-VIDs are assigned to groups of users or to all users subscribing to a particular DSL service. C-VIDs are not used in this case.

When Mobile WiMAX is used to provide access to DSL services over a wireless infrastructure, the WiMAX network emulates the behavior of the DSL access network and provides Ethernet connectivity between the customer interface and the V interface in the same way as the Ethernet-based DSL aggregation network. Figure 2.22 shows the combined network architecture of DSL and Mobile WiMAX for the case of interconnecting over the V interface.

Ethernet connectivity across the V interface supports IP services as well as VPLS for business customers. This internetworking scenario requires Ethernet services support in the Mobile WiMAX access network.

Details of the Ethernet connectivity across the V reference point are defined in [137]. The V reference point does not comprise a control protocol part but only describes a data path. The R3 interface of the simple ETH architecture of Mobile WiMAX specifies only the control part but does not make any statements on the data path. This allows interconnection of the simple ETH Mobile WiMAX network on the R3 reference point directly with the V interface of the DSL network.

Due to the plain L2 connectivity across the Mobile WiMAX network, authentication of the DSL user is still performed by the DSL within the PPPoE protocol. The access line and its customer side termination are implicitly identified by the DHCP option 82 fields when IPoETH is deployed. User credentials as well as the authorization and accounting information for IP services are handled by the AAA server in the DSL NSP.

The WiMAX security framework is applied to establish secure Ethernet connections over the IEEE 802.16 radio interface based on device or user identities with the credentials stored

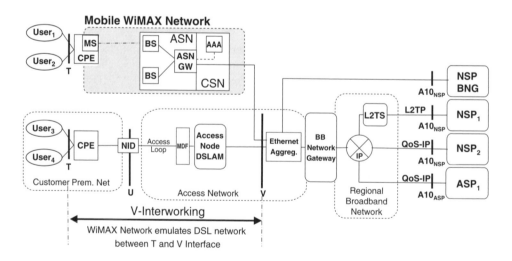

Figure 2.22   WiMAX integration with DSL access networks over the V interface

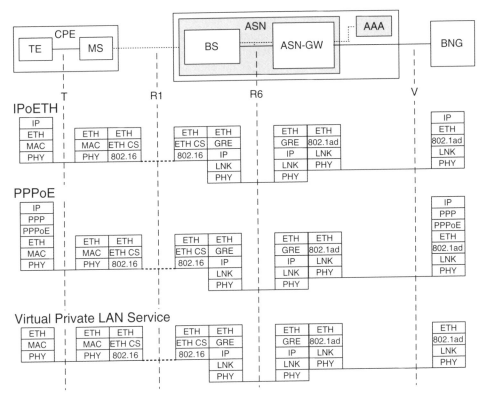

Figure 2.23 Protocol layering for V interworking

in the MS. The identity of the CPE unit is used to manage the provisioning and accounting of the Ethernet connectivity across the WiMAX network according to the profiles stored in the WiMAX AAA server. Essentially, there is no difference between device authentication and user authentication in the DSL interworking case. The identity is always pointing to a particular MS embedded in a CPE, and it makes no difference to the authentication procedure whether the identity is hard coded or soft coded in the MS.

Subscriber lines or particular services are segregated by stacked VLANs according to IEEE 802.1ad on the V interface, allowing aggregation of a group of subscribers into one VLAN as well as separation of subscriber ports to realize distinct Ethernet connections for VPLS for business customers.

Figure 2.23 depicts the protocol layering on the data path when Mobile WiMAX is deployed for wireless DSL access networks. The V interface can be directly mapped onto the R3 data path without any further network functions being needed in the CSN when the R3 interface does not deploy MIP for management of the connection.

2.5 Mobile WiMAX and 3GPP SAE/LTE

In response to the high interest of operators in a radio interface and network supporting broadband IP services, 3GPP started its activities at the end of 2004 on the long-term evolution

(LTE) radio interface together with the enhancement of the network architecture called system architecture evolution (SAE). The first complete set of specifications for LTE/SAE was published in spring 2009 as part of 3GPP Release 8.

This section provides a comparison between the network architectures of Mobile WiMAX and 3GPP SAE. After a short introduction to SAE with its network reference architecture, network entities and reference points, the main differences to the Mobile WiMAX architecture are pointed out. In addition to the technical differences in the architectures, the section concludes with a comparison of the network evolution paths for increasing coverage, throughput and business models.

2.5.1 Introduction to SAE

SAE is the name of the network architecture of the LTE mobile communication standard developed by 3GPP. The network is an evolution of the GPRS core network with a simplified, all IP-based architecture for higher throughput and lower latency radio access networks. In particular, it supports multiple, heterogeneous radio access networks including non-3GPP technologies in addition to legacy 3GPP systems, including mobility between the access technologies. The architecture for 3GPP accesses and non-3GPP accesses is specified in [139] and [140], respectively.

2.5.1.1 Reference Architecture

Figure 2.24 shows the architecture reference model of SAE for the non-roaming case.

The SGSN and the reference points S3, S4 and S12 represent support for legacy 3GPP radio access systems.

In addition to the non-roaming architecture, the roaming architecture distinguishes two scenarios. Either home routed traffic is supported by connecting the serving gateway in the visited network over a S8 reference point to the PDN gateway in the home network. Or, alternatively, a local breakout scenario is applied with access to the IP services in the visited

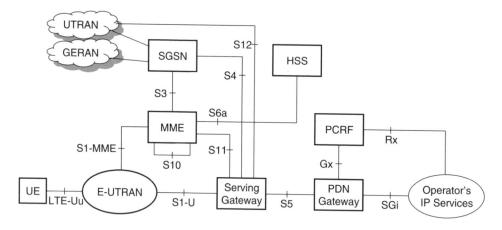

Figure 2.24   Non-roaming architecture for 3GPP accesses

network and a connection to the PCRF of the home network over an interface labeled S9 with the PCRF in the visited network. In all roaming cases, the HSS function in the home network provides the authentication and security framework.

2.5.1.2 Network Entities

The evolved packet core (EPC) is the main component of the SAE. The EPC serves as equivalent to GPRS networks and introduces three new network entities called the mobility management entity (MME), serving gateway and packet data network gateway (PDN gateway), in addition to the home subscriber server (HSS) and policy and charging rules function (PCRF) already present in the previous versions of the GPRS.

The subcomponents of the EPC are:

- **HSS**: This consists of the HLR (Home Location Register), which contains the user identification (subscriber identity), addressing (phone number) and all related user profile and subscription information, and the AuC (Authentication Center), which generates the keys for mutual authentication between network and terminal (UE) and the ciphering information for encryption of user data over the air.

  The HSS in 3GPP is comparable to the authentication and authorization part of the AAA server in Mobile WiMAX, although the actual authentication and authorization procedures are conceptually different between 3GPP and WiMAX.
- **MME**: This deals with all subscriber and session-related control plane functions in the access network. It interacts with the HSS for authentication of the subscriber, and the retrieval and setup of the subscriber profile and ciphering keys. Based on the subscriber profile, it chooses the serving gateway for the initial attachment and intra-LTE handovers involving a relocation of the serving gateway, and it is involved in the bearer activation and deactivation process. It is responsible for the tracking and paging of idle mode terminals. In addition, it provides control plane functions for the mobility between GERAN/UTRAN and E-UTRAN, and access to the control plane signaling for lawful intercept.

  The MME in 3GPP overlaps with the control plane functions of the ASN-GW in Mobile WiMAX, although there are major differences due to the policy enforcement located in the ASN-GW.
- **Serving gateway**: This is the anchor of the data path for all mobility intra-3GPP access systems, i.e. intra E-UTRAN mobility as well as mobility between E-UTRAN and legacy systems based on UTRAN or GERAN. It handles data delivery for idle mode terminals and is involved in replication of user traffic in the case of lawful intercept.

  The serving gateway in 3GPP has similar functions to the data path portion of the ASN-GW in Mobile WiMAX.
- **PDN gateway**: This is the mobility anchor and termination of the bearers set up between the terminals and the packet data network providing the IP services. It comprises the policy enforcement function as well as packet filtering functions for deep packet inspection and accounting and charging. The PDN gateway also acts as anchor for the mobility between 3GPP and non-3GPP access networks.

  From a functional perspective, there exist some similarities between the PDN gateway in 3GPP and the HA or LMA function in Mobile WiMAX. Functional differences are mainly caused by the different locations of policy enforcement. For WiMAX networks, the default policy enforcement point is the ASN-GW.

- **PCRF**: This function integrates the policy decision function and the charging rules function by linking the subscriber profile with the QoS requirements of applications. It authorizes QoS resources and instructs the PDN gateway on how to proceed with the data traffic. The function was introduced in 3GPP in earlier releases to enable flow-based QoS control and charging, and was made mandatory for LTE/SAE.

In a Mobile WiMAX deployment with default AAA-based QoS control, the 3GPP PCRF has no direct equivalent. However, WiMAX can optionally deploy the 3GPP PCC framework accompanied by a PCRF. There are some differences as the WiMAX PCC specifications are based on the 3GPP Release 7 standards together with a number of WiMAX-specific extensions and conceptual differences like mobility support.

2.5.1.3 Reference Points

The reference points in LTE/SAE are:

- The **S1-MME** reference point represents the control plane protocol between E-UTRAN and MME.
- The **S1-U** reference point between E-UTRAN and the serving gateway defines the per bearer user plane tunneling and inter base station (eNodeB) path switching during handover.
- The **S3** reference point between MME and SGSN defines the user and bearer signaling information exchange for access network mobility across GERAN, UTRAN and E-UTRAN.
- The **S4** reference point between SGSN and the serving gateway provides control and mobility support between the GPRS access network and the anchoring function in the serving gateway. S4 also carries user plane tunneling, when direct tunneling between the UTRAN and the 3GPP anchor function in the serving gateway is not established.
- The **S5** reference point between the serving gateway and the PDN gateway provides user plane tunneling and tunnel management functions for the relocation of the serving gateway due to UE mobility, or for connection to a non-collocated PDN gateway according to the requested services.
- The **S6a** reference point forms the interface for user authentication and authorization between the MME and the HSS.
- The **S8** reference point is used in roaming scenarios as equivalent to the S5 reference point for the interconnection between the serving gateway in the visited network and the PDN gateway in the home network (not shown in Figure 2.24).
- The **S9** reference point provides an interface for the exchange of QoS policy and charging control information between the PCRF in the home network and the PCRF in the visited network, when services of the visited network are provided (3GPP nomenclature: local breakout function). The reference point is not depicted in Figure 2.24.
- The **S10** reference point provides an interface between the adjacent MME for relocation of the MME and inter-MME information exchanges.
- The **S11** reference point between the MME and the serving gateway supports the standardized split of control and user plane functions into two separate functional entities to allow flexible strategies for upgrading the radio access networks.

- The **S12** reference point carries user plane tunneling between UTRAN and the serving gateway in the case of a direct tunnel. In cases where S12 is not deployed, user plane tunneling is carried over S4.
- The **Gx** reference point supports the transfer of the QoS policy and charging control information from the PCRF to the enforcement function in the PDN gateway.
- The **SGi** reference point provides connectivity from the PDN gateway to service networks of the same or other operators, e.g. for delivering IMS services. It corresponds to Gi for 3GPP networks.
- The **Rx** reference point provides a control interface between the PCRF and the AF (Application Function) in the service network for the transport of application level session information [141]. Rx can identically be used for a WiMAX deployment with PCC support.

2.5.2 Architectural Differences

Despite the similarities between LTE/SAE and Mobile WiMAX, there are a number of essential differences in the network architectures, which are illustrated in this section.

2.5.2.1 Interoperability

The LTE/SAE architecture is specified on a functional base with reference points defined between functional entities. The specification makes no assumptions on how the functional entities can be combined for implementation.

The specification of Mobile WiMAX focuses on interfaces between business entities (R3, R5) and interfaces between real network elements (R4, R6, R8). The structure of the reference points in Mobile WiMAX is aligned to real interconnections, with multiple protocols and functions running in parallel across the same interface. In addition, certification and interoperability testing are part of the specification work of the WiMAX Forum.

2.5.2.2 Network Sharing

The LTE/SAE architecture comprises all functions of a network providing mobile telecommunication services in a uniform manner without any operational boundaries inside the network.

The Mobile WiMAX network architecture assumes a split of the service provisioning into different operator roles of NAP, NSP and ASP following the established models in the fixed broadband Internet market. By splitting the access network operation into NAP and NSP portions, Mobile WiMAX introduced a standardized model for sharing the radio access infrastructure, which is of special interest for regulators as well as for service providers with limited access to spectrum or less focus on the operation of their own wireless access network.

The differentiation between NAP and NSP introduced in Mobile WiMAX led to the location of the policy enforcement and accounting functions in the ASN-GW to allow the access operator to maintain control over the assignment of the radio resources. Policy control became a part of the subscriber information provided to the NAP, when the subscriber enters the network, and authentication, authorization, policy control and accounting are integrated into a single AAA interface.

2.5.2.3 Core Network Requirements

The LTE/SAE architecture assumes that the service provider operates an enhanced 3GPP core network with HSS and PCRF as mandatory functions.

Mobile WiMAX has a less demanding approach to allow a broader variety of service providers to leverage a Mobile WiMAX access infrastructure. At the low end, an access router and a RADIUS server are sufficient to start a Mobile WiMAX service offering, eventually enhanced by a HA to enable wide area mobility. Nevertheless, Mobile WiMAX has the possibility to upgrade to network functionality in the core comparable to LTE/SAE, if such complexity is required for the particular service offering.

2.5.2.4 Trust Architecture

The LTE/SAE trust architecture is designed with the SIM card as the core of the trust architecture. It is assumed that every terminal contains a UICC for the authentication and generation of the ciphering keys.

While Mobile WiMAX is also able to handle UICC as subscriber credential, the trust architecture is built around EAP as the universal authentication protocol supporting any kind of credentials, from simple username/password combinations up to highly complex schemes based on certificates. The architecture is designed such that the security of the access network is independent of the kind of subscriber credentials used by the NSP. Service providers are able to differentiate their subscriber security level without any effect on the complexity in the radio access network, e.g. embedded devices like a digital camera with instant uploading of pictures may be secured by other means than a mobile phone.

2.5.3 Deployment Differences

The design choices made for the mobile network architectures affect the deployment options of the technologies. It makes a huge difference whether the architecture is aimed at adding another radio access layer to fully established mobile networks, like LTE/SAE, or is aimed at allowing operations to start new businesses from scratch, cover new regions, open the pipe to the Internet or address new applications.

2.5.3.1 Entry Burden

Starting a new business is always risky. The higher the initial investment, the riskier the business, in particular when entering new business domains with no guarantee that the venture will be successful. Setting up a new kind of network requires careful planning and the possibility to test-drive the new infrastructure before making the final decision.

The Mobile WiMAX network architecture was designed to enable small deployments with stepwise upgradability to nationwide networks. The initial network consists of a couple of base stations, an ASN-GW and an AAA server. It does not yet provide all the possible functions, but it allows services to standard terminals.

LTE/SAE has a much higher entry burden. More complex network equipment is necessary to establish an initial deployment.

{
1.
2.
3.
4.

800
1200 = 6000
1600
800

Practice (2)

- Read the abstracts of the two articles.
- Identify the common features of a good abstract. Tick in the table below:

Features	Abstract 1	Abstract 2
Topic	✓	✓
Purpose		
Scope and questions	✓	
Methods	✓	
Research results	✓	✓
Conclusions		
Recommendations		
Others		

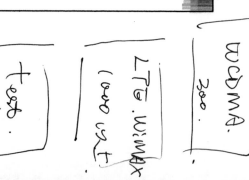

2.5.3.2 Area Growth

Depending on the equipment, a usual ASN-GW may serve up to about 100 base stations, already sufficient to cover a midsize town or a larger region. If larger areas are to be covered, more base stations and more ASN-GWs can be added. Each ASN-GW may build an ASN and defines its own coverage area, or several ASN-GWs can act as a cluster within the same ASN. Regardless of the deployment scenario, the R4 interface between the ASN-GWs enables seamless mobility of terminals across the ASNs without influencing the connection between the NAP and NSP. The R3 interface allows reallocation of the anchor of a terminal to an ASN-GW nearer to its physical location, which makes sense, when an appropriate backbone is in place.

LTE/SAE follows a similar path when extending the coverage area. More base stations and more MME and serving gateways are installed. As one difference, splitting the access controller into MME and serving gateway usually requires two new network elements instead of a single element as in the case of the ASN-GW. Furthermore, LTE/SAE does not support the direct interconnection of serving gateways to forward bearers to the initial (anchor) gateway. Handover between gateways always causes handovers on the S5 reference point.

2.5.3.3 Capacity Growth

Growing numbers of users as well as increased bandwidth demand in a coverage area require more and more transfer capacity over time. Capacity is primarily limited by the design of the base stations, and even with much more advanced technologies the increase of the throughput and the spectral efficiency have reached their limits. When the technologies are near to the physical limits of radio transmission, more capacity requires either more spectrum or more base stations. Regulatory constraints usually restrict access to more spectrum, so increasing the number of cells in a coverage area is usually the way to go. More base stations require more forwarding capacity in the access gateways.

To allow the upgrade of user plane capacity without the need to replace the control plane server when mainly user plane throughput has to be increased, LTE/SAE adopted the separation of the control plane server and the user plane server of the access gateway into the MME and serving gateway. The separation sounds beneficial for cost-effective upgrades of the throughput, but has issues in real deployments. One aspect is the introduction of an interface between two highly complex network elements, which not only causes additional complexity in the network elements, but also requires more implementation effort and interoperability testing, in particular when multiple vendors have to interwork.

Replacing the serving gateway to increase throughput may become cumbersome after a couple of years as standardization progresses, and an upgrade to the latest release may be required for the MME before being able to install a new serving gateway. The difficulties in upgrading a multi-vendor combination of MME and serving gateway may outpace the efforts to replace integrated access gateways, when the MME and serving gateway are kept together in one network element.

After lengthy discussions during the initial design phase of the network architecture, Mobile WiMAX took the latter path, finally following the experiences of the fast-growing Internet; in IP routers, separation of the control plane and user plane is rarely seen.

2.5.3.4 Business Growth

So far the mobile communication business has followed the model of a subscriber with one device for accessing the network services of the mobile operator. When the mobile communication business reaches saturation because nearly everybody has a mobile subscription and competition reduces the usage fees of the established mobile services, operators can either generate new revenues by selling new services or reduce the cost of network operation.

In addition to what a traditional mobile network like LTE/SAE is able to offer, Mobile WiMAX provides the following options to extend business and increase profitability:

- **Network sharing**: By sharing the radio access infrastructure among multiple operators, a single infrastructure can be leveraged for multiple operators without affecting the interface between service provider and subscriber.
- **Wholesale business**: Due to usage of standard ISP equipment for the service provider's core network, many more entities can establish their own network service, e.g. governmental or community organizations, enterprises, public security.
- **Low-cost subscriber identities to allow for multiple devices per subscriber**: Also known as the $50+$ billion devices challenge, this allows any EAP method to be used for authentication of subscribers with RADIUS servers, providing the lowest cost per identity.
- **Access paid by application provider**: The provisioning and accounting model supports charging of the application provider instead of the user of the terminal. It allows bundling of terminals with paid services offered by application providers directly to customers.

3

Subscription Handling and Security

In this chapter, in addition to developing the overall security model in WiMAX networks and numerous related methods, the focus will be on a small collection of data describing a 'subscription'. This will especially allow for elaboration of the quite important role that the concept of a subscription plays in WiMAX security, as well as of subscription handling in relation to security. Motivated by this, network access security in Mobile WiMAX will be analyzed, with the focus on understanding the building blocks that are provided by the standards and how they impact deployment decisions. One additional building block that is of key importance for the overall business case of WiMAX operators and that reflects the different business models compared to existing mobile cellular networks is over-the-air provisioning of a subscription, which will be presented from both a technical perspective and looking at its business importance.

3.1 The Meaning of a Subscription

From a network operator's perspective, and this especially relates to the WiMAX NSP operator as – per WiMAX specifications – the one being solely responsible for providing the 'home' network services to its customers, the most valuable resource of its business today is of course its customer base. The set of a customer's profile and policy settings that are based on an actual contract allow control of the use of the services offered through the operator's network facilities to the customer. In technical terms, everything is distilled into this small but important collection of data. It actually represents the subscription across the operator's network entities. It brings the 'subscriber' to life and is actively involved when granting access to the network resources and services that the subscriber has 'subscribed to', independent of whether a subscription relates to a human user, a device or a machine.

From a human user's perspective, the subscription is analogous to a specific contractual relationship between this person and a particular operator. Specifically, it describes the set of network services that can be used, covers a basic set of technical configuration parameters like the set of devices allowed to be used with the subscription, relates service use correctly to the user's account, and contains all data required to actually use the service.

Many experts, after dealing with the problem of how to best define a subscription, for the sake of simplicity end up with merely a one-to-one relation between the subscription and a

Deploying Mobile WiMAX   Max Riegel, Dirk Kroeselberg, Aik Chindapol and Domagoj Premec
© 2009 John Wiley & Sons, Ltd

specific billable account in the operator's system. The simple logic behind this is that there is always an account through which the bills can be settled after the subscriber has consumed services that the operator is charging for. However, as shown by the above discussion, this certainly is an oversimplified approach that also does not accommodate the end customer's view well. Also, it is important to note that a billable account might better be related to the contract that a user agreed with the operator, as it can serve as more than just a single subscription.

Throughout this book, we refer to a subscription from a primarily technical view mainly as the set of permanent or semi-permanent data that is required in both the WiMAX device and the network. This is to make the overall system work for both sides as soon as the subscriber – being the owner of the subscription insofar as it is the entity finally receiving the bill, just to make use of the oversimplified view for the sake of simplicity here – starts connecting to a network.

Let us consider some different types of subscriptions without going too deeply into technical details:

- Human user subscription stored on a universal subscriber identity module (USIM) smartcard inserted into a mobile phone (and therefore not bound to the actual device, with exceptions like an operator-subsidized device that the subscription is locked to). This example is listed at the top due to the fact that the number of GSM/UMTS mobile phone customers surpassed 2 billion in 2006, according to the GSMA organization (www.gsmworld.com).
- Device subscription in a DSL modem. The device can be used by several users and the operator, according to today's deployed technology, has no means of identifying human users 'behind' the CPE device, besides the contracted one paying the bill.
- Human user with several identities mapping to the same subscription. A subscription is always bound to a specific identity, but can also accommodate several identities under the same subscription (e.g. business/private). Alternatively, there can be independent subscriptions in this case, possibly mapped to the same user account in the operator's backend system as soon as it comes to billing.
- Human user operating several devices with different identities, mapping to the same subscription (e.g. laptop, smart phone, voicemail). Here, a single subscription is used across several devices and several such devices could be active at the same time.

Section 4.3.1 elaborates further on this and provides the specifics of a WiMAX subscription in the context of secure network access.

Leaving aside a pure WiMAX perspective here, the network that a subscription is being used to access does not necessarily have to be a WiMAX network. It might be a 3GPP2-defined cellular or a WLAN network. Also the home network and therefore the subscription itself does not necessarily have to be a WiMAX one. An example would be a UMTS subscriber with a handset equipped with an additional WiMAX radio that is roaming through a WiMAX network in an area without UMTS coverage, to get access to its home operator's services. Many more examples are readily available in this area.

Finally, as part of this motivation for aspects related to a subscription, user experience is one of the most important factors as soon as it comes to the practical handling of subscription data. Today, most users are experienced with handling subscriptions in two different areas that are in the process of moving closer together – or trying to dig into each other's domains – both in the operator's infrastructure and on the device side, and WiMAX might play an especially interesting role here.

The first example considers mobile phone networks like GSM/UMTS or CDMA. These systems are often seen as a 'walled garden' model, where many areas are under the tight control of the operators, communities of operators are well established to provide for global roaming, and also subscription handling and security typically follow centralized models. Clearly, the global success of these systems can partially be related to the centralized subscription handling where, after one-time installation, the end user has nothing to worry about besides entering a PIN code when switching on the mobile. Especially in the GSM case, the user does not even need to worry about this when traveling the world. It will be sufficient to just switch on the phone when leaving the plane for most cases to get a basic service (especially for those who do not care too much about their bill).

In summary, the effort for the end user regarding handling the subscription is minimal, which has led to broad adoption. However, the range of services used in practice is quite limited today, although the subscription in principle could be used as a central one for a broad range of operator-provided services.

Looking at the second example, the Internet in contrast is considered a very open and heterogeneous community with a diverse services landscape offered by a wide range of ASPs. These ASPs operate largely independently of each other. This has led to maximum diversity in the services offered due to an environment that allows rapid realization of new ideas. The result is of course beneficial to the end user, who is now able to choose among a broad variety of novel offers. On the other hand, strictly concerning security issues, the open model also led to a fascinating variety of user accounts, i.e. subscriptions, where most standard Internet users are required to handle dozens of usernames and passwords, thereby creating a huge burden for the end user. In addition, it also created huge amounts of work on identity management solutions that are trying to improve the given situation, most of them struggling for wide adoption.

Considering WiMAX, the above discussion shows that subscription is about the subscriber, and the subscriber can be seen as the precious resource of an individual operator. Business environments and the operator's or service provider's role in different markets vary a lot. A WiMAX subscription is expected to primarily be a ticket to the wireless access and required network services like IP mobility or QoS support, but with much less involvement in the applications space.

However, the WiMAX model for subscription handling and security has been developed to accommodate a wide range of deployment choices and is clearly not limited to one specific ecosystem.

3.2 A Network Reference Model for Security

In the standard WiMAX network reference model, the access service network (ASN) includes the access functions for network access. It consists of WiMAX base stations (BSs) and introduces the ASN gateway (ASN-GW) as a central controller unit that – depending on the actual dimensioning of the ASN – may control everything from a single to a few hundred base stations. So the ASN roughly corresponds to the radio access network of cellular systems, although optimized for broadband data instead of cellular voice.

Security functionality specific to the ASN includes – besides the network-side termination of the 802.16 radio interface security – centralized handling of access security in the ASN-GW that acts as a network access server (NAS) and has security-related procedures for ASN-based mobility. This potentially spans a rather large geographical area including a large number of BSs.

The connectivity service network (CSN) provides IP connectivity and contains network entities like the tunnel endpoint for all user traffic between ASN and CSN that may also be the anchor for IP mobility, DHCP server, AAA infrastructure and subscriber management functions. It corresponds to what is typically known as the core network.

For security, the CSN is mainly responsible for authenticating and authorizing all WiMAX subscribers. This of course includes all identity management-related procedures.

One decision that arose when developing the WiMAX specifications has been to base the architecture on reference points instead of interfaces. A reference point defines a generic connection between high-level building blocks like ASN and CSN and typically groups a number of different control protocols and data together that are exchanged across the reference point. At the highest level of abstraction, R3 connects ASN and CSN, or a NAP to a NSP. R5 as the important reference point for roaming between NSP operators connects two (visited and home) CSNs and may be considered as a subset of R3. Within an ASN, important reference points are R6, connecting the BS and ASN gateway, and R4, between two ASN gateways.

All these reference points are responsible for supporting a certain set of security-related control signaling, and have their specific security requirements and mechanisms for protection (referred to as 'reference point security').

As soon as we start focusing on security aspects and consider the related functions in WiMAX, a slightly more detailed version of the network reference model can be drawn, as shown in Figure 3.1.

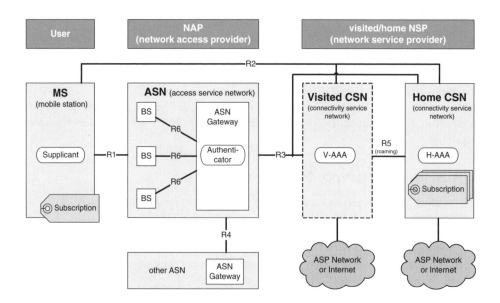

Figure 3.1 Network reference model for security

For securing access, network access authentication in WiMAX builds on the Extensible Authentication Protocol (EAP) framework [3] and the RADIUS protocol [6]. It uses the common three-party EAP/AAA model for authentication and authorization, involving the supplicant (Supplicant) in the device (MS), the authenticator in the ASN that plays the role of a network access server, AAA proxies in the visited CSN (V-AAA) when the device is roaming, and the AAA server in the home CSN (AAA or H-AAA when roaming).

The EAP itself only specifies a generic 'container' and a well-defined state machine that is capable of exchanging the actual authentication protocol messages. Such instances of authentication protocols are called EAP methods and are represented by R2 in the reference model. This architecture results in a potentially large variety of EAP methods that can be used for authenticated access to WiMAX networks (although only a small set of EAP methods is relevant for existing deployments).

Depending on the EAP method, further decisions by WiMAX operators are necessary for the actual method used for performing authentication in their network. This is always mutual according to the WiMAX specification, and may be based on:

- shared secrets (or passwords as a special case) that can be stored in the MS or on a removable smartcard (SIM);
- public/private key pairs and certificates; or
- hybrid solutions using TLS for server authentication and passwords or shared secrets for subscriber authentication, similar to the common https Internet model for accessing a secure website.

WiMAX supports a number of different authentication modes, such as user or device authentication. Also, both can be combined into a single authentication run during network entry.

EAP is carried between supplicant and authenticator across the R1 reference point by the 802.16 MAC layer and across the R6 reference point by a WiMAX-defined protocol. The base station acts only as a relay here, without taking part in the actual EAP message exchange.

To exchange EAP messages during access authentication between the authenticator and the (home) AAA server holding the actual subscription, RADIUS or Diameter [11] are used across R3, and across R5 in roaming cases.

3.3 Subscription versus Device Authentication

For secure access to network resources across a wireless link, an initial authentication and authorization phase – in WiMAX based on EAP methods – needs to be executed between the device attempting access and an authentication server in the core network responsible for granting such access. Such authentication is based on security credentials that are available prior to entering the network, or the network will request the device to enter a procedure for initial provisioning that results in such credentials, which are referred to as over-the-air (OTA) provisioning, see Section 3.6.

Assuming the former case, a subscription needs to be available at the device side as well as in the authentication server in the network. In the device, security credentials, as part of the subscription, are securely stored, e.g. in tamper-proof storage within the device or in a tamper-proof removable module inserted in the device.

3.3.1 Subscription Types and Security Credentials

Common wireless and cellular access technologies typically perform cryptographic authentication for a subscription that directly maps to a human user, but do not authenticate the device itself. Examples of security credentials for wireless user subscriptions are:

- SIM and USIM in GSM/UMTS cellular networks that store symmetric keys and perform cryptographic algorithms for authentication within a removable smartcard module (UICC).
- Username/password accounts commonly used in WLAN environments.

The main difference between WiMAX and these user-centric mechanisms is that the WiMAX device also can be authenticated based on cryptographic mechanisms. Otherwise, the common forms of subscription security credentials like username/password, shared secret, smartcard based (e.g. SIM), or public/private key based are all possible in WiMAX network access. As an example, existing deployments of Mobile WiMAX broadband access in South Korea and in the United States are using different approaches like a WiMAX-SIM smartcard-based subscription, or a public/private key pair and certificate in the device. The latter does not differentiate between the subscription and the actual device because it is bound to the device. This indicates that the actual choice of how to realize WiMAX subscriptions, regarding what is actually handed out to the subscriber, vary a lot based on specific operator requirements and processes to hand out such subscription credentials to the actual user or device. On the one hand, this may be considered an advantage, and the existing specifications allow this flexibility on purpose. On the other hand, it limits flexibility regarding the choice of devices. Not every device will support all types of subscription credentials (Figure 3.2).

Subscription authentication typically ensures – from the operator's perspective – that the user holding a WiMAX subscription, while using network resources and mobile services possibly based on WiMAX access, can be uniquely and securely related to the subscription. Simply speaking, the operator needs to make sure the bill gets paid.

In contrast, device authentication serves a different purpose unless the subscription is a pure device subscription, e.g. in a CPE device, and no user authentication is additionally performed. The main advantage of authenticating the device in addition to a subscription installed on the device is that the verifying operator learns the cryptographically verified device identity. This can subsequently be used to support mechanisms that for example ensure the device is not a stolen one, or even help to identify misbehaving devices that might try to access the network by

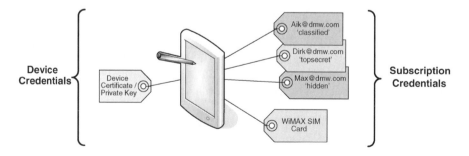

Figure 3.2   Security credentials in a WiMAX device

spoofing the identity of another device. Further use cases may include methods based on the device authentication that would allow the operator to verify that a device entering the network comes with appropriate certification for a specific WiMAX specification release, firmware version, etc. However, the latter use case typically tends to result in considerable complexity and conflicting practical aspects and can be considered hard to realize in practice.

A clear advantage of devices that are already shipped with preinstalled means for performing device authentication is that initial provisioning, i.e. creating a subscription on first use of the device with a selected operator, becomes more secure and convenient for both the operator and the user. An overview of how device authentication is used in OTA provisioning of WiMAX devices is provided in Section 3.6. Finally, there are countries where national regulation mandates mobile network operators to provide means to securely identify a mobile device when entering their network. This does not necessarily require cryptographic device authentication, but would surely benefit from the resulting authenticated device identity.

In Table 3.1, the relation between the type of authentication performed during network entry and the subscription installed in the device is summarized. The table takes into account what is specified for WiMAX networks according to [13].

It can be determined that for WiMAX networks, subscription authentication is always mandatory for gaining unrestricted access to the network, and device authentication may optionally be performed depending on the policy applied by the WiMAX NSP and on the device's capabilities.

In cases where no valid subscription is installed on a device, e.g. for new devices that are taken into use, device authentication will be performed to 'bootstrap' the process of secure initial provisioning to install and activate a new subscription for a specific WiMAX NSP (see also Section 3.6).

Let us take a quick look to see where other network access technologies stand today to get a feeling of what the actual difference of WiMAX is compared to state-of-the-art mechanisms. Examples are:

- 3GPP networks, where the subscription stored on a removable USIM card typically maps to a human user. Device authentication happens only implicitly by verification of the mobile terminal's identity (IMEI), but no cryptographic mechanisms for verifying the IMEI are used and in turn most of the security relies on secure storage of the device identity in the mobile phone, so it cannot be modified in the device prior to being sent to the network. Also, a user can plug the USIM card into different mobile terminals, so the user's subscription and identity are not necessarily bound to any specific mobile device.

Table 3.1 Relation of authentication and subscription type

Subscription type installed	Authentication performed	
	Subscription authentication	Device authentication
User subscription	Mandatory	Optional
Device subscription	Mandatory	Optional but implicitly performed through subscription authentication
No subscription	Not available	Mandatory to secure initial provisioning

- WLAN networks, where only subscription authentication is available for the Wi-Fi Forum's WPA and WPA2 profiles. The WLAN device identity – the MAC address – cannot be cryptographically verified and in fact it is typically very easy to change or spoof the MAC address of a WLAN device.
- DSL or fixed WiMAX access [4], where the subscription relates to the device only. In such environments, any user is usually authenticated through an independent step after the device has gained access to the network, typically based on username/password combinations.

One may question the effort that the WiMAX Forum took to develop cryptographically strong mechanisms to authenticate devices. The motivation for investing additional effort here becomes clearer when looking at the expected business models for Mobile WiMAX, where device authentication is clearly more advantageous compared to existing mobile cellular networks like 3GPP. Devices like standard off-the-shelf laptop computers can no longer be expected to be under full operator control. Hence, it is clearly beneficial to introduce cryptographically protected device authentication procedures that are based on digital certificates imprinted into the device during manufacture.

Certainly, the overall security level achieved by this largely depends on how well a certain number of key questions are answered, like secure storage of keys and trusted certificates on the device, and the PKI (Public Key Infrastructure) system's effectiveness in handling aspects like certificate revocation in a possibly very mobile environment.

The WiMAX Forum put significant effort into creating an appropriate PKI environment including appropriate profiles for digital certificates, root certificate authorities for device- and server-side certificates, and revocation mechanisms. Every device produced by a manufacturer that complies with the WiMAX Forum official specifications and processes comes with a device certificate and root certificates imprinted during manufacture and is therefore fully prepared to perform device authentication (see Section 3.4 for further details).

3.3.2 Authenticated Network Access

As motivated by the above section, Mobile WiMAX networks strongly rely on the IETF-defined EAP/AAA model for providing secure access to network resources for authorized subscribers. The same framework is already known and might be considered a well-established technology for securing access to WLAN infrastructure – although in many cases this is still only true as far as implementations are available in WLAN equipment. It is not true when looking at what is actually being used in reality. A simple check is to look at the authentication method used in most WLAN equipment. Most likely, it will be based on simple pre-shared keying (PSK) instead of EAP.

However, there are clear reasons why this is different for WiMAX. Many typical WLAN use cases do not include central authentication infrastructure as being available in large mobile operator's networks. The use cases often are still called 'fly-by access', with common examples being Internet access during a hotel stay or just for a brief stop at a coffee shop or an airport. Also there is no need to deploy EAP-based authentication requiring more costly infrastructure in home scenarios compared to simple PSK authentication. All these use cases might benefit, but typically do not require a permanent subscription for secure access that would justify EAP usage. Also, when traveling worldwide, as heterogeneous as the experienced cultures are operators providing WLAN access and ISP services. It is still a rare case where a WLAN 'subscription' can be used on a different international business trip.

In contrast to this, WiMAX understands a subscription as a central prerequisite for being able to access network resources and services. Even more important, WiMAX networks are expected to consider roaming a critical success factor. Clearly, EAP-based authentication complemented by an AAA protocol with full support for roaming between networks is the choice here.

Now let us take a more detailed look at the procedure for authenticating a mobile or fixed subscriber and network during initial network entry, while trying to elaborate the main differences in the role of EAP in WLAN access authentication.

Standard EAP as defined by the IETF can be performed in the simplest case as a two-party protocol, just involving a supplicant or EAP peer (typically located in the end user device) and an EAP server that is integrated into the authenticator. An example might be a laptop computer running the EAP peer and the WLAN access point implementing authenticator and EAP server functions.

WiMAX is using EAP as a three-party protocol. Here, the authenticator becomes an EAP pass-through that is only relaying EAP messages between the EAP peer and server. The EAP server can reside at a different location and actually in a different network. The general advantages are that the authenticator in pass-through mode becomes independent of the actual EAP method being used, so no changes to deployed equipment in the access are required in case there are changes in the actual EAP methods. These only impact the end devices and the EAP server. Also, of course, any authentication termination in a central authentication server, whether or not roaming plays a role, clearly requires such a three-party model.

Across the wireless link, EAP is encapsulated in PKMv2 at the wireless MAC layer as per [1] that is transferred across the wireless physical (PHY) link as shown in Figure 3.3. So the wireless interface specifications need to offer specific payloads for exchanging EAP messages. The examples here are [104] to carry EAP over the WLAN radio interface and of course [1] for WiMAX.

Whereas, for WLAN, authenticators can directly reside in the access points, a major architectural aspect in Mobile WiMAX networks is that the authenticator function resides in the ASN-GW that centrally controls the related procedures for BSs. It only acts as a simple relay for EAP (not considering special cases where BS and ASN-GW functionality are

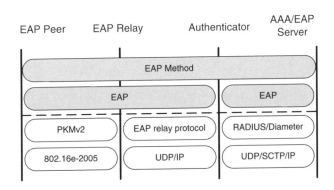

Figure 3.3 WiMAX protocol stack for EAP and EAP methods

integrated into the same physical equipment), and is after successful termination of an EAP method just provided with the appropriate security keys to protect the wireless link. Considering the fact that EAP messages traverse all the way from the MS to the AAA server in the home network and that an EAP method typically requires several round trips for a successful authentication, this clearly provides a significant performance improvement, because EAP procedures are only performed for changing the ASN-GW, in contrast to every BS change.

An EAP method runs end to end between the supplicant in the MS and the EAP/AAA server and is transparent to the authenticator in the access network and to any intermediate proxy. This holds especially for the security aspects: only the supplicant and AAA server run the required security algorithms and possess the required security credentials. Hence, a decision on the EAP method to be used is up to the operator of the AAA server (as long as the device supports the selected method). It does not depend on any intermediate AAA instance like the local access or any visited network. User authentication in WiMAX always terminates in the AAA server of the user's home network that owns the subscription.

WiMAX devices are shipped with a device certificate securely imprinted by the device vendor. Verification of these certificates, if used during network access authentication, becomes possible by using a common root authority (WiMAX root CA) for WiMAX devices that is established by the WiMAX Forum.

Through such public key infrastructure and with the specifications requiring network equipment to support adequate EAP methods, it becomes optional for operators to perform both device and subscription authentication during network entry. If both device and user authentication are to be performed, tunneled EAP methods like EAP-TTLS are used that establish an outer protected tunnel between the EAP peer and server based on the device certificate and then run subscription authentication through this tunnel by using a so-called inner method. Details regarding different possible EAP methods are given in Section 3.3.3.1. However, for this section it is sufficient to understand that with the selection of an appropriate EAP method, it is possible to perform either subscription authentication or device authentication (e.g. in the case of initial provisioning as explained in Section 3.6), or both, by running a single EAP method.

For the remainder of this section, we will put together the basic steps that are performed to allow a MS to securely establish a WiMAX connection with the network.

As shown in Figure 3.4, access starts with activation of the wireless link that includes, for example, all the required steps for selecting a WiMAX network among the locally available ones. The EAP is performed subsequently, hence happening in an early phase of the network entry procedure where for instance the MS does not yet own an IP address. The AAA server responsible for performing EAP authentication (and hence acting as EAP server for this MS) is selected by the network based on identity information provided by the MS itself: the network access identifier (NAI).

More details regarding the NAI as one of the central identities used in WiMAX are given in Section 3.7. However, within the scope of this section this identifier, if being used for a WiMAX subscription, typically consists of a user identity and a realm part denoting the user's home operator.

If the EAP method protocol run was successful, i.e. both EAP peer and AAA server successfully verify authentication information provided by the other end, a resulting master session key (MSK) is transferred to the authenticator in the ASN-GW. The MSK is used as the basic session key for all subsequently derived keys that are required to protect the different

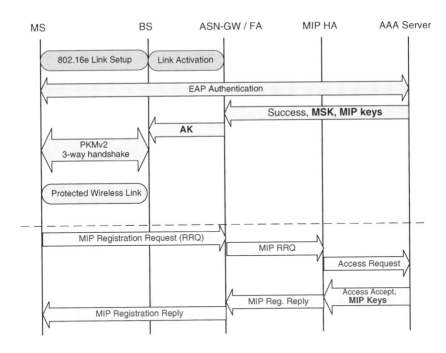

Figure 3.4   Basic steps for secure WiMAX network access

logical control and data channels established across the wireless interface for this MS. Most of these keys are derived in the WiMAX BS. However, the ASN-GW performs one key derivation step from the MSK itself and transfers the resulting AK key to the BS across the R6 interface.

Security for the wireless link is actually established by a three-way handshake specified in [1] as part of the PKMv2 protocol at the wireless MAC layer.

After successful completion of the PKMv2 handshake, the wireless link is protected. This in turn allows the MS to start setting up IP connectivity with the WiMAX network. The example in Figure 3.4 assumes that the Mobile IP (MIP) is used, so the MS sends a MIP registration request (RRQ) to the network that is passed on by the ASN-GW's foreign agent (FA) to the assigned Mobile IP home agent (MIP HA).

Protection of the MIP signaling messages shown in Figure 3.4 between the MS, the ASN-GW and the HA is performed at the MIP level. This is based on mobility session keys derived from the EAP authentication phase that are securely distributed in the WiMAX network. So as a security mechanism that is rather specific to WiMAX, there is a coupling of the initial EAP authentication procedure with the security required as part of subsequent steps of the overall network attachment procedure in WiMAX.

3.3.3 EAP and EAP Methods

As already briefly discussed in the above overview section, network access authentication in WiMAX is performed through EAP, and EAP is used as a three-party protocol by defining the roles of an EAP peer, the authenticator in the ASN-GW and the EAP server that is part of the

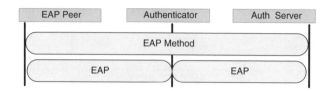

Figure 3.5   EAP three-party model with pass-through authenticator

AAA server located in the subscriber's home network (Figure 3.5). The BS as a pure EAP relay is not considered here. It is not actively involved in the EAP procedure for network access and only comes into play as soon as EAP has been successfully performed. The BS receives keys resulting from the EAP procedure from the ASN-GW across the R6 reference point, to perform the PKMv2 handshake and protection of the air interface.

In this section we will take a closer look at the EAP methods that are considered for WiMAX networks, or might be considered in the future.

3.3.3.1 EAP Methods in Mobile WiMAX

EAP methods have been around for quite a while already, and up to today an impressive number of such protocols for authenticated network access have been defined [29]. However, methods of practical relevance are rare, and for many others it would be very difficult to find a publicly available specification describing the method.

Simple authentication schemes have been defined by the EAP specification itself to allow the support of legacy frameworks in the fixed access area using the Point-to-Point Protocol (PPP) for access. These include for example CHAP [23] support that is directly built into the EAP as the 'MD5-Challenge' type.

The EAP specification itself requires any key generating EAP method to be able to derive two resulting 64-byte (or longer) session keys, MSK and EMSK, where the MSK's sole use is to act as root session key for protecting the wireless link. EMSK usage is marked – purely following the EAP specification – as being reserved for future use. WiMAX defines such EMSK usage with deriving session keys for protecting Mobile IP from EMSK. Hence, only EAP methods defining EMSK derivation in addition to MSK (the original EAP RFC 2284 that is deprecated by RFC 3748 did not require EMSK generation, and older EAP methods comply with the old RFC only) are applicable to WiMAX networks – and these are a limited number of the more recent methods only.

Initially, the approach taken by the WiMAX Forum was not to introduce a set of selected EAP methods as the default ones for WiMAX deployment; the idea was to only limit the choice of EAP methods by technical requirements like the ability to derive EMSK.

Later on, this approach was deprecated in favor of pushing for interoperability between different vendors and implementations, and especially to ensure a basic support that can be assumed for any WiMAX MS regarding EAP authentication. Support of EAP methods in devices is typically considered the major limiting factor, in contrast to the broader network-side support offered by a typical AAA server implementing a range of different EAP methods.

The default methods requiring support in WiMAX networks and devices are the following ones:

- EAP-TTLS/MS-CHAPv2 [26] [27]
- EAP-AKA [28]
- EAP-TLS [24].

To be more precise about the requirements for WiMAX implementations, EAP-TLS, which is mainly intended for authenticating the device itself but can also be used for subscription authentication if differentiation between device and subscription is not needed, is implemented in all WiMAX devices and networks. For EAP-TTLS and EAP-AKA, which are the default choices for authenticating the actual WiMAX subscription, a MS has to support at least one of these two methods. The network in contrast is recommended to support both.

Nevertheless, WiMAX allows – assuming support in the MS – an operator to choose an EAP method best matching its specific networking and security requirements. This is especially true due to the fact that the EAP method is fully transparent to the authenticator in the ASN and intermediate AAA proxies in any visited CSN. Therefore, any specific deployment choice in this area does not require the operator to update all possible access networks and it especially does not impact roaming.

But why does WiMAX already define three default EAP methods, when for interoperability the best thing would be just to have one of them?

To motivate our further considerations regarding different deployment scenarios, a general differentiation can be made based on the type of security credentials that are supported:

- **Symmetric**: Pre-shared keys (PSKs) where the owner of the subscription and the AAA server share a common secret (typically a value of at least 128 bits). EAP methods supporting PSK-based authentication – although this cannot be generalized – are typically considered very efficient but require that a common secret known by the operator's AAA is already provisioned in the MS prior to running the EAP method.

  The commonly used passwords also fall into this category, as passwords are merely a weak form of PSKs with lower entropy (and hence with either decreased security or posing special requirements on the EAP method like protection against password guessing attacks).

  PSK methods like EAP-AKA (which, however, has a strong relation to cellular systems) are the default choice for subscription authentication, especially if not combined with device authentication.

- **Asymmetric**: Public/private key pairs and certificates (in short: public key). EAP methods based on public key credentials like EAP-TLS are typically heavier regarding computational impact, data transferred OTA and required number of round trips, but are in principle more flexible in scenarios where the existence of a PSK between the MS and AAA server cannot be assumed. This, however, only holds at a general level as the actual process of initially provisioning the security credentials needs to be considered at the same time, and may play a more significant role.

  Public key-based methods include EAP-TLS, which is the default choice for device authentication. They can also be used for subscription authentication, e.g. if the subscription is bound to a specific device (the certificate on the device serving for both subscription and device authentication).

- **Hybrid methods**: EAP methods using both public key credentials and PSKs combine the advantages of both approaches, but still suffer from being heavier choices than pure PSK methods. A common combination here is – as in the widespread Internet model when accessing secure websites via https - to authenticate the subscriber by using a PSK and the AAA server via public key credentials.

WiMAX makes use of a hybrid method with EAP-TTLS/MS-CHAPv2 that allows the combination of subscription and device authentication.

Now let us take a closer look at the possible deployment scenarios for the selected default methods, to understand which method is best applied under specific circumstances.

3.3.3.2 EAP-TTLS

For standard WiMAX network operation, subscription authentication must be performed to gain network access. In addition, device authentication might also be executed depending on the operator's policy. The MS itself may authenticate the network (actually from a security perspective this is always recommended; however, the specifications do not force a MS to do so).

EAP-TTLS is defined here as the default method that has to be supported by any WiMAX network AAA server. This method allows the home network's AAA server to authenticate itself against the MS using a server certificate (see Section 3.4 for details regarding the WiMAX PKI aspects). Subscription authentication is performed using a simple challenge–response authentication exchange called MS-CHAPv2 that is sent through the protected tunnel established by EAP-TTLS. Hence, EAP-TTLS is called a 'tunneled method': it first establishes a protected channel and then allows for running an 'inner method' through this tunnel. Actually any EAP method can be performed through EAP-TTLS. Using other exchanges than MS-CHAPv2 is possible and is a valid option for deployment-specific choices as long as support on the MS side for such combinations can be assured.

When additional device authentication is required by the operator, this can also be done via the same EAP-TTLS procedure by the MS authenticating the (typically only server-authenticated) tunnel with its own device certificate.

3.3.3.3 EAP-AKA

A WiMAX MS is mandated to support either EAP-TTLS or EAP-AKA. The main difference between EAP-AKA and EAP-TTLS is that AKA – the authentication protocol inherited from 3GPP UMTS networks and put into an EAP method – is purely based on symmetric keys. No TLS tunnel needs to be established to protect the MS-CHAPv2 exchange which on its own would be too weak for authentication across an open wireless link. Additional device authentication is not possible with EAP-AKA that does not support public key-based authentication or certificates. The latter may still be possible as soon as EAP-AKA is run as the inner method through EAP-TTLS, but this would first require devices to implement support for such a combination.

Clearly, EAP-AKA is more efficient than EAP-TTLS with an inner method, e.g. in terms of the number of round trips.

Also, due to its heritage, EAP-AKA is the set method for interworking between 3GPP and WiMAX networks and neatly supports the use of USIM cards as removable smartcard tokens carrying the credentials for a WiMAX subscription.

3.3.3.4 EAP-TLS

Being one of the first standardized methods, EAP-TLS [24] has received a number of enhancements that are documented in [25]. The new standard actually deprecates the original protocol version that had been an experimental RFC, from the IETF specifications perspective. However, it takes time to update implementations – and this is especially true for devices coming from a large number of different manufacturers. WiMAX is still based on the original EAP-TLS protocol for device authentication (although technically it would also be possible to perform subscription authentication for a user through EAP-TLS). There are minor differences between the specifications that would introduce backward-compatibility problems with existing WiMAX equipment when merely replacing the old by the new RFC.

Of course, it is possible for an operator to choose [25] the original EAP-TLS method as a deployment-specific option. The main obstacle that can be expected for the realization of such a choice, again, is device support.

EAP-TLS supports WiMAX deployments specifically for use cases related to the initial provisioning of 'blank' devices; that is, devices that are not yet provisioned and only hold a generic device certificate as a security credential that is not bound to any subscription at the time of purchase. Also, reprovisioning in cases where an MS loses its provisioned subscription data makes use of EAP-TLS.

3.3.3.5 EAP Message Flow

Besides the exchanged EAP method messages, a general EAP message flow for entering the WiMAX network consists of a few additional EAP messages to initiate the method, and to communicate success or failure of authentication. Also, the 802.16 link layer adds to this by defining an additional EAP-Start message (similar to the WLAN link layer defining an EAPOL-Start message, but only used for re-authentication in WiMAX as we describe later in this section).

For initial network entry as shown in Figure 3.6, the authenticator in the ASN starts the EAP exchange by sending an EAP-Request/Identity message across R6 that is relayed by the BS to the MS requesting access (1). This message indicates one proposed EAP method. In (2) the MS responds to this either by sending an NAI as identity information or by sending a NAK message indicating that the proposed method is not supported.

Assuming the former case, the exchange continues in a sequence of EAP-Request/Response pairs to carry the actual EAP method messages (3). The number of Request/Response pairs depends on the EAP method itself. It is interesting to note here that for EAP methods it is always the AAA server sending the first message.

After successful completion of the EAP method (i.e. the AAA server and MS authenticated each other and the AAA server authorized the MS network entry), the AAA server sends an EAP-Success indication to the Authenticator. The AAA message carrying the EAP-Success also carries the required keys for setting up PKMv2 security across the wireless link and Mobile IP security, if required (4).

The EAP-Success message is finally relayed across R6 and the wireless link to the MS (5).

3.3.3.6 EAP Method Negotiation

For EAP it is important to negotiate which method is to be used for secure network entry. This will typically depend on the (rather limited) set of methods supported by the device, support

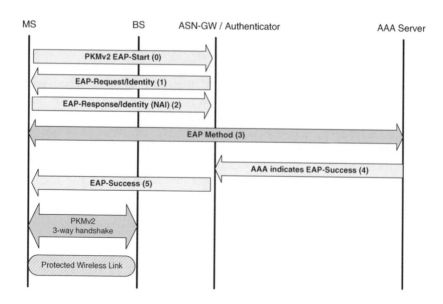

Figure 3.6 EAP flow with identities

offered in the AAA server, whether device authentication is required, and on possible policy settings related to a specific subscription. Looking at a single operator with good control over the device models handed out to subscribers, this may be a straightforward choice and can efficiently be based on preconfiguration, especially on the device side.

However, considering WiMAX devices that may not have any a priori relationship with an operator prior to being initially provisioned, negotiation of an adequate EAP method for network attachment is important. It is also necessary to limit the number of default EAP methods for improving interoperability, which motivates the choice of a small number of default methods in the WiMAX specifications.

In EAP itself only very basic support for negotiating the EAP method is available. On receipt of the method indicated by the authenticator, the MS may not support this method. In this case it sends a NAK message back to the network that can contain a list of methods actually supported by the device. On receiving this list, the AAA server can verify whether one of the indicated methods matches the AAA's internal policy for granting network entry, and chooses this method. With no match, EAP authentication will of course fail, and the AAA server will send an EAP-Failure indication instead of EAP-Success.

Although EAP itself is limited in supporting method negotiation, in practice WiMAX is not expected to suffer from interoperability problems in this area when assuming broad support for the small set of default EAP methods. In the above use case for off-the-shelf WiMAX devices with no relationship or preprovisioning for any operator (like a standard laptop with added WiMAX support), EAP-TLS will be used for secure access to a WiMAX network that subsequently runs an initial OTA provisioning procedure, which is discussed in more detail in Section 3.6. Further EAP methods can be selected as part of this procedure to minimize possible interoperability problems.

If it comes to the question of which mode of authentication – whether device plus subscription or subscription only – is to be performed, there is no need for additional negotiation prior to executing the EAP method. This is related to the class of EAP methods being used for combined device and user authentication, the so-called 'tunneled' EAP methods like EAP-TTLS. Here, for subscription authentication the 'outer' TLS tunnel is established with only the AAA server supplying public key credentials and thereby authenticating itself to the MS (unilateral authentication). Through this secure tunnel, an 'inner method' – MS-CHAPv2 in the simplest case, but also a full EAP method like EAP-AKA can be performed – is run that actually performs user authentication. In contrast, if combined device and user authentication are required, in addition to what is described above, the MS uses its own device credentials to make the first step in tunnel establishment a mutually authenticated one, by adding this device authentication.

Returning to negotiation, it is therefore sufficient to negotiate a single (tunneled) EAP method here. Whether both device and subscription authentication or subscription-only authentication is subsequently performed will depend on just the provided credentials to the method, but not on method negotiation itself.

3.3.3.7 Re-authentication

Step (0) in Figure 3.6 shows an EAP-Start message that is not defined as part of the EAP itself, but is an additional trigger provided by the PKMv2 protocol responsible for carrying EAP messages between the MS and the authenticator. WiMAX does not send EAP-Start messages for initial network entry due to the fact that EAP-Start would be unprotected when being sent prior to any established secure channel across the wireless interface. A secure channel is only established after the successful execution of an EAP method and after completing the PKMv2 handshake. Unprotected EAP-Start messages will be ignored and dropped by a WiMAX BS.

However, the EAP is also performed for re-authentication. Such re-authentication typically happens when the lifetime of the initial EAP authentication session and the security keys generated as part of it time out. It actually allows wireless security to be renewed and fresh session keys to be generated without requiring the MS to leave the network and perform a new network entry.

To allow the MS to trigger re-authentication when it gets close to the end of the key lifetime for the preceding EAP run, the MS can send an EAP-Start message to the authenticator to trigger an EAP-Request/Identity message, starting a new EAP exchange for rekeying.

In practice, there are no fixed lifetimes for an EAP session in WiMAX access. However, a reasonable default value for EAP-derived keys and in turn for the re-authentication frequency is one hour. This is also in line with what is recommended in [1] as a default value, with an additional minimum threshold of one minute and a maximum threshold of one day.

It is interesting to know that there is no single lifetime value for an EAP session, due to the different entities and possibly also network operators involved in the procedure of establishing and maintaining an EAP session for WiMAX access. In fact, the AAA server will set a lifetime for the keys derived from an EAP authentication run (and hence for the EAP session). This might be based on policy internal to the AAA server, or the AAA server can communicate a timer value to the authenticator in the ASN. The ASN itself will run its own timer, which can be set to any value provided by the AAA server. However, the ASN, depending on local security policy, may also choose to set a shorter lifetime.

In addition to the timers in the network, the MS sets its own timer (timers set by the AAA server of the authenticator are not communicated to or synchronized with the MS, unless there is some preconfigured value). Finally, this leads to the situation where the involved entity (MS, authenticator or AAA server), setting the smallest timer for the lifetime of the EAP session, will finally trigger re-authentication in advance of a timeout, if the MS is still in the network.

Without such re-authentication, the network will terminate all MS connectivity and state after the EAP session lifetime expires.

3.3.3.8 More EAP Methods

The EAP methods selected as the default choices for Mobile WiMAX are well-established ones. However, there are many more, and a well-established technology does not automatically mean it should always remain state of the art.

The IETF EMU (EAP Method Update) effort, for example, is chartered to develop properly standardized new EAP methods. Among the available results are

- an updated specification for EAP-TLS [25] that provides a number of improvements over the original EAP-TLS method; and
- an EAP-GPSK providing a lightweight method that is based on PSK credentials [30].

For Mobile WiMAX, the newer version of EAP-TLS has not yet been adopted as a replacement of [24] for the reasons identified in Section 3.3.3.4.

Considering the fact that EAP-TTLS belongs to the class of heavyweight methods for both the MS and the AAA server, EAP methods supporting only PSK-based credentials are the more efficient choice for most scenarios where there is no requirement for device authentication, and subscriber authentication is the only authentication performed. This is based on the fact that public key-based credentials require more computational resources at the respective ends. Impact on battery lifetime in the MS might be neglected, regarding the frequency of EAP authentication and assuming the device has sufficient computational power to perform public key operations and validate server certificates without increasing network entry time. However, the computational impact at the AAA server side might be significant regarding the possibly large number of active parallel subscribers in a typical mobile network deployment.

Another quite important criterion is certainly the number of required round trips (pairs of Request/Response messages) for successful execution of the EAP method. EAP authentication is commonly considered to account for a significant part of the overall delay in initial network entry, which is the time experienced by a subscriber prior to being able to actually use a service. This is due to the fact that EAP messages run end to end between the MS and the AAA server in the network. And this holds especially for roaming scenarios where different network operators in different geographical locations are involved and need to route the EAP messages.

To confirm the above, PSK methods like EAP-AKA or EAP-GPSK offer a clear advantage regarding the number of round trips required:

- EAP-TTLS with MS-CHAPv2, four round trips;
- EAP-AKA, one round trip; and
- EAP-GPSK, two round trips.

The numbers show the round trips between the MS and the AAA server for a successful authentication, excluding EAP-Identity messages and EAP-Success.

3.3.3.9 Deployment Options for Authentication Endpoints

If we consider Figure 3.1 and termination of the EAP in the network, the most simple deployment model for Mobile WiMAX becomes a quite simple one. Independent of whether subscription-only or both device and subscription authentication is performed, this authentication always ends in the home NSP 'owning' the subscription. And taking the actual specifications for WiMAX networks as per [116] into account, it is this deployment model that is supported. The current specifications do not allow termination of an EAP method exchange for an existing subscription in the visited NSP.

This is trivial for a deployment without roaming where there is only one NSP, namely the home one. However, as soon as a V-NSP is involved, the V-NSP will also operate a WiMAX-compliant AAA server and might have specific interests in being involved in authorization decisions. So why has it not been involved in the authentication and authorization process for network entry? Certainly, subscription authentication can only be performed with the help of the home AAA server; no V-NSP would have access to the subscription credentials.

But why not at least split device and user authentication and keep the former one local? One of the reasons has certainly been that, at the time of developing the initial WiMAX network specifications, the majority of involved service providers did not see a need for terminating device authentication locally in the visited access network when, at the same time, the subscription authentication required is performed by the home network anyway. In addition, to make different policies for allowing network entry useful handles for possibly limiting access to specific types of services, some effort would be required to develop appropriate mechanisms for negotiating an authorization policy between the different involved business entities in the network (being a selection from different NAPs, visited and home NSP, broker networks).

Also, when focusing on the EAP from a technical perspective, splitting device and user authentication between visited and home network would require some additional considerations regarding the supported use of the EAP. Let us elaborate a little on this to better motivate the different technical approaches and options that were considered for adding cryptographic device authentication to WiMAX and to understand different deployment options and their impacts.

Technically, when device and user authentication are to be performed, WiMAX network access does not allow two subsequent and independent EAP conversations to be performed for one single network entry (i.e. for the same MS). Also, running two such EAP conversations independent of each other would introduce security problems related to potential attackers stepping in between the two exchanges (man-in-the-middle), when no cryptographic binding of the two EAP steps is done. For the same reason, the so-called 'chaining' of EAP methods within the EAP, i.e. performing two subsequent EAP methods within a single EAP conversation, is not allowed (see [3] for further details regarding the security properties).

The solution adopted for WiMAX is to make use of a tunneled EAP method with the default choice being EAP-TTLS with MS-CHAPv2 as the inner method for user authentication. Any tunneled EAP method counts as a single EAP method when looking at the EAP itself, so it is

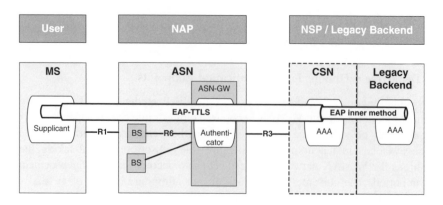

Figure 3.7   Example use of EAP tunneled and inner method

different to method chaining. However, being one EAP method, this terminates in the same (in fact, the home NSP's) AAA server.

It is interesting to consider here the property of tunneled EAP methods where the inner method can be extracted by the AAA server terminating the outer tunnel, and can be forwarded to another backend AAA server, like the one that is already part of a backend authentication infrastructure and connected to the WiMAX CSN. An example configuration of this is shown in Figure 3.7. So, in principle, it would be feasible in WiMAX to make the AAA server in the V-NSP or even an AAA server in the NAP the one terminating the outer TLS tunnel of EAP-TTLS and thereby performing the device authentication, while relaying the inner method to the home NSP's AAA server.

EAP-TTLS/MS-CHAPv2 does not support such a deployment option due to the fact that all keys derived in WiMAX for further applications like Mobile IP from the EAP method must be generated in the home network's AAA server itself. Unfortunately, MS-CHAPv2 does not support the required key generation for WiMAX, so the only option in practice is to terminate the full TTLS exchange in the home AAA. Also, forwarding plain MS-CHAPv2 without the protection of the outer TLS tunnel would probably not match the security requirements of operators.

When combining EAP-TTLS with other inner EAP methods capable of generating appropriate keys like EAP-AKA or EAP-GPSK, scenarios like the one shown in Figure 3.7 become possible. It would also be possible to perform device authentication in the local ASN, although this is not yet supported by the WiMAX network specifications following Release 1.5 [116].

Another technical approach to perform device authentication locally has been discussed in the context of IEEE and the WiMAX Forum. This had been worked on in draft versions of [13] but later on was dropped due to a lack of operator requirements at the time. The solution here is to add support and negotiation capabilities for two subsequent EAP conversations at the 802.16 MAC layer that cryptographically binds the second round to the first one to counter the above-mentioned man-in-the-middle attacks. In fact, the second EAP conversation is protected by keys derived from the first one OTA. The final session key MSK for subsequent protection of the wireless link is blended from keys generated by both EAP conversations.

The additional flexibility provided by this approach, also known under the term 'double-EAP', compared to the use of tunneled EAP methods, lies in:

- freedom in the choice of EAP method (for tunneled EAP methods, besides the fact that there are only a few of them, only the inner method can vary);
- full support of independent locations for the AAA servers terminating device and subscription authentication;
- cryptographic binding of the two EAP conversations. (In contrast, [26] is subject to the related class of attacks and requires further efforts, e.g. regarding local configuration of devices, to minimize the impact; an updated version v1 of this protocol that added support to cover this was not adopted by WiMAX, although it is possible to use it.)

However, it is also important to note that this approach requires a slightly higher effort in implementation, which is of course always tied to cost. So, as usual, there is a tradeoff and in this case the decision has been in favor of the more limited but available tunneled methods. Double-EAP would clearly require support in WiMAX chipsets, so without full support from the standards and commitment by chipset vendors it cannot be used. Both are currently not available.

An overview of how double-EAP could support device and subscription authentication is given in Figure 3.8. The first EAP conversation for authenticating the device starts with the usual identity exchange (1) and (2), followed by the EAP method (3). On successful completion, the wireless link can already be protected with an integrity key called EIK that is derived from the device authentication EAP conversation.

Across the protected wireless link, the MS triggers the second EAP conversation by sending an EAP-Start message (6) to the authenticator that in turn starts the EAP conversation (7) to (9) for subscription authentication.

Secure binding of the two EAP conversations then occurs with the authenticator using both keys MSK-1 and MSK-2 (10) and merging them into the MSK to derive all further keys for establishing the secured wireless connection. This final step is the same as in normal network entry with a single EAP method, the only difference being the way to create the MSK.

The advantage of creating the cryptographic binding in the access network is that backend AAA servers for device and subscription can be deployed fully independently of each other. In such a deployment scenario and assuming that the AAA server handling device authentication is in the local access or visited network, it may be useful to communicate to the home network information on whether device authentication has been successfully performed. However, this could easily be performed by adding this information to the AAA messages exchanged between the authenticator and the AAA server during network entry.

3.3.4 WiMAX-SIM Cards

In 3GPP mobile cellular networks, subscription data and security credentials are distributed on a smartcard token called the universal integrated circuit card (UICC) that reaches the subscriber via out-of-band means like being picked up in an operator's shop by the new subscriber. So the actual model for initial provisioning and activation of a subscription is different to the OTA provisioning model developed for WiMAX networks as discussed in Section 3.6. The latter rather covers the dynamic creation and installation of a subscription in an unprovisioned MS through a special network entry procedure supporting such provisioning 'on-the-fly'.

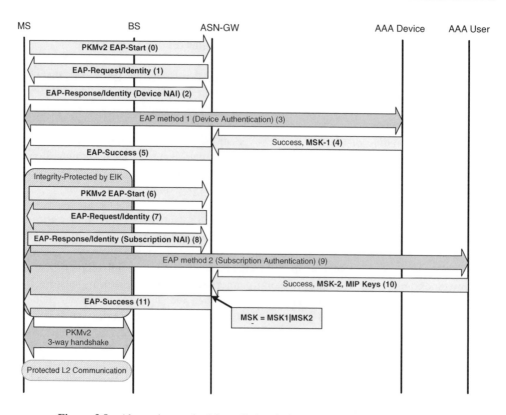

Figure 3.8 Alternative method for splitting device and subscription AAA servers

The concept of a smartcard token, or UICC, carrying a universal subscriber identity module (USIM) application in a UMTS mobile phone is unchanged in WiMAX. This holds with the exception that for WiMAX a more general identity format for the subscriber identity based on the NAI has to be supported instead of the international mobile subscriber identity (IMSI) format used in 3GPP networks. Logically, none of the physical properties of the UICC and the required support for UICC tokens in the terminal are modified. However, instead of the logical USIM application that makes up a UMTS subscription and also includes appropriate authentication algorithms for network entry, WiMAX uses a WiMAX-SIM application on the card [101]. This new application, while mostly reusing USIM structures and specifications, standardizes the additional data files on the card that are required to carry all subscription information specifically related to WiMAX.

Let us now take a look at the deployment aspects regarding UICC-based subscription provisioning versus OTA initial provisioning, where the underlying assumption is that a UICC-based subscription cannot be provisioned OTA for security reasons. Using the WiMAX-SIM is certainly of major interest to operators that have already implemented the well-established and widely used processes for issuing and delivering UICC-based subscriptions to customers. Alternatively, subscriptions can be created and provisioned on-the-fly through dynamic OTA provisioning of the MS and the use of portal web pages displayed to the user through the web browser on the MS. The latter example rather reflects the much more dynamic as well as temporary provisioning model that is common nowadays in WLAN environments, but is also considered less secure than a UICC-based initial provisioning of a subscription.

Aspects of a UICC-based subscription that should be considered for any deployment decision are:

- The higher level of protection applied to the subscription credentials where the long-term secret keys are already put into the tamper-proof hardware by the operator and never leave the tamper-proof and PIN-protected hardware token for the whole lifetime of the subscription.
- The fact that the subscription and all related configuration information can easily be transferred to a different terminal provides a clear benefit to the user, e.g. when changing to a new MS.
- In general, the UICC also carries the authentication algorithm that is – by putting it on the UICC – made independent of the terminal used by the subscriber. Therefore, the operator is in full control of the actual authentication algorithms based on the fact that the same operator hands out the UICC cards and also deploys the AAA server for network access authentication. In WiMAX, there is a fundamental difference to 3GPP deployments regarding this security property, as the algorithms are part of the EAP method which is typically implemented by the terminal. So if, for WiMAX, the subscription were stored on the UICC, the authentication method would still be performed by the MS. Certainly, it is technically possible and recommended to realize the authentication and key derivation algorithms of the actual EAP method directly on the UICC. However, this would be a deployment-specific choice as no generic standardized interfaces are available to split the EAP method functionality between the terminal and the UICC. EAP-AKA stands as an exception: although the method specification itself does not provide guidance regarding which parts are to be realized in a UICC, such a split implementation is rather straightforward. The UICC by default supports the AKA authentication algorithm of 3GPP and implementations can largely be based on specification work done in the ETSI Smart Cards Project (SCP) like [102]. As a result, for a deployment with UICC-based subscription credentials the EAP-AKA (or alternatively EAP-SIM) methods would be the most obvious and technically viable choice.

For the last of the above aspects, it is important to note that it is possible (and actually such a solution has been rolled out by at least one major WiMAX operator) not just to perform the cryptographic algorithms of the authentication and key derivation procedures for network entry in the UICC; instead the full EAP method like EAP-AKA is implemented in the UICC as a deployment-specific choice. Appropriate specifications to describe the required file structure on the UICC application and some basic interaction between the terminal and UICC module are available from ETSI, see for example [103], and the realization of the full EAP method on the smartcard may help the operator to overcome interoperability issues between different MS implementations. In WiMAX, this is also supported as an option. However, it still remains a deployment-specific realization option as there is no standard available that fully describes the interaction between an off-the-shelf WiMAX-enabled terminal and the EAP-enabled WiMAX-SIM application on the UICC. Here, especially the interaction and shared responsibility between the EAP capabilities of the MS and those of the WiMAX-SIM (where currently only EAP-AKA is covered by existing specifications) require additional alignment between the UICC and the MS manufacturers which would likely need to be encouraged by operators looking for such a solution. One of the technical aspects to be solved in such a case is that, for example, EAP-AKA would run on the UICC for subscription authentication, but EAP-TLS would still have to be performed in the terminal for OTA provisioning, for instance. Here, implementations need to ensure a smooth interaction between the EAP methods that are implemented in different modules. One has to keep in mind that the logic of when to trigger a particular EAP method still needs to remain in the MS. A valid use case for this would be where the owner of a WiMAX-enabled laptop has a WiMAX-SIM-based subscription from the home operator: for example, when a user staying in a different country wants to set up an alternative subscription via OTA provisioning with a local WiMAX operator to avoid international roaming fees.

The decoupling of the mobile user equipment (UE) and the (U)SIM card has likely been a relevant factor for the success of 3GPP cellular networks. However, whereas a WiMAX-SIM makes a lot of sense for certain deployments, it is also clearly limited or even not suitable for other deployments. Let us consider the example of a user switching on a WiMAX device without a valid subscription being installed in a certain location and needing to get broadband data access as quickly as possible. A common case would be staying in a hotel during a trip abroad. The device will attach to the locally available ASN and then offer a list of possible NSP networks where the user can set up a new subscription via OTA provisioning. Here, the dynamically created subscription would clearly not be UICC based and the actual provisioning model simply would not match those that UICCs are used for.

3.4 Certificates and the WiMAX Public Key Infrastructure

To briefly recapture the above discussion regarding device and subscription authentication, the former is independent of whether a WiMAX device has already been initiated with a subscription, or is an off-the-shelf one without any credentials for, or prior knowledge of, any WiMAX network.

From a technology viewpoint, there is nothing new in the WiMAX Forum's defined device authentication methods. These are based on well-known X.509 version 3 certificates [31] as specified by the IETF [32] with a WiMAX-specific profiling on top [14] and [15]. The WiMAX profiling serves to optimize usage for the given case of authenticating devices based on their MAC address.

Considering the overall use case, however, the concept of device authentication is novel insofar as it would be challenging to find a similar established system in a comparable mobile network infrastructure. Under the central control of the WiMAX Forum, root-level certificate authorities (CAs) have been created and are operated to serve as the roots of trust for a WiMAX-wide public key infrastructure (PKI). This covers both operators' AAA servers and end user devices. And the most important enabler for making the whole system work in practice is clearly that all WiMAX devices will be shipped with a preinstalled device certificate. At this point it becomes necessary to clarify the definition of a 'WiMAX device' within the scope of this section: that is, the set of devices approved by the WiMAX Forum certification process that will likely include all devices produced by major device manufacturers. It is important to point out here that the term 'certification' is overloaded. When speaking about security, it refers to digital certificates and device or network authentication. When speaking about WiMAX devices, it refers to those devices that have successfully completed a set of standardized testing procedures and are approved as being 'certified' by the WiMAX Forum.

These two areas are completely separate, and the overlapping meanings of certification often lead to confusion.

Also, while referring to the WiMAX endpoint as a 'device' throughout this section, the piece of hardware that the device certificate is actually bound to is not necessarily a full device as seen from the end user's perspective. When considering the case where a laptop computer is using WiMAX for wireless access to broadband data services, the actual WiMAX device is rather the WiMAX modem that may, for example, be externally attached as a USB stick.

A PKI for all certified WiMAX devices is a huge undertaking, regarding the overall resulting system and infrastructure in terms of the number of devices and the number of involved business entities, although the effort for an individual operator or device manufacturer might be neglected. This is especially true because real-life PKI systems for very large numbers of end devices are scarce in terms of deployment. Consider the example of mobile phones where even state-of-the-art authentication is based on pre-shared secrets and PKI methods may only be used as optional additions (see e.g. [40]) that are rarely available in practice. As another example of the attempt to create a 'global PKI', the OMA-defined digital rights management system [41] might be considered. So, of course, the WiMAX approach still needs to prove its feasibility and the validity of its use case.

This section will not attempt to make too many predictions regarding the above general considerations, on the future scale of deployments making use of this PKI system and of device authentication. It will instead focus on identifying the main use cases for device authentication. With the ability to securely 'bootstrap' devices that have not yet been initiated, this is at least one good reason why the investment effort does exist.

3.4.1 A Brief Overview: Digital Certificates and PKI

One of the main benefits of PKI-based systems and security credentials including a public/private key pair for endpoints authenticating to a network, and vice versa, is that there is no need to share secret information between the device's credential store and the authentication server of the network. See also Section 3.3.3 for the security credentials used in EAP methods. It is therefore in theory much better suited for scenarios where it cannot be assumed that the communicating parties have had any prior and secure communication with each other, like the one required for setting up a subscription – including a shared secret to be used for access authentication later on. In practice, PKI approaches so far have often been deprecated, although the technology must be considered well understood. This is mainly related to high infrastructure costs and to the complexity of contractual issues including, for instance, operational cost or liability issues between the involved stakeholders, rather than being due to the higher computational load in the device and network for the specific cryptographic operations. The number of contractual issues becomes potentially higher as soon as third parties provide the root of trust for such systems.

This section will not focus on the theoretical and cryptographic aspects of the underlying security mechanisms. However, the interested reader is encouraged to consider background literature like [76].

In PKI-based authentication, each participating device owns a public and a private key. The private key needs to be stored securely and security is based on its secrecy. So it is good practice to store such keys in tamper-proof hardware such as a dedicated security chip or smartcard.

In contrast, the public key is public information and is required by other devices to set up a security context with the owner of this public key.

Without additional measures and without cryptographically binding the identity of the public key owner to the key itself, it would not be possible for other participants to verify that the communication partner 'at the other end' is the valid owner of this public key. Hence, public keys are certified by applying a digital signature over the public key, the identity of the owner and a set of additional parameters, all bundled together in a standardized format – an X.509 certificate [31]. So basically the certificate securely binds the certified entity's public key and identity.

These signatures and the resulting certificates are issued by certificate authorities (CAs) that, prior to issuing a certificate, perform a registration procedure and verify the identity of the future owner of the certificate. They are also responsible for ensuring that each participant of a PKI is assigned a unique identity within the given PKI.

To allow an entity participating in the system to verify the signature of a certificate, the signing CA has its own public/private key pair and signs the public key (self-signing), thereby becoming the root of trust or root CA in the system. The resulting CA certificate can then be accessed by all participating entities (recall it is public information). However, to spoil the whole system an attacker would just self-sign its own certificate and distribute this as the root certificate. To prevent such attacks, it is therefore important that all devices store trusted certificates in a certificate store protected against unauthorized modification and only keep certificates of trusted CAs in there.

Now let us discuss delegating CA rights to subordinate CAs. Instead of signing the public key of a device, a CA can also sign the public key of a subordinate CA that in turn will issue and sign device certificates. This allows delegation of the right to issue certificates. leading to CA hierarchies. The top-level CA for a PKI (the one signing itself) is the root of trust, and all subordinate CAs signed by the root CA can issue certificates themselves.

3.4.2 WiMAX PKI Hierarchy

With this motivation of the fundamental steps in a PKI system and looking again at WiMAX, Figure 3.9 shows the hierarchy that is supported by the WiMAX PKI:

- Certificates are issued for both a WiMAX device to authenticate to the network and for AAA servers deployed by operators to authenticate to devices entering the network.
- For both device and server certificates, the WiMAX Forum established root CAs as the roots of trust. For security reasons, device and server hierarchies are rooted by separate CAs, so there is no single CA covering both.
- As the second level of the hierarchy, it is possible for device manufacturers to set up their own manufacturer CAs with a certificate signed by one of the device root CAs. This is the common approach for device manufacturers of larger volumes, allowing them to keep the process of issuing device certificates under their own control instead of being tied to an external CA service. For smaller operators, in contrast it might be more efficient to rely on an external party to receive device certificates. The WiMAX Forum is in control of related processes like setting the common policies and requirements for the root CAs or any subordinate CAs, but does not operate the root CAs for WiMAX devices itself. Contracted CA operators like

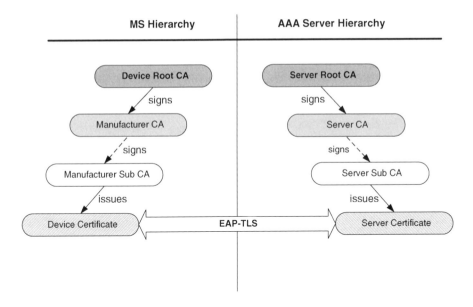

Figure 3.9   CA hierarchy of the WiMAX PKI for device authentication

Verisign and Motorola offer such a service to manufacturers not interested in setting up a manufacturer CA on their own.

- For server certificates, it is possible in a similar way to create server CAs that are approved and signed by the server root CA. An example would be a large operator running its own, already established PKI that the operator wants to make use of to generate certificates for the operator's WiMAX AAA servers.
- An optional third level in the hierarchy allows for setting up subordinate CAs under the control of (and signed by) a manufacturer CA or a server CA, respectively. This is certainly not required or practical in all environments; however there are very good examples why to deploy subordinate CAs. One of them includes a large device manufacturer installing a subordinate CA under its manufacturer CA for each factory. This moves the control over creation of device certificates right into the factory itself, whereas in contrast central delivery of certificates to the factories would put a risk on the whole production process. The actual risk is that timely delivery of certificates cannot be guaranteed. This also provides the reason why it would not be acceptable for large device manufacturers to rely on device certificates centrally issued by a WiMAX device root CA operator.
- The device itself will be provisioned with a device certificate during the production process. In addition, it has to be provisioned with all trusted root certificates for server authentication including at least the server root certificate and certificates from the server CAs of selected operators – the latter being optional but beneficial for efficient operation later on.

In a similar way, each AAA server will be provisioned with its own server certificate issued by either the server root, or a subordinate second- or third-level server CA. In addition, for being able to verify device certificates received during network access authentication a AAA server needs to store the device root CA certificate, and additional manufacturer CA certificates as required, for efficient operation.

The above considerations provide general guidance about how to choose the appropriate hierarchy for either a device manufacturer or an operator. Large manufacturers likely have PKI-related processes already as part of their production chain, including experience with operating so-called trust centers that provide CA services to internal and external customers. They will also benefit from setting up and certifying a CA for WiMAX devices, or will even consider this the only acceptable option. Creating another hierarchy level by setting up local manufacturer sub-CAs allows local control of the creation of certificates. However, it also has to be considered that every additional level of hierarchy in a PKI system puts an extra burden in terms of computational effort and storage capacity on the devices that will have to store and work through longer certificate chains to verify certificates.

Smaller manufacturers may benefit by getting device certificates simply through the WiMAX Forum's approved root CA operators, as the cost per device might be much lower than setting up and operating an own manufacturer CA.

Interestingly, based on common numbers as of today, the cost per device and certificate is even lower for large manufacturers relying on these external facilities for large volumes. The interested reader is referred to [38] for up-to-date WiMAX X.509 certificate and pricing information.

So the key factors for deciding whether to set up an own manufacturer CA do not include cost per device, but rather smooth integration into the existing production process and high reliability and availability in the process of issuing certificates, including all legal aspects in case external partners are involved. Consider a factory being forced to stop production because it is waiting for new certificates to arrive.

For operators, the main factor for deciding how to provision their AAA servers with WiMAX server certificates is whether an internal trust center can be used that is capable of hosting appropriate CA services – in compliance with the WiMAX-defined certificate profiles – or whether it is more convenient to rely on external services in this area. Also, it is important to consider whether integration into existing operation and maintenance (OAM) procedures in the operator's network can efficiently take place to minimize operational cost. The latter aspect, however, may be of less importance to AAA servers as their number is typically relatively low, even for large networks.

3.4.3 WiMAX Certificate Profiles

The WiMAX Forum has published certificate profile specifications for both device certificates [14] and server certificates [15]. Such profiles are required, and are common also in other organizations like 3GPP due to the fact that standard X.509 certificates and their Internet profiling done by the IETF Public Key Infrastructure (X.509) effort allow for quite a number of options to be selected.

One example is that, in WiMAX, certificates are signed by using the SHA-256 algorithm with RSA [33], whereas many PKI systems today are still using the older SHA-1 version. Also, the latest commonly agreed security requirements are addressed by allowing a 1024-bit key size for the RSA algorithm in device certificates only for devices that are not expected to be in use after 2010. All other devices are required to come with device certificates using 2048-bit RSA keys. From a security point of view this is motivated by a recommendation for cryptographic key sizes published by the US National Institute of Standards and Technology

[37]. These are considered secure for a period until 2030 based on estimating the progress of computing performance and scientific progress leading to reduced strength of cryptographic algorithms.

Considering the continuous advances in the computing power of small devices, putting higher requirements on such devices by doubling the key size does not impact the overall system too much – this might rather be relevant to authentication servers being required to perform many such operations per second. However, doubling key sizes increases the size of the certificate. An average X.509 certificate with 1024-bit RSA keys and DER encoding means that a piece of data in the order of 600 to 800 bytes has to be stored and exchanged, not considering large names included in the identifiers etc. Moving to a 2048-bit key size may push the certificate length up to more than 1 kilobyte. Especially across wireless transport and also when carried within an EAP method, this may cause issues like fragmentation and negative impact on the wireless throughput.

The larger key size of 2048 bits is – in contrast to device certificates – already the default for CA certificates. For example, the self-signed certificates for the initial WiMAX Forum device and server root CAs that were created during December 2007 are 2048 bits.

3.4.4 Certificate Revocation

One of the important but operationally rather costly features in a PKI system is revocation. Once a certificate of a device, a server or a CA – listing, respectively, the least to most critical regarding overall system security – is issued and published, the entity owning the certificate and the corresponding private key is able to authenticate itself to any other participant of the same system. If the security of such an entity is compromised, e.g. by revealing the private key, without mechanisms to inform all participants about the security breach and removing the broken end from the system by invalidating it, i.e. revoking the corresponding certificate, the security of the whole system fails. In general, this is not just about an entity being hacked and the private key being opened up, but more about using device authentication and certificates as a means to easily exclude specific devices that might lack authorization, for whatever reason, from the system.

This then takes us into the realm of use cases, and a potential one in this area is stolen devices. In mobile phone networks stolen devices are a considerable problem (see e.g. [74] for numbers from the UK) and in the 3GPP world, for example, technical means have been created to address this area. Devices are shipped with a unique ID called IMEI (International Mobile Equipment Identity). In the world of 3GPP operators and the GSMA organization as the largest association of GSM and 3GSM operators, a central registry, the EIR, was created to allow stolen devices to be tracked based on their IMEI number [75]. Assuming the system is widely used by operators and that reporting of stolen mobiles works well, an operator can check the IMEI of a device attempting to enter the network against the EIR and deny access as soon as the mobile is listed as stolen. Exceptions might be – based on country-specific regulation – requests to access emergency services.

However, as discussed above, there is no cryptographic protection applied to the IMEI when transmitted to the mobile phone network. The IMEI is of course protected against manipulation in the device itself, but for significant numbers of mobile phones today these mechanisms are opened up by hackers in a relatively short period, allowing them to replace the IMEI number in stolen phones and thus circumvent the EIR effort.

In WiMAX, device authentication technically allows a similar system, or even an operator-specific database for checking against identities of stolen devices, to be based on the device certificates. The main difference to the above GSMA example is that the device identity is signed by a CA and embedded in the certificate. Also, if authentication includes verification of the device certificate and verification of the possession of the corresponding private key, the device's identity can be cryptographically verified by the network as part of the network entry procedure. Of course this also depends on the security within the device itself, but it is considered harder to hack a device and replace both the device certificate and private key for device authentication, instead of just showing a fake identity to the network when the network cannot base verification on digitally signed certificates.

Summarizing this use case, device certificates have the potential to improve the operator's means to fight against stolen devices. But this of course assumes large-scale cooperation between operators and creation of a central registry similar to the EIR in the GSM world – a system that the WiMAX Forum may leverage in the future.

Returning to revocation – which, besides ensuring overall system security, is required to realize use cases such as the above one based on device certificates – two rather different mechanisms are specified in the PKI area and have also been adopted as part of the WiMAX PKI system, namely certificate revocation lists (CRLs) [32] and online verification using OCSP [34, 35]. Both mechanisms are required to be supported by any CA that is part of the WiMAX PKI [18, 19].

The basic difference between these two revocation models is that the first one can work offline besides downloading the latest CRL from time to time, whereas the second one requires online checking but can at the same time be optimized to match the requirements of wireless systems. However, they both share the same idea of allowing the receiving end to learn whether a given certificate of a communication partner is still valid, or has been revoked and therefore can no longer be trusted.

Simply speaking, CRLs are files containing the list of revoked certificates under the control of a specific CA. The list itself must be protected against manipulation, hence it is signed by the CA itself and any entity participating in the PKI system can verify the validity of a CRL issued by the root CA by using the trusted root CA's certificate. Due to the length of a single certificate, CRLs only contain the unique serial number of all the revoked certificates. Depending on the number of entries in a CRL, the related files can still grow large.

Access to these lists typically happens by the participating entities downloading the CRL file from a known, publicly available directory. A common interval for updating CRL information is every few days to once per week. In summary, CRL handling in the PKI system is relatively simple, but the drawback is the need to frequently download possibly large files to end devices, which might negatively impact the overall performance of a wireless link like WiMAX. And as it would probably be hard to convince customers to pay for such downloads, the cost falls on the operator. Also, there is a tradeoff between the frequency of CRL downloads and the timeliness of CRL information.

WiMAX mandates all participating CAs to timely issue corresponding CRLs, i.e. either in case of a revocation event or at least every three months for the top-level CAs.

The alternative to CRLs is an online check to verify a specific certificate to a dedicated server (OCSP responder) in the network that is aware of the latest revocation information available for the PKI system. OCSP is the commonly used protocol for such online checking and the OCSP responder in the WiMAX PKI is related to a specific CA that is responsible for signing the

OCSP responder – as it would sign a CRL - to allow devices to verify that it is one of the trusted entities in the PKI. OCSP responses are cryptographically signed to make statements about the revocation status verifiable, and an additional message exchange is required for the standard OCSP communication, so this also does not come without cost. In addition, it needs to be ensured that an OCSP server is available at the time of certificate verification, whereas the CRL can be downloaded offline and is then locally available for verification. Regarding freshness of revocation data, revocation checking by using CRLs can never be as accurate as online checking and revocation status information sourced from a CRL that is already a few days old might be considered less secure.

One interesting fact regarding the use of OCSP in WiMAX is that during certificate verification as part of the network entry procedure, a WiMAX device is just running the EAP with the network infrastructure but otherwise does not yet have any IP connectivity. Hence, it would not be able to reach an OCSP server at this point in time. WiMAX is therefore using OCSP optimizations for network access [35] and allows OCSP to check the revocation status of the server certificate. This is performed as part of the EAP-TLS messages exchanged during network entry with device authentication [36].

3.4.5 Challenges

In general, PKI technology as used in WiMAX is a well-established and well-studied one. As a result, the challenges also are well understood.

Typically, a major part of the overall effort to set up and operate a PKI system is related to operational procedures and processes between the involved business entities. The WiMAX Forum created documents for both device [16] and server [17] certificates to define a framework for these procedures and describe the basic policy and security requirements, e.g. for operating WiMAX CAs. The documents describe the responsibilities of the different entities in the WiMAX PKI while focusing on the root CA aspect.

Interestingly, [16] limits the usage of device certificates to the access to WiMAX servers, i.e. to be used in EAP device authentication with a WiMAX CSN operator's AAA server. In general, the availability of device certificates in combination with a huge PKI covering these certificates would be a valuable enabler for adding security in other use cases. An obvious example would be to reuse the existing infrastructure for secure access to web servers where TLS is common and the underlying authentication protocol is the same as the one used through EAP-TLS or EAP-TTLS in WiMAX access. It would be sufficient to add the WiMAX device root CA certificate and appropriate manufacturer CA certificates, as required, to the list of trusted CA certificates in the web server to enable it to verify device certificates. Authentication of the web server by a device would not need changes, assuming that the web server certificate is issued by one of the root CAs commonly installed in the device browser's trusted certificate store.

However, the use of device certificates is currently limited strictly to the basic WiMAX network operation itself, because it is commonly considered bad practice in security to use the same certificate and private key across conceptually different applications. So would it make sense to use the existing WiMAX PKI and add application-specific certificates? The idea looks obvious at first glance. Looking at it in more detail, it becomes clear that for any application or class of applications, a new root CA would need to be set up – or likely two, one for device certificates and one for servers. Also, depending on the application class, new policy and requirements documents would need to be created and agreed upon to match application-specific needs. In summary, the effort would not be minor, and there may be other already existing roots for applications that make use of PKI-based security. So, as a general recommendation it is more reasonable to make use of existing PKI approaches for specific applications, instead of considering WiMAX device certificates, or the PKI created for issuing these.

Besides operational considerations, there are also technical challenges for the WiMAX PKI. One of them is related to the availability of a trusted time source in the participating entities. When verifying a certificate, it is necessary to check whether the lifetime (better: 'valid through') that is part of every certificate has not yet expired. Otherwise the certificate would need to be treated as invalid. Verification of time information of course requires accurate and trusted time information to be available in the device. This is especially true because a common trusted time source that is accessible via online means cannot be assumed – and would not be accessible at the time of server certificate verification during network entry to a device. Also, time information cannot be received from the network's AAA server due to the fact that this is the one still to be verified whether it is trusted.

Typical fixed and mobile devices like CPEs, mobile phones or laptop computers have their own time sources available in most cases. This holds with a few exceptions: some mobile phones in the United States receive time information directly from the network and do not allow the user to override it, even in cases like international roaming where the network-based time is not necessarily accurate. After all, the question of whether an internal time source can be a trusted one has to be answered by looking at the devices' internal security methods.

The general understanding, however, is also that an internal time source cannot be assumed for all WiMAX devices, especially considering those that are modems only, like USB sticks. For such devices, verification of server certificates can only provide limited insurance that the device is talking to a trusted network – the certificate or even a corresponding CRL might be an old one, so the certificate might be outdated or even revoked at the time of verification. And technically there are no means to overcome this issue as even OCSP-based online revocation is based on time stamps. Still, it can be considered a rather unlikely event that both the network's authentication server and the revocation infrastructure are malfunctioning.

3.4.6 Use Cases

Finally, after studying the means that are in place for allowing device authentication based on certificates in WiMAX, let us continue the investigation of use cases started at the beginning of this section.

Currently, the main motivation for making use of device certificates and the related device authentication procedure is to perform OTA provisioning, as described in Section 3.6. In the case of initial activation of a WiMAX device, there is no security association like a common username and password in place between the device owner trying to create a new WiMAX subscription and a network operator. In this case, the device certificate is used to establish the initial security between the device and the operator targeted at setting up a new subscription. Technically this happens by performing EAP-TLS together with the device certificate used for client authentication, with a secured WiMAX wireless link and keys generated from the EAP method. The EMSK key is used as the basis to derive a session key for securing subsequent OTA message exchanges between the device and the OTA provisioning server that dynamically gets this key from the AAA server. Hence, the existence of certificates even in non-initialized WiMAX devices allows the device owner to verify that the targeted network is the one it claims to be. In addition, it allows the network operator to verify the device's identity (and possibly bind it to the freshly created subscription) and to see that it is at least an approved WiMAX device registering with this new subscription.

Regarding the different scenarios for activation provisioning, there is the case where devices are already prepared for activation with a specific network operator; the network operator might even have sold the device through its own distribution channel. For such cases, the operator could have already prepared a subscription. As soon as the actual user initiates an OTA provisioning procedure, the operator can learn the authenticated MAC address by verifying the device certificate as part of the EAP-TLS handshake in device authentication. This in turn permits identification of the subscription prepared for this device via the authenticated MAC address. So the operator can create a final subscription based on the specific and known properties of this device.

This example might be interesting to operators because the diversity of potential WiMAX-enabled devices is expected to be large, see also Table 3.2 below that is explained in more

Table 3.2 DevType MO listing different classes of WiMAX devices

Type of device	DevType string for population	Terminal equipment node
PC card – single mode	SingleModePCCard	Required
PC card – multi mode	MultiModePCCard	Required
Express card – single mode	SingleModeExpressPCCard	Required
Express card – multi mode	MultiModeExpressPCCard	Required
USB card – single mode	SingleModeUSBCard	Required
USB card – multi mode	MultiModeUSBCard	Required
Basic modem	BasicModem	Optional
SOHO modem	SOHOModem	Optional
Personal media player (PMP)	PMP	Optional
Multi mode PMP	MultiModePMP	Optional
UMPC	UMPC	Optional
Laptop	Laptop	Optional
Internet tablet	InternetTablet	Optional
Single-mode handset	SingleModeHandset	Optional
Multi-mode handset	MultiModeHandset	Optional
PDA	PDA	Optional
Gaming device	GamingDev	Optional
Video phone	VideoPhone	Optional
Machine to machine	M2M	Optional
Digital camera	Digital Camera	Optional
Digital camcorder	Digital Camcorder	Optional
Wearable device	WearableDev	Optional
Multi-mode messaging device	MultiModeMsgDev	Optional
Electronic book	EBook	Optional
Navigation device	NavigationDev	Optional
In-vehicle entertainment device	InVehicleEntDev	Optional
Home media gateway	HomeMediaGW	Optional
Music player	MusicPlayer	Optional

detail in section 3.6. Mechanisms for the dynamic exchange of device capabilities had been considered as part of the overall effort on device certificates by the WiMAX Forum. However, approaches to add such capability information to the device certificate itself have been deprecated. Certificates are considered an inappropriate means for carrying such information, one reason for this being the inflexibility of a certificate to ever change the signed content again during the device's lifetime, although device capabilities might well change – the idea is not to allow for dynamic reprovisioning of a device OTA with a new certificate.

Device authentication can also be performed for normal network entry, i.e. when a subscription already exists on the device and is used for entering the WiMAX network. Running device authentication during initial network entry in addition to subscription authentication (e.g. by performing EAP-TTLS with EAP-AKA as the inner method) might allow the network operator to limit subscriptions to specific devices. The operator's AAA server learns the device's MAC address also during subscription-only authentication, but this is not an authenticated one.

OTA activation of subscriptions certainly is the most relevant use case for WiMAX operators supporting device authentication. However, another interesting scenario happens to arise from the area of emergency services, where WiMAX devices with commercial VoIP functionality installed are required to allow their users to make emergency calls.

In general, this area is based strongly on local regulations that vary between different countries with different national laws. Support for emergency calls with a subscription installed in the device is commonly required by law. An area of particular interest is the capability to make emergency calls without a subscription being available. This is typically referred to as unauthenticated mode. Taking mobile phones and traditional cellular voice services into account, the requirements for unauthenticated calls vary greatly between countries, e.g. a number of European countries require all cellular mobile phones to allow their owners to make an emergency call without a SIM card inserted (i.e. with a blank phone without any subscription data available in GSM/UMTS networks).

In WiMAX networks unauthenticated emergency calls can be supported and require the operator subject to such regulatory requirements to allow network entry without a valid subscription being installed in the MS. The solution for this is based on device authentication. Further details are given in Section 4.7.

3.5 Security Design Considerations in the WiMAX Network Architecture

There are a number of basic design choices that were made when developing the WiMAX network reference model. In general, the idea was to enable a broad range of deployment models starting from very small ones with possibly only a single BS, but going up to country-wide networks of large operators, supporting very large numbers of subscribers and providing all the necessary means for full mobility and roaming.

The latter class of business models where all equipment is part of a single operator's network matches today's cellular networks like those deploying GSM/UMTS technology or CDMA. WiMAX, in contrast, is designed to allow for more dynamic business models by decoupling the access part and the core part of the network. So operators can define their role in the WiMAX ecosystem as being a WiMAX NAP by covering the access part and running an ASN, choose to act as a NSP by deploying a CSN, or combine both in cases where they run a full WiMAX network. The WiMAX network architecture allows ASN operators to interact with more than a single CSN by offering NAP sharing, where such deployments might be comparable to existing business models in the fixed broadband access (e.g. DSL) area rather than in mobile phone networks. In the DSL area, a common model is where the 'last-mile' infrastructure to the customer is owned by one operator, and other operators that are rather acting as the ISP lease this infrastructure but do not operate their own lines.

To flexibly support all the different possible deployment models and operator roles, several mobility and handover schemes were developed so that it becomes possible by choosing the appropriate options to optimize the WiMAX network efficiently for a specific deployment with particular business relations to partnering operators.

Mechanisms for providing subscription handling, exchanging subscription-related identity information and securing the overall network communication are designed to match the inherent requirements for flexibility. The following will consider the major security design choices to support efficient and secure operation within a WiMAX network, and between operators, from a deployment perspective.

3.5.1 Authenticator Mobility Support

For network access security, the authenticator in the ASN-GW is the central function in an ASN; Section 3.3.2 provided a general overview of the ASN and authenticator roles in securing network access.

Let us now look deeper into network operation when it comes to mobile devices handing over between BSs. A MS that performs initial network entry triggers a BS to select an ASN-GW for the new MS and then relay security-related information between the MS and the ASN-GW. The assumption here is that the BS is configured with information about which ASN-GW to choose.

As discussed in Section 3.3.2, the EAP exchange is relayed through the BS because the authenticator function is centrally located in the ASN-GW. The design decision to split security responsibilities in the ASN between the BS as the termination point for the air interface security and the ASN-GW as the central controller for authenticated access in the ASN is not primarily related to security considerations, but implements an important performance optimization for mobility.

From a security viewpoint, it is reasonable not to put the authenticator functionality that is critical to the ASN's overall security level into each BS. Consider the fact here that a BS is often located in a rather exposed physical environment and therefore is clearly an easier target for attacks, compared to a central controller for the access network that will typically be located in a physically secure environment. To provide an extreme example here, BS functionality might be implemented in a femto CPE device where the operator is very unlikely to be in control of any physical location.

However, the main reason for putting the authenticator functionality into the ASN-GW is for mobility. Authentication via the EAP, depending on the EAP methods' number of round trips, requires end-to-end communication between the MS and the backend AAA server in the home operator's CSN. And especially in a roaming situation this might involve numerous entities that the communication has to traverse. Speed is important, particularly for initial network access, as a long delay after switching on a WiMAX device clearly affects end user experience adversely. Also, such end-to-end communication increases the overall load on the network and interoperator interfaces.

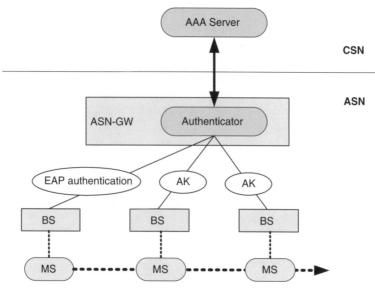

MS moves between base stations

Figure 3.10   MS movement within an authenticator domain

Keeping this in mind, it made a lot of sense in the design of the WiMAX network architecture to include the authenticator function in the central ASN-GW instead of putting one into each BS, as depicted in Figure 3.10. Otherwise, any handover from one to another BS would require a rerun of EAP authentication with the new BS and establishing a new security context for wireless link protection. Instead, with the authenticator in the ASN-GW, EAP authentication only has to be performed once during initial network entry and does not need to be executed again as re-authentication before the end of the EAP session lifetime. So any MS connected to the ASN can move between BSs without the need to rerun the EAP and with minimal delay for setting up a fresh security context with the target BS. Keys within the network will be bootstrapped by the authenticator directly into the target BS from the active MSK key – the authenticator and the MS will generate a new AK key (see Section 3.3.2) from MSK for every new base station and this is kept an ASN-internal operation at the network side.

Of course, this model can only work as long as the MS hands over between BSs that are connected to the same authenticator. Any other authenticator would not be in possession of the current security context for the EAP session of the MS. So an authenticator spans a set of BSs and these will choose the authenticator for an MS entering the network.

Let us introduce a few definitions at this point. The authenticator and the set of all BSs covered by this authenticator are called the 'authenticator domain'. A typical example of an authenticator domain is the set of all BSs connected to the same ASN-GW. An MS that enters the network will be assigned the authenticator of the authenticator domain that the BS belongs to. This authenticator will be called the 'anchor authenticator' for the MS EAP session, and the anchor authenticator will not change during handover until another anchor authenticator is chosen within an EAP re-authentication – a procedure that is called 'authenticator relocation'.

Also, to describe an EAP session within the scope of WiMAX networks, the term 'WiMAX session' is used. A WiMAX session spans the MS session established by EAP/AAA-based authentication and authorization during initial network entry, and ends at the time of network exit. It continues to live across and is not terminated or altered by an EAP re-authentication.

Note that an authenticator domain does not limit flexibility in the design of a WiMAX ASN: a BS can be part of more than one authenticator domain – although in such cases it needs to be configured with appropriate logic to choose the correct authenticator domain for a MS entering the network.

3.5.2 Anchor Authenticator and Relocation

As soon as a MS that is connected to the WiMAX ASN hands over to a BS not covered by the domain of the current anchor authenticator, i.e. the BS is not connected to the same ASN-GW, the new ASN-GW responsible for the new BS will come into play. In this situation, shown in Figure 3.11, the WiMAX architecture allows for two different ways of how to continue service to the MS.

The first option is the simpler one: an EAP re-authentication is triggered by the network, and the MS establishes a new security context with the new authenticator, therefore moving the anchor authenticator from the original to the new ASN-GW. The benefit of this method is that the MS can now be fully handled by the direct authenticator covering the actual BS that the MS is connected to. However, it requires the EAP to be performed and therefore impacts handover performance. Even if the EAP methods and their implementation in both the AAA server and the MS support (like most state-of-the-art EAP methods do) a fast re-authentication exchange, a message exchange with the CSN is still required.

As the second option does not require such interaction with the CSN, it is possible to keep the original anchor authenticator. This means that the new BS receives the required security context for wireless link protection including the BS-specific AK key from the anchor authenticator of the original authenticator domain, but this is requested by the new BS through

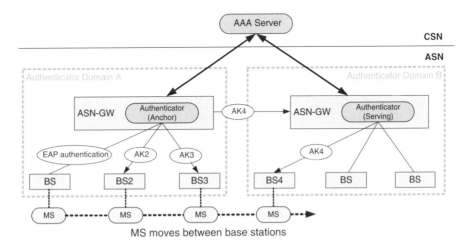

Figure 3.11 Handover to a BS covered by a new ASN-GW

the authenticator of the authenticator domain that the BS belongs to (AK4 in Figure 3.11). This new authenticator is the one located in the *serving* ASN-GW. In contrast to performing a re-authentication, this second option avoids the delay introduced by rerunning the EAP. It does, however, slightly increase signaling complexity in the network for the MS as related messages now need to travel, as depicted in Figure 3.11, from the BS across the R6 reference point to and from the serving ASN-GW and across R4 to and from the ASN-GW holding the anchor authenticator.

Due to the fact that this process can be repeated several times and add more intermediate ASN-GWs as hops for the MS-related signaling, the communication chain may become too costly, and the network can decide to actually trigger the first option and perform authenticator relocation. This means performing re-authentication to pull the anchor authenticator into the serving ASN-GW.

3.5.3 Identity Hiding

During the development of the WiMAX network specifications, a substantial effort was made to address the area of identity hiding. This is all related to the identity information of WiMAX subscribers passed across the network reference points and the wireless interface, and, although the resulting architecture for identity hiding can be considered a reasonable tradeoff between addressing privacy requirements and avoiding negative impacts on network operation, there are several reasons why the effort was worthwhile.

One of them is wireless security, which is of course in the interest of the subscriber. It is commonly understood that the threat level regarding identity privacy and identity theft in wireless communications is a serious one. This is due to the fact that it is easy for an attacker to just scan the whole radio communication in the current location and record what is going on. Lists showing the identities of all the subscribers of an operator who happened to be in a certain location, such as a central place in a downtown area, are not things that any operator would like to reveal to the broader public or, even worse, to competitors.

To provide an example to show that other wireless technologies have been taking this seriously, the interested reader is referred to UMTS networks and the related 3GPP specifications [39]. There, the IMSI as the central subscription identity is sent over the air interface only once for the initial procedure to bring the subscription on the USIM card into action. For all subsequent communication across the wireless link, the IMSI is replaced by a temporary identity (TMSI) that does not allow attackers to identify, or track, a specific mobile by recording OTA communication.

But we must also consider the specifics of WiMAX network design and operator requirements. As discussed above, in contrast to existing architectures for mobile networks WiMAX introduced the possibility of splitting the access and the 'ISP' part of the network into ASN and CSN that can be run by different operators acting as NAP and NSP. The subscriptions, however, are owned by the NSP and are an extremely valuable resource for its business. So the NSP operator's business interests are in conflict here: on the one hand, there is a need to partner with a number of NAP operators to offer wider coverage to its subscribers; on the other hand, the NSP operator might have no interest in sharing the identities of all its subscribers with the NAP. Imagine that a NAP can also operate its own NSP business in addition to sharing NAP services with other NSPs and therefore might have a good reason to learn about the 'partnering' NSP's customers and try to pull them into its own network.

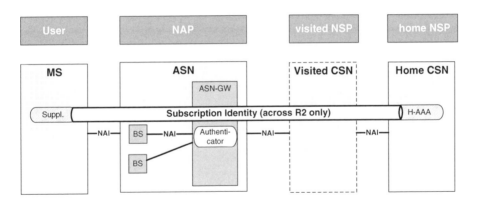

Figure 3.12   Identity hiding across NAP and visited NSP

This scenario provides the basic reason why WiMAX protects subscriber identities, in addition to hiding them OTA, from the NAP and any involved visited NSP network.

The central identity tied to WiMAX network operation is the NAI used as outer identity in EAP messages during the network entry procedure. However, the NAI is also used in mobility and in AAA signaling in the network. As discussed in Section 3.7, an AAA session ID is used in addition to handle all session-related information related to the same subscription.

In contrast, the permanent identity of the subscription is only used in the EAP method itself, as depicted in Figure 3.12. When choosing an appropriate EAP method like EAP-TTLS with the inner EAP method for subscription authentication, this in fact means that only the EAP method endpoints – the MS itself and the home CSN AAA server – can see the permanent subscription identity. All other network operation is based on the identity transferred as outer NAI in the EAP. So, especially the ASN network and any visited CSN, i.e. the NAP and all visited CSN operators, cannot see any subscription identity of the home CSN subscribers.

Note that in contrast to 3GPP there is no need in WiMAX to convey the permanent subscription identity at any time to the NAP or visited CSN operator.

WiMAX enhances this by defining the concept of a pseudo-NAI. The NAI is composed of a 'username' part, followed by the '@realm' part (see Section 3.3.3 for further details). In contrast to the permanent subscription identity that, although not necessarily being in NAI format, may carry the real subscriber name in the username part, WiMAX allows the exchange of a unique random value as the username part instead of the real username. So the major NAI property for correct network operation – uniqueness within the home operator's realm – is fulfilled without providing any subscription-related information. Each pseudonym value is chosen by the MS itself for an EAP run. To allow for additional protection against traceability of an MS-related session across different EAP re-authentications (i.e. for a longer period to record the subscriber's movement), it is possible to change to a new pseudonym for every re-authentication. Such new pseudonyms cannot be linked to former ones by scanning the wireless communication.

Independent of this generic pseudo-NAI concept of WiMAX, there are EAP methods that inherently support identity hiding as part of their design. The best example here is EAP-AKA, which implements the identity hiding scheme of 3GPP. However, implementing either one of those mechanisms needs careful consideration to avoid conflicts between the WiMAX feature and method-specific mechanisms. To provide one example, in EAP-AKA the AAA server is responsible for generating the pseudonyms used in the NAI, whereas WiMAX generates them in the MS.

The pseudo-NAI concept only covers the network procedures related to the subscription identity, but not the MS itself with the MAC address as the device identity. In fact, an attacker could track the MAC address of the device instead of focusing on the NAI to learn about the MS behavior. There are important differences of course between the subscription and the MS. For example, the subscriber can easily exchange one MS against another without an attacker noticing it by tracking NAI information. However, it is understood that to provide complete identity privacy on the wireless link, MAC address privacy would also be required.

While developing the WiMAX network identity hiding scheme it was actually considered whether identity hiding should be extended to also cover device identity. However, the effort for hiding the permanent MAC address of the WiMAX device would have been several times higher than that required to hide the subscription identity. This is based on the fact that the MAC address is an inherent part of the WiMAX wireless MAC layer and ASN signaling across the R4 and R6 reference points. The expected effort to apply changes in this area would have exceeded the benefits of protecting device identities in the initial WiMAX Forum specifications for a Release 1 system.

Changing the MAC address can in principle be considered as an identity spoofing attack against the network and is expected to create problems for proper network operation in many places of the WiMAX ASN. The WiMAX network specifications take such spoofing scenarios into account. Methods are available as an optional functionality for Release 1 equipment and are mandated for any ASN-GW or BS implementing the Release 1.5 network specifications. The actual mechanisms to identify and tackle misbehaving devices that are either simply misconfigured, or spoofing the MAC address on purpose, include:

- ASN support for parallel new network entry with the same MAC address as that of an already existing WiMAX session. This is allowed to proceed until the new entrant either proves to have a valid subscription or is denied entry. This first-level protection cannot detect all cases of MAC spoofing, but it disables all critical denial-of-service attacks that otherwise may become possible for an attacking device misusing the radio link signaling with a wrong MAC address. Also, if the claimed MAC address of the new entrant is the wrong one, any attack would be easily traceable by the operator later on, back to the actual subscriber using the misbehaving device.
- Device authentication in addition to the above measure as a second level of protection, which, if enforced by the operator, allows cryptographic verification of the device's MAC address and provides strong protection against MAC address spoofing.

Interestingly, the extensions that were made in the ASN-internal signaling across the R6 interface to protect against MAC address spoofing support several devices using the same MAC address for a limited period of time. In fact, the ASN became more independent of the MAC address as the central identifier of the session information related to a MS being attached to the network. And returning to the original aspect of MAC address hiding, a further extension of the same concept to also cover R4 and all the WiMAX-defined ASN internal messages would make the ASN signaling independent of the MAC address and in turn would be able to support MAC address hiding.

Furthermore, in combination with the appropriate extensions in the specification of the wireless interface, this would lead to full support of MAC address hiding in WiMAX. For WiMAX systems based on the WiMAX Forum Release 1.5 specifications for both the network and the radio interface profile, this is not yet possible. Also, it would not make much sense for a vendor or operator to try to realize this in any proprietary way. However, bearing in mind that

(1) the actual effort regarding extensions within the WiMAX ASN would be quite limited,
(2) the CSN entities do not depend on the MAC address, and
(3) the upcoming next generation of WiMAX radio specifications as per 802.16 m can be expected to include methods for enabling MAC address hiding,

WiMAX is likely to provide relatively strong features in this area in the mid-term future.

To return once more to the status quo, an example is provided in Figure 3.13 of a roaming scenario to show which identities can be seen (and traced) on the wireless link, in the NAP and in the NSP. The example assumes that the MS is moving from ASN A to ASN C via ASN B. ASN A and ASN B are connected to the subscriber's home CSN A via the visited CSN B. ASN C is connected to the home CSN A via the visited CSN C. The MS is assumed to perform re-authentication on entering a new ASN.

Independent of this specific example, by eavesdropping on the communication across the wireless link it would be possible to learn the pseudo-NAIs 1 to 3 independently of each other, including the MAC address of the device. If the device and MAC address remain the same, it would be possible to relate the three different pseudo-NAIs to the same device. However, an attacker would not be able to prove that the same subscriber has been using the device all the time.

The NAP covering ASNs A to C would also be able to correlate the pseudo-NAIs via the MAC address. This would not be possible when the ASNs belong to different NAPs. In contrast, the visited CSN B would see, besides the MAC address, the pseudo-NAIs 1 and 2 due to the fact

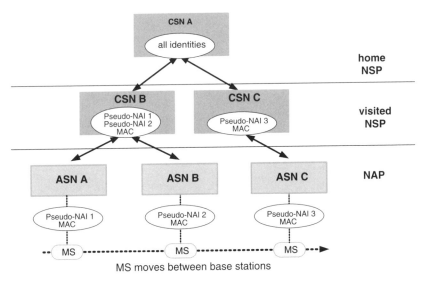

Figure 3.13   Identities as seen by NAP, NSP and on the wireless link

that it learns the NAI-MAC binding from ASN A and ASN B. CSN C can only see pseudo-NAI 3 in addition to the MAC address.

Only the home CSN of the subscriber and the MS would be able to see the real subscription identity. So the example shows that, although there is a limited possibility for an attacker to track the different pseudo-NAIs across different WiMAX access networks and correlate them by using the MAC address, the permanent subscription identity is protected.

Summarizing the above, the motivation for subscription identity hiding lies mainly in the unprotected wireless interface that is an easy target for eavesdropping. It also comes from limited trust between the different and possibly diverse interacting business entities in the WiMAX ecosystem. The recommendation that can be given here is clearly to make use of the pseudo-NAI mechanism. Independent of whether there is full trust between operators or not, the former motivation provides a sufficient reason for this.

3.6 'Bootstrapping' a Subscription OTA

The above sections focused on what happens in terms of security and identities for initial network access, re-authentication and during handover between different BSs of even ASN-GWs. But what happens when there is no subscription available on a WiMAX device, i.e. the device has not yet been provisioned and activated?

3.6.1 Subscription Creation and Device Models

A 'blank' WiMAX device without any valid WiMAX subscription and security credentials being installed cannot access any WiMAX network without performing additional steps. For any normal WiMAX network entry, appropriate subscription information including a valid identity and security credentials is necessary – recall that any WiMAX network that complies with the WiMAX Forum specifications will only allow network access after establishing a secure channel across the wireless link.

WiMAX devices and networks are expected to offer broad support for OTA provisioning and activation, besides device management that can be performed by the same means, and this has been a key feature for operators during development of the WiMAX specifications.

First, let us look at other existing wireless technologies to determine why WiMAX is different here, and put a stronger focus on dynamic OTA provisioning from the beginning.

Starting with WLAN, as of today it can at least be assumed that the majority of WLAN usage still happens without any subscription being available in advance, and also without a subscription being created after use. There are networks of WLAN hotspots that are centrally covered by larger operators (often with a strong background in cellular systems), and where subscribers can get connected without an initial 'provisioning' step. Also, there are some larger roaming federations. Anyway, most WLAN users typically buy a one-shot access by purchasing a scratch-card, registering with a portal page through the device's web browser, etc. More permanent subscriptions are often based on simple username/password login information, where the user is finally responsible for keeping such subscription data secure – something which can become particularly inconvenient when different hotspots are covered by different, independent operators.

At the opposite end, in large mobile phone networks access is typically not possible without a valid subscription being available in the mobile phone, leaving aside here the possible exception of an emergency call. For example, in the GSM world, putting a blank mobile phone into operation would not make a lot of sense as long as no SIM card is inserted. Here, it is not the phone storing the subscription data but the smartcard with a SIM or USIM application that includes the subscription information and securely locks away the subscription keys. Initial provisioning (although there might be more accurate descriptions of the actual steps) basically requires the user to physically acquire the SIM card and then insert it manually into the mobile phone or data card, to activate the subscription and thus to actually enter and use the network.

For WiMAX, it is important to look at the operator's models for getting users into their networks and this especially includes considering the expected distribution chain and types of WiMAX devices, and the level of control that the operator has over the WiMAX devices actually used in combination with a subscription.

As for types, WiMAX radio modules can be part of quite different equipment, e.g. fixed CPE-like devices, laptop computers, handhelds, smart phones, or even consumer electronics like digital cameras or portable music players. And when considering the diversity of devices and the expected ones that will support WiMAX first (i.e. CPE devices, and laptops or handhelds) in larger volumes, it becomes clear that these devices are typically not under the control of an operator's common distribution channels. So here is one of the main motivations for deploying WiMAX networks with OTA provisioning facilities from the beginning: unlike today's mobile phones, a significant number of WiMAX devices will be bought as 'blank' devices regarding the availability of a WiMAX subscription, and need to be provisioned first, before any WiMAX network service can be used. Such devices are rather comparable to the situation found today for WLAN: a laptop supports WLAN radio (and may also support WiMAX) but does not come with any a priori relation with an operator to make use of WLAN services. The important difference here is that WLAN usage is easily possible without creating a subscription. This is clearly different for WiMAX due to the regulated spectrum.

However, it is equally possible that WiMAX devices will be sold through the operator's channels and will therefore be under the operator's control, making it a completely different provisioning model compared to the above one. Such a model equates to today's models of cellular operators as well as DSL or cable operators and allows for subsidizing devices, or controlling device lock mechanisms where a subsidized device is most likely bound to the operator's network that handed out the device as part of the new subscription.

To reflect the different provisioning schemes as much as possible, WiMAX devices are categorized into two basic 'models':

- Model A devices are those that are controlled and possibly subsidized by the operator.
- Model B devices are those that are sold through general retail channels. Such devices can be further categorized into two groups: those being completely unrelated to any operator (model B_1) and those coming with some initial configuration prepared for provisioning with a specific operator's network as soon as the user decides to put the WiMAX into use (model B_2).

Important applications of OTA provisioning include:

- Setting up a new subscription on a blank device or a device that has already been prepared to establish a subscription with a specific operator
- Modifying the settings of an existing subscription, such as registering a new device that is used with the subscription.

- Modifying the device's configuration information. Common examples for the latter are updates of the operator's preconfigured network information for network discovery in the device (list of preferred NAPs, roaming partners, etc.) or removing a device lock that was set earlier.

3.6.2 OTA Provisioning Model and Protocols

Simply speaking, the WiMAX dynamic OTA provisioning defined by [20] allows the MS to connect to a provisioning server in the CSN operator's network in a well-defined way. The provisioning server itself deploys connections with the backend subscription management facilities of the operator that runs entities like the central subscription database, subscription policy and management servers, or portal servers for interaction with the actual user and with the AAA server. However, these operator-internal interfaces are subject to specific vendor implementations and are, besides some triggers and parameters exchanged between the provisioning server and the AAA server, not standardized by the WiMAX Forum. During provisioning, the ASN only plays a passive role as it is not directly involved in the process of provisioning that is fully covered by the NSP, not the NAP.

For the communication between a WiMAX device and the provisioning server of an operator, two standard protocols are supported (Figure 3.14). Both were developed by different organizations and had to be extended by all the required provisioning parameters specific to a WiMAX device:

- OMA-DM [42] as developed by the Open Mobile Alliance (OMA) with the WiMAX-specific profiling defined in [21].
- TR-69 [43] as developed by the Broadband (former DSL) Forum with the WiMAX specific profiling defined in [22].

For many mobile devices like mobile phones today, the OMA-DM protocol suite is used for provisioning. It defines exchanges between the provisioning client in the device and the provisioning server that are based on XML messages. The actual information to be provisioned into the device is described as a set of management objects (MOs) that are logically organized

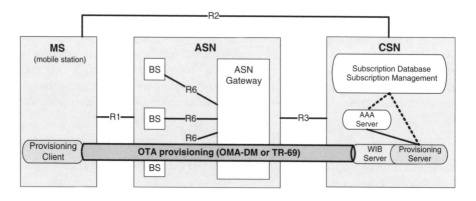

Figure 3.14 WiMAX OTA provisioning network reference model

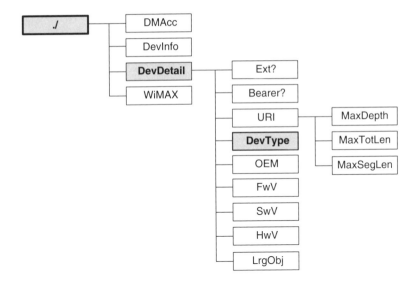

Figure 3.15   OMA-DM management tree for WiMAX devices: DevType MO

in a tree structure called the management tree. For WiMAX, its own management tree is defined to cover all the WiMAX-specific MOs. This is depicted in Figure 3.15, where the first level below the root of the tree is completely shown, and the second level zooms into the DevDetail MO and shows an example path toward the DevType MO.

This example is not meant to provide a full overview of the WiMAX-defined MO, but shows the specific path through the WiMAX management tree to let us have a look at the DevType MO. Interestingly, this provides some rough insight into the envisioned classes of WiMAX devices (Table 3.2).

Conceptually, the TR-69 method used in WiMAX OTA provisioning is rather similar to OMA-DM considering the XML-based protocol using SOAP-encoded messages between the WiMAX device and the provisioning server. It also organizes the data to be provisioned in a tree-like structure called a parameter management tree that consists of a set of parameters as the tree's leaves.

The actual parameters in the management tree describing a generic Internet gateway device (IGD) are modeled by the Broadband Forum in [44] and the WiMAX-specific profiling of [22] extends this DSL IGD model for WiMAX devices that are expected to be CPE-like devices in most cases. Although the class of devices implementing TR-69 comes with a number of different properties compared to Mobile WiMAX devices using OMA-DM, a substantial subset of the data is the same between the methods, like profiles for EAP usage and parameters specific to the wireless link.

3.6.3 Activation and Initial Network Entry

The network entry procedure of a WiMAX device requires a subscription with valid subscription credentials to be available to the device, otherwise it would not be able to successfully complete the EAP and establish a secure link to the network's resources across the wireless interface.

Now let us consider a device like a WiMAX-enabled laptop that has been bought from a general electronics store and does not have any such credentials that would allow connecting to a specific operator's network. Here, a special method for initial activation of the device's WiMAX capabilities with a selected operator becomes necessary, otherwise only out-of-band provisioning like walking to an operator's store or creating a subscription while visiting an operator's web portal would be possible.

As part of a standard network entry procedure, the device will perform the four general steps of:

1. NAP discovery (finding out about the available access networks).
2. NSP discovery (finding out about the available NSPs that are also the potential targets for setting up a new subscription).
3. NSP selection (optionally choosing an NSP in case of roaming as the visited one).
4. Performing network entry (including EAP).

To enable dynamic OTA provisioning for initial activation of general retail devices for a specific operator, WiMAX makes exclusive use of the EAP-TLS method. A device looking for WiMAX networks without any viable subscription installed will therefore perform the following steps, resulting in a modified network entry procedure as shown in Figure 3.16:

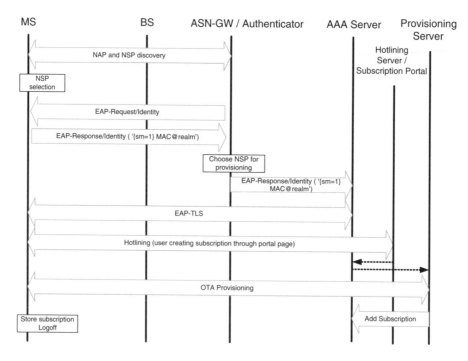

Figure 3.16   OTA activation provisioning for a new WiMAX subscription

1. After NAP discovery and selection, gather information about the available networks (NSPs) that provide for OTA activation provisioning. Here, it is important to consider once more that a NAP as the pure access provider might operate with NAP sharing enabled: it is interconnected and partnering with more than one NSP network that makes use of the NAP's access. It would be close to impossible to preconfigure any WiMAX device with all such – possibly changing – business relations in advance, so the NAP's BSs are able to transmit a list of possible CSNs. By enabling this mechanism the NAP operator is able to advertise the available NSPs to choose from to create a subscription. Technically, the list of available NSPs is sent to the MS as part of the MAC management messages exchanged during a network entry.

2. Given that a list of possible NSPs is advertised, the device can be preconfigured with an initial policy for choosing a specific NSP and can select to enter activation provisioning with this NSP. That is, although not equipped with a valid subscription from the beginning, a device might be already prepared for a 'preferred' operator, mapping to a device model B_2. Alternatively, a device model B_1 would display the received NSP list to the user and allow manual selection of the operator to create a subscription.

3. Now, the actual process of entering the network for provisioning can begin. For this the device will trigger the network at the lower layer and the ASN-GW will send an EAP-Identity Request and the device will reply by sending an EAP-Identity Response. The important difference of the initial provisioning case from standard network entry is the NAI that is sent by the device as part of this response. The NAI will carry the device's MAC address as the username part and will either come without a realm at all (leaving the decision for choosing an NSP for provisioning the device to the NAP's default policy) or show the NSP realm that is requested for provisioning as the realm of this NAI.

   As an example, a blank device with the MAC address 'FE0B5D979383' receiving the information from the selected NAP that either 'wo_besteffort.de' or 'wo_bestqos.de' are available as NSPs and selecting the first NSP, would put 'FE0B5D979383@wo_besteffort. de' as the NAI in the EAP-Identity Response to the ASN. The ASN-GW would forward such a EAP message to the AAA server of the selected NSP 'besteffort.de' to handle the request and start the process of provisioning the MS new subscriber.

   At this point, one essential piece of information would still be missing in the network. This is the actual indication to the network that this network entry is requesting provisioning and is not a normal network entry for an existing subscription. Without providing such an indication, the AAA server would not be able to resolve the received identity information in the NAI to an existing and active subscription, and could only respond by rejecting the entry attempt.

   To solve this, WiMAX uses a WiMAX-specific NAI 'decoration' to indicate OTA provisioning requests. Being a decoration of the NAI this is actually a set of characters added to the NAI in front of the standard 'username@realm' content. The actual decoration used is '{sm=1}', which states that the device is requesting to enter the network in service mode 1, which indicates OTA provisioning. Another service mode 2 is defined for emergency mode network entry.

   As a result, the actual NAI used by the device in the above example is '{sm=1} FE0B5D979383@wo_besteffort.de'.

4. Subsequently, the device will perform EAP-TLS with the selected NSP to establish a secure connection OTA and with the device certificate and server certificate of the NSP's AAA

server for authentication. The EAP-TLS exchange does not provide much information or assurance to the NSP network from a security point of view, except a verified MAC address of the requesting device that might be used by the operator later on as part of a new subscription, or it can help to relate the requesting device to a set of model B devices that had been prepared for provisioning with the operator. However, it allows the device to authenticate the network within the WiMAX PKI system and hence gain a certain level of assurance that it is talking to the expected operator.

5. After successful network entry with EAP-TLS in provisioning mode, the device will be granted limited IP connectivity to the NSP's network. It can now enter a hotlining phase where the device connects to a server and allows the user to interact (typically by displaying an interactive page in the device's browser) with the operator's subscription portal, study the terms and conditions, and finally create the subscription. However, hotlining is an optional mechanism that is described by the WiMAX specifications but might be replaced by similar operator-specific mechanisms in the provisioning phase.

6. The device will now use its limited IP connectivity to contact the provisioning server using – depending on the device type as discussed above – either the OMA-DM or TR-69 protocol.

   Here, it is interesting to look at the security applied to the actual provisioning communication between the device and the provisioning server, as these two entities are found together in a possibly quite dynamic environment, meaning they cannot be expected to have any common ground in advance regarding security – like a pre-shared key for protecting the exchanged messages. However, it is important that the provisioning process especially operates at a very high security level, otherwise the resulting new subscription will not be fully trustworthy on either side over the full subscription lifetime.

   WiMAX solves the above by deriving a fresh session key from the initial EAP-TLS exchange. This resulting key is generated directly in the WiMAX device in parallel to further derived keys like the MSK used as the basis for air interface protection. At the network side, the provisioning server needs to talk to the AAA server to receive the same key. By this method, both the device and the provisioning server share a key that is then used to secure the OTA provisioning exchanges at the application layer.

7. After successful provisioning of the device as well as creation and activation of the new WiMAX subscription in the operator's infrastructure, the device will leave the network and the service mode session for activation provisioning is terminated.

From now on, the device can use the new subscription to perform standard network entry with the operator as its home NSP.

3.6.4 Reprovisioning

An important topic for operators is to keep service and maintenance cost reasonable, so it is beneficial to both operator and customer to be able to dynamically provide customers with subscriptions or make it fast and easy to modify the subscription details according to a customer's needs.

It is also beneficial to provide facilities for reprovisioning in cases where a subscription in a WiMAX device fails to work for reasons connected with corrupted data, or the device selecting the wrong subscription for a network in cases where several are installed in parallel. Under these circumstances, it is essential for the operator to implement a dynamic mechanism for reprovisioning the device with the correct subscription and to fix the problem by minimizing the risk of losing a frustrated customer to a competitor, or to another technology offering wireless broadband access. Also, it is reasonable to avoid situations where affected customers will be forced to call the operator and possibly enter manual procedures to replace the corrupted subscription with a new one.

WiMAX devices can behave in a specific way to allow reprovisioning in cases when the subscription used for network entry does not work for the reasons mentioned above.

After a failed network entry attempt (i.e. on receiving an EAP-Failure message from the NAP) the WiMAX device can in fact retry network entry a second and a third time. However, if the subscription still does not work for network entry after this, WiMAX devices are expected to fall back and try network entry with EAP-TLS, requesting provisioning. The provisioning server of the home NSP will then be able to reprovision the device with the correct subscription details, which will usually fix the problem.

Discussion within the WiMAX Forum identified a side effect impacting security on implementing an automatic fallback to provisioning mode in WiMAX devices, which would potentially allow a rogue operator to force devices into provisioning mode and to modify their configuration data. This would obviously allow for a larger range of attack scenarios. The likeliness of such attacks is actually considered very low due to the fact that the device will verify the network server's certificate during the EAP-TLS exchange and a rogue network will fail to successfully complete EAP-TLS. Nevertheless, this serves as a good example of why it is important to make use of the WiMAX PKI in a secure way. This includes secure storage of trusted root certificates in the device, making use of certificate revocation checking, especially at the device side by using either up-to-date CRLs or online verification via OCSP, and proper verification of the server's certificate and certification chain.

3.7 Identities in Mobile WiMAX

For proper network operation, identities can be considered the central binding between the different functions and processes running within the operator's network, between different operators and between the network and the WiMAX devices. They are used across all sections of the network. The validity is partially just hop by hop between adjacent network elements, but many identities are used in WiMAX and have an end-to-end significance in the overall system.

Identities are among the most important pieces of information for all network-level services, but are paramount for the sound operation of:

- the overall system's security;
- subscriber-related accounting as well as interoperator settlement; and
- interaction of the WiMAX network with application services like location or VoIP.

Whereas the application-related aspects will be considered in more detail in Chapter 4, this section highlights a selection of identities that are of special importance to network operation and therefore need specific consideration for any WiMAX network deployment or MS implementation.

Table 3.3 Selected identities and their significance

Identity	MS	ASN	Visited CSN	Home CSN
MAC address	X	X	[X]	[X]
NAI	X	X	X	X
Subscription ID	X	[X]	[X]	X
NAP-ID	X	X	X	X
NSP-ID	[X]	X	X	X
WiMAX-Session-ID	—	X	[X]	X
CUI	—	X	[X]	X

3.7.1 Overview of WiMAX Identities

The identities considered in this section are carried within control messages, and as a rough categorization control messages are exchanged either between the MS and the WiMAX network, or within the WiMAX network only. Where the WiMAX network splits into ASN and possibly several involved CSNs, identities can be meaningful to only one, several or all of the NAP or NSP operators.

A basic choice of identities in WiMAX networks is presented in Table 3.3, where an 'X' indicates that the identity is always required in the respective part of the system, and a '[X]' indicates that the identity is not central and may only be used in some cases.

A good example for the former case is the MAC address of the MS that is used as a central identifier in the message headers of both the wireless link MAC layer and the WiMAX-specific layer of the R6 signaling between the BS and ASN-GW. To provide an example for the latter case, the MAC address of a device will be used in the CSN for device authentication where the AAA server matches the signed identity from the device certificate with the identity provided by the MS to the ASN. It is always sent across R3 to the AAA server, but only used for specific cases like the certificate-based device authentication, or where the device identity is part of the subscription profile. Another example would be the WiMAX-Session-ID that does not appear at all on the wireless link, but is central to the ASN and home CSN operation. Any intermediate visited CSN may use the WiMAX-Session-ID, but can also just forward it transparently in the control signaling of R3.

Of course there are many more identities. However, a selection of those of specific importance and used across the whole network or having relevance for many functions in parallel will be presented in more detail.

In the following sections we will take a closer look at the roles of the MAC address as the device identity, and at the NAI and subscription identities as the main piece of identity information in a WiMAX subscription. Also, a set of network identities will be highlighted, including operator identities and temporary identities for binding subscriber-related sessions.

3.7.2 Device Identity

To start with a more detailed consideration regarding the individual identities, let us look at the one identifying the WiMAX device (or the interface card, depending on the actual device type). As for WLAN networking, in WiMAX the device's unique MAC address is used. It is important to note that a MAC address identifies the device's wireless adapter rather than the device itself. This might be a one-to-one relation in cases where the wireless chipset is an integral part of the device. However, by adding an external wireless adapter, devices like a laptop can easily change their MAC address shown to the network. So in general, the MAC address is rather of limited use to a network operator for identification purposes. This holds especially for devices with an open platform, like general-purpose laptops, or handhelds.

Depending on the actual operating system used by the device, even this might in general allow the MAC address value of a networking adapter to be easily changed. However, on recalling the discussion of device certificates in Section 3.4, it becomes clear that such a feature conflicts with WiMAX operation and would exclude a WiMAX device from attaching to any network that makes use of device authentication. The MAC address is part of the signed device certificate which is verified by the network during device authentication. So, changing the MAC address would create a mismatch with the device identity stored in the certificate, and would – depending on the operator's policy – likely result in rejection by a network enforcing device authentication.

The MAC address presented by the device during initial network entry is bound to the wireless security session created after EAP authentication and is verified by the ASN, so a device cannot change the MAC address value in mid session. In combination with device authentication during network entry, this allows both the local access and the home operator to rely on the MAC information provided by the device. One example could be to limit the use of a specific subscription to a specific MAC address only. However, the WiMAX Forum specifications do not consider such cases in detail, so it is completely up to the operator's local policy whether to introduce such bindings for a subscription based on the device identity or whether to completely separate the subscription and device.

Regarding the visibility of the MAC address, it is transferred without being encrypted OTA, and is mainly used by ASN entities like the BS and ASN-GW as part of the MS-related session information. In addition, the MAC address is passed by the ASN-GW to the home CSN AAA server, making it visible for the home CSN and any intermediate visited CSN operator. As a result, and in contrast to the NAI, no identity hiding is applied to the MAC address in WiMAX (see also Section 3.5.3).

3.7.3 Subscription Identity

The NAI [5] is acting as the central identity used for handling network entry but also has relevance for controlling already-established sessions for a subscriber. By default, the NAI as the subscription-related identity carries the permanent subscription identity and is composed of a 'username' part followed by a '@realm' identifier to identify the home operator as the issuer of this subscription. This is for example the case in WLAN deployments that are using EAP-based access authentication (typically indicated in the WLAN driver configuration by an 'Enterprise' security setting).

Interestingly, the identity information sent in the EAP from the MS to the authenticator via the EAP-Identity Response message is always carried in the clear across the wireless link for initial network entry. Due to the fact that eavesdropping on the wireless link is a relatively simple task not requiring any special equipment, and therefore also tracking all identities transmitted in a specific geographical location is simple, it is in the interest of both the subscriber and the operator to hide the permanent subscription identities from being readable by unauthorized third parties. As we have already detailed in Section 3.5.3, WiMAX supports an identity hiding scheme that allows a pseudo-identity value to be put in the username part of the NAI, instead of transmitting the permanent subscription identity.

The important thing here is that the NAI as discussed above is often referred to as the 'outer NAI', the one carried outside the actual EAP method. An EAP method itself typically carries an additional identity that is referred to as the 'inner NAI', or inner identity, because it is – depending on the EAP method used – not necessarily carried in NAI format. The inner identity is used within the EAP method to allow the AAA server to uniquely identify the subscription. But also, depending on the security properties of the EAP method, this inner identity might be transmitted in encrypted form, so it can only be seen by the MS itself and by the AAA server of the home operator. The inner identity is specific to the EAP method and has no additional meaning in WiMAX other than for networks like WLAN, so the focus in this section is on the outer NAI.

As a third option with specific relevance to device authentication, the username part of the NAI can carry the MAC address of the WiMAX device that matches the one signed in the device certificate.

Options on how to compose a NAI are shown in Table 3.4. From this we can determine that specific combinations are only relevant to specific EAP methods. In detail:

- EAP-TLS will use the device's MAC address as this is the identity meaningful for device authentication. Also, in most practical use cases for EAP-TLS authentication the device will not be able to present a meaningful subscription identity due to the lack of a meaningful subscription (as for initial provisioning).
- EAP-AKA does not provide support for device authentication as used in WiMAX, so we will only see permanent subscription identities or pseudo-identities in a NAI used for EAP-AKA. Interestingly, EAP-AKA has its own built-in identity hiding scheme that may be used in combination with the WiMAX pseudo-identity mechanism (see Section 3.5.3).
- EAP-TTLS provides for device authentication and for protected transport of subscription authentication within a TLS tunnel, by using an inner method (like MS-CHAPv2, or EAP-AKA). So the 'inner' identity will be hidden by the protected TLS tunnel in any case, and the outer NAI presented for EAP-TTLS mainly serves to identify the home operator via the realm part. Here, it is up to the deployment to choose whether the NAI username part carries the device's MAC address, the permanent subscription identity or a pseudo-identity. For device authentication, the device's MAC address is already carried in the device certificate, and the subscription identity is provided to the network in the inner method. So there is no strong technical reason for choosing a specific one from these three options.

However, it is clearly not recommended to put the permanent subscription identity in the EAP-TTLS outer NAI without a strong reason for doing so, since this would bypass any identity hiding across the wireless link.

Table 3.4 Options for NAI composition

NAI username part	NAI realm part
Permanent subscription Identity Pseudo-identity Device MAC address	@homeoperator.com

3.7.4 *Identities within the Network*

There are many identity values that are part of the control messaging carried within WiMAX networks across the different reference points and interfaces. Here, we will only highlight a few of the more interesting ones.

3.7.4.1 NAP-Identity (NAP-ID) and NSP-Identity (NSP-ID)

The WiMAX network and operator itself is identified by either a NAP-ID or a NSP-ID, depending on what type of network the operator has deployed. Both are 24-bit values. In those cases where an operator owns an ASN and a CSN, the NAP-ID and NSP-ID may be the same. Depending on the actual deployment, an operator can use one or more NAP-IDs, with the restriction that the NAP-ID has to be the same within a single ASN.

To allow a MS to learn about the available NAPs in its location, the NAP-ID is advertised by each BS owned by a NAP. This is transmitted as part of the base station ID (BS-ID) over the air interface to support WiMAX devices in NAP discovery.

The NSP-ID identifies a specific NSP operator. It is typically found as the realm part of the NAI of all the NSP's subscribers. Also, the ASN may send a list of NSP-IDs to a device as part of the initial network entry procedure, to allow the device to select a visited NSP for roaming, or to learn about the available NSPs if initial OTA provisioning is requested (see also Section 3.8.3 for NAI decoration with NSP-IDs).

WiMAX operators will act as part of a global ecosystem. Hence, the NSP-ID and the NAP-ID identifying a specific operator are expected to be globally unique. Such values require a global registry for number assignment. For WiMAX operators, there are two ways to get such unique operator IDs:

1. Requesting NSP-ID or NAP-ID assignment from the central broadband operator ID space [46] maintained by the IEEE Registration Authority.
2. Computing the NSP-ID or NAP-ID from a globally unique mobile country/mobile network code (MCC/MNC) as used by GSM/UMTS operators. The exact method for the derivation, allocation and administration is described in the ITU-T Recommendation E.212, and mapped to the broadband operator ID space maintained by the IEEE as defined by the IEEE Registration Authority.

3.7.4.2 WiMAX-Session-ID

The identifiers for the network itself are static ones: that is, they do not have any relation to specific devices, or different sessions of a device. For handling sessions of a specific device and subscription with a WiMAX network, a set of temporary identifiers are used within the network and are exchanged within the AAA signaling messages (see also Section 3.8). A very important identity is the WiMAX-Session-ID that is used to bind all AAA control signaling between the different network entities serving a specific MS and subscriber into a single session context, as described in Section 3.5.1. Together with the home network responsible for this subscription, the WiMAX-Session-ID helps to uniquely identify all related session context from network entry until the MS finally exits the network between the involved ASN, hCSN and any vCSN for roaming. It is a value that is unique to WiMAX as it does not exist in the standard IETF AAA-related specifications. The session ID is mostly required to bind AAA signaling together that otherwise would be fragmented, for example, due to a handover or relocation of the AAA client in the ASN for a specific MS. Assigned by the home CSN AAA server at the time of network entry of a MS, it is part of every AAA message exchanged in the network that is related to this WiMAX session.

The WiMAX-Session-ID is used across different network functions and especially helps to bind these together. As one example, a subscriber can have multiple accounting sessions and use more than one IP address during the overall WiMAX session. Also, parts of the network-internal control messaging might only be subject to one specific ASN-GW–CSN pair, so in the case of handover to a new ASN-GW, new 'sessions' are created between the CSN and the new serving ASN-GW. In such cases, the WiMAX-Session-ID serves as the link between all these individual 'sessions' and especially allows the home operators to finally relate everything to the same subscription. Also, for handover between different ASN-GW entities, the WiMAX-Session-ID can be seen as the important piece of information making the AAA 'session' mobile, which is not provided by the standard RADIUS or Diameter protocols.

The WiMAX-Session-ID is designed to span a WiMAX session. However, after the device finally detaches from the network, this ID will lose its relevance, and the device will be assigned a fresh WiMAX-Session-ID for every new initial network entry.

This serves well to provide proper session control in the WiMAX network. However, there are specific applications, or network functions, that come with clear requirements to be able to track devices or subscribers across more than one WiMAX session. As long as the same home operator's AAA server is in control of all WiMAX-Session-ID assignments for a specific subscriber, the home operator can relate the different IDs of the same subscriber to each other. But other network entities cannot be assumed to be able to do so, especially when considering that many deployments are expected to use the pseudo-identity in the NAI instead of the permanent subscription identity. For such entities, the only way to be able to bind several independent WiMAX sessions to the same subscriber would be to resolve the current WiMAX-Session-ID to the subscription identity. This would need to take place through the home operator's AAA server – a service that is currently not supported by any of the WiMAX specifications (although it would technically be feasible by using deployment-specific extensions to existing AAA signaling messages).

3.7.4.3 Chargeable User Identity

To address this WiMAX-specific inability for network entities including application servers to relate different WiMAX sessions to the same subscription due to the potential use of pseudo-identities and identity hiding that especially impacts visited NSPs, the chargeable user identity (CUI) can also be used. The CUI in principle serves the same purpose as the WiMAX-Session-ID. However, the major difference is that it can be kept constant by the AAA server across several WiMAX sessions (the actual value to be used is under the control of the home operator's AAA server). So the home operator can individually configure CUI usage to match the actual needs of different network entities that talk to the AAA server via AAA control signaling. The CUI for example can be kept constant:

- for a specific number of WiMAX sessions of a subscription;
- per device used with a subscription; or
- over a specific period of time.

It all depends on what the home operator is willing to communicate through CUI usage. But making the CUI a semi-permanent or even a permanent identifier for a subscription still would not reveal the permanent subscription identity to a visited NSP. In summary, the CUI provides additional means to an operator to address business needs, or regulatory and legal requirements, without being required to reveal important identity information about the actual subscriber.

One major example where the CUI comes in handy is to allow operators to fulfill lawful interception requirements. WiMAX becomes challenging in this area due to the identity hiding scheme, especially in cases where interception takes place in a visited network that is operated in a different region under legislation different to the home operator's region. Without the permanent subscription identity being available in the ASN or visited CSN, it is only possible to relate the different WiMAX sessions of the same subscriber through the device's MAC address. However, the device used for a subscription can change, so the MAC address cannot be considered a very reliable piece of information.

To bypass this, the respective operators can agree that the home operator should keep the CUI value constant for a specific period of time to allow easier local tracking of the WiMAX sessions. Further measures could improve this further. For example, changes of the CUI value could be communicated through the standard AAA control path to visited and access operators with the AAA server embedding the old CUI value within the new CUI. However, such improvements would be subject to deployment-specific agreements and are not part of the specifications. Nevertheless, the home operator would still be in full control of the policy and could apply different mechanisms to different roaming partners or partnering NAPs, or could even distinguish different policies on a per subscriber basis.

3.8 AAA Protocols and Routing in WiMAX

The main control protocols used across the R3 reference point between the ASN and CSN are RADIUS [6] or Diameter [11] for AAA signaling, and Mobile IP. Mobile IP might not be required across R3. This depends on the actual mobility mode of a specific MS and the mobility mode supported in general by the ASN, which might be a simple IP-only deployment. The AAA protocol remains the only part that is always in use across R3 as soon as a subscriber enters the network.

But why is there a need to allow for two protocol choices instead of just focusing on a single one? Actually, considering the protocols' history, the answer becomes clear when looking at the natural transition process in deployments from the older RADIUS to the newer and more flexible Diameter protocol and at the fact that existing AAA infrastructure is often based on RADIUS, but Diameter is also common in specific areas. So, as usual, the market is not uniform regarding AAA protocol deployments. Examples of mostly RADIUS-based infrastructure are in the area of DSL fixed access, or CDMA cellular networks. Diameter deployments are common in service-related interfaces like the 3GPP-defined IMS system or for Diameter credit control-based prepaid charging services.

The network architecture for WiMAX is based solely on the RADIUS protocol in the initial Release 1 [13], but Diameter support is available in parallel in Release 1.5 [116], so both protocols are fully supported as AAA solutions in a WiMAX network and the choice is up to the specific deployment. Although it would not make much sense to deploy both protocols in parallel for the same purpose, AAA signaling in a WiMAX network serves different applications and so it may actually make sense to run both protocols at the same time across the R3 reference point.

This is due to the fact that the AAA protocol in WiMAX is used for a number of different main purposes, or 'applications', including:

- Authentication and authorization (the first two 'As' in AAA) for network entry and re-entry of a device.
- Accounting (the third 'A' in AAA).
- AAA support for IP mobility services and DHCP keying.
- QoS set up for the subscriber's sessions.
- Exchange of location information related to a WiMAX device.

The focus in this section will be on the first item in the above list. All others are discussed in detail as part of Chapter 4.

3.8.1 RADIUS and Diameter Overview

The main purpose of the UDP-based RADIUS protocol is to exchange messages for performing AAA as well as to exchange a broad range of configuration data between a so-called network access server (NAS) acting as the AAA client and an AAA server in the network's backend infrastructure. A basic functionality in the WiMAX network regarding AAA functionality is that the AAA client can send a request message to the server with information about a network entry request, and the server, being responsible for authenticating and authorizing the request, answers with a response message and passes back the required configuration information to make this happen. For WiMAX, the NAS functionality maps to the ASN-GW in the WiMAX network architecture that implements the AAA client functionality (unless the ASN is deployed as an integrated base station (IBS)).

In addition to the AAA client and server roles, there can be AAA proxies in the AAA communication path. These are by default assumed in WiMAX in cases where the ASN is connected with the home CSN through one or more intermediate visited CSNs. Also, so-called AAA broker networks might be used by individual operators, or in large interconnections of operators, for centrally handling settlement of the subscriber's service usage via AAA signaling. The latter category, however, is not considered in the basic network specifications of WiMAX and harmonization of existing infrastructure in this area with WiMAX-specific requirements and data formats is dealt with in a dedicated global roaming work group (GRWG) of the WiMAX Forum. For further reading regarding roaming models, see for example [105].

Due to the nature of RADIUS passing request/response messages between client and server across a possible chain of proxies, RADIUS is understood as a hop-by-hop protocol where there is not necessarily an end-to-end relation between the client and the server - except for the trivial case where the client and server are talking directly to each other. This impacts the security model of RADIUS and also Diameter, which is a hop-by-hop one, meaning that the messages of both protocols are only secured between two adjacent RADIUS nodes. As a result, in the AAA infrastructure each individual proxy must be a trusted one from the perspective of the AAA client and the backend server. Actual security is achieved in RADIUS through a simple integrated authentication and message integrity mechanism. The primary method for protecting Diameter is TLS, with IPsec being a generic alternative for both protocols.

The hop-by-hop communication style also impacts routing of RADIUS messages that travel in the direction from client to server. There is no general need for an AAA client to know the final AAA server's address, but the messages can be passed to an AAA proxy that will handle further forwarding to the next AAA hop to the destination server.

For subscriber authentication, the plain RADIUS protocol has integral support for password-based authentication mechanisms like PAP and CHAP that are used in network access via the PPP. Both are simple challenge/response methods that would fail to match the required security level when used across any wireless link without additional protection. Hence, in WiMAX the EAP-based authentication is used instead through RADIUS. Here, EAP messages are carried by RADIUS [7], as shown in Figure 3.3.

RADIUS messages carry the authentication, authorization and configuration information to be exchanged as a set of attributes. So a RADIUS message typically consists of the message header indicating the actual message type at the beginning and also carrying an identifier value that allows for matching corresponding request and response pairs (called a RADIUS transaction). The header is then followed by a list of attributes that are encoded in TLV (Type/Length/Value) format to carry the actual content of the message.

Here, extensibility comes into play: RADIUS defines a set of basic attributes but any RADIUS deployment can add an additional set of attributes for deployment-specific use that does not impact standard RADIUS operation. Hence, in case an AAA server parsing a RADIUS request message comes across an attribute that it does not recognize, the attribute will typically be ignored and just skipped to allow for backward compatibility in implementations.

WiMAX is using this mechanism with its own set of WiMAX-specific attributes, as we will see below.

The Diameter protocol is the successor of the rather old RADIUS protocol developed in the IETF community to reflect and address as much implementation experience as possible from RADIUS deployments.

Specific improvements in general are in the area of the underlying transport, where Diameter can be based on either the TCP or the newer SCTP, which can be beneficial over RADIUS, especially for accounting, where packet loss typically has a direct impact on the accuracy of the resulting settlements either between operators or directly with the customer. Also, Diameter has a tighter specification in several areas like failover handling, capability negotiation or security, leading to significant simplification in making implementations interoperable and establishing links between network operators, which can be considered beneficial for roaming.

Diameter also implements additional functionality over RADIUS in the base specification, such as server-initiated messages to the AAA client. These are only possible by implementing separate additional specifications in RADIUS deployments. A specific reference can be made to the dynamic authorization extensions for RADIUS [8] that define the optional messages that allow a RADIUS server to initiate communication to the AAA client, with one example being a disconnect message to terminate the related WiMAX session. To improve interoperability, this extension has been made an integral part of WiMAX networks. For Diameter, similar functionality is available already as part of the baseline protocol.

Diameter exchanges data in the form of AVPs (Attribute Value Pairs) instead of the RADIUS attributes. The Diameter concept for supporting deployment-specific needs is quite different from RADIUS. Whereas RADIUS just allows for adding individual attributes, Diameter introduces the concept of applications. A basic set of AVPs is part of the Diameter protocol specification itself. However, additional applications like the one for carrying EAP messages across Diameter [9] can be defined in separate documents and AVPs of different applications can be combined with the basic AVPs in Diameter messages. So the basic idea is that an implementation starts with the basic Diameter protocol and adds the documents to the implementation that are required to match the overall system's requirements.

To improve interoperability, Diameter provides for dynamic routing based on the set of supported applications between the Diameter entities, so a Diameter message can be dynamically directed to a server best matching the required functionality. Without any match, at least this can be determined by the involved entities in a controlled fashion and without ending up in unexpected and fatal errors. Diameter applications are registered with the Internet Assigned Numbers Authority (IANA) to provide a central registry of existing applications [45]. For interoperability, Diameter also adopts the concept of a criticality flag called the 'm-bit' that can be set for individual AVPs. When set, it tells the receiving end that understanding this AVP is required and the AVP cannot just be skipped in case it is unknown. So this introduces an important mechanism to allow the efficient handling of backward compatibility or a partial equipment upgrade affecting the AAA interface.

In contrast to RADIUS, Diameter introduces a session concept. So the sum of messages exchanged that are related to a specific 'activity' like authenticating and authorizing a subscriber for network access is bound by an identifier called the Diameter session identifier. It is important to note here that this identifier does not overlap with, and cannot replace, the WiMAX-Session-ID that in contrast binds together the whole session of a MS across the different entities of the WiMAX network.

Sessions and the related session state are typically only maintained by the Diameter client and the Diameter server. However, this brings us to the Diameter concept of intermediate entities, which logically formalizes a number of common flavors that are all summarized in RADIUS deployments under the ambiguous notion of a 'proxy'.

In Diameter, intermediaries are called agents, and a Diameter agent can come in different flavors including:

- proxy agents and all agents that may take their own policy decisions in addition to forwarding messages and, based on local policy, may also send their own reject messages;
- relay agents that are simple message-forwarding entities and only analyze routing-related AVPs in the messages but are otherwise stateless; and
- redirect agents that indicate to a client to directly connect to another AAA server and therefore redirect the client's requests. So a redirect agent is typically not in the message path for redirected messages, but may, depending on the implementation and configuration, act at the same time as relay or proxy for other types of Diameter messages.

In addition to the above types of Diameter agents, translation agents can be deployed. These perform mapping between RADIUS and Diameter in cases where networks operate mixed deployments, or different network operators have different AAA protocols deployed and need to interconnect despite the protocol mismatch. This latter category may have some significance because the WiMAX specifications do not mandate the use of one or the other protocol. However, the technically less complex and therefore more obvious approach for operators is to consider dual-stack WiMAX equipment just supporting both protocol options.

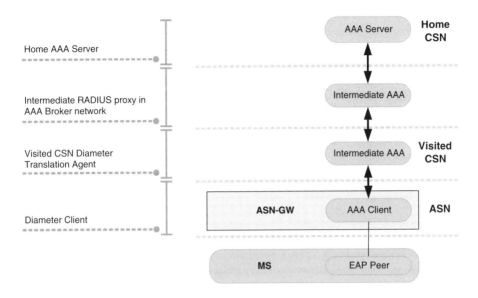

Figure 3.17 Example AAA path for WiMAX deployments

The AAA protocols might operate independently of each other in cases where RADIUS is used for different applications than Diameter in the same network. The case where network access is based on RADIUS but prepaid charging uses Diameter-based messaging might serve as a practical example here. However, as soon as different WiMAX operators need to interact, translation of RADIUS attributes into Diameter AVPs, and vice versa, becomes a valid scenario. A possible example is shown in Figure 3.17, where a Diameter-only ASN and visited CSN serve a roaming subscriber. AAA communication passes through the visited operator's translation agent that converts all messages to the RADIUS-based home network of the subscriber into RADIUS. Also, the RADIUS transactions pass through an intermediate broker network that interconnects the visited and the home operator.

Recalling the concept of a Diameter session, agents are typically considered stateless regarding a session and rather hold a transaction state to allow for mapping a response to the related request message, for being able to route back to the clients. However, an agent can also be implemented to act in a stateful way by tracking a Diameter session and maintaining related state information in the agent.

3.8.2 Making WiMAX Speak AAA

As discussed in the previous section, both AAA protocols allow extension of the basic functionality without modifying the base protocol itself. In general, to improve interoperability between different access systems controlled by the same backend infrastructure, it is worthwhile to limit additions to the base protocol functionality to a reasonable and well-defined set. The development of the WiMAX network specifications, however, serves as a good example that additional specific functionality is clearly required for creating a better optimized network architecture. This is in contrast to smaller systems that may cover a single focused use case only. Still, it is important to consider such interworking with other existing systems when designing the new pieces, to avoid conflicts later on.

The WiMAX AAA architecture for the initial release of network specifications [13] is based fully on RADIUS. So although in Release 1.5 Diameter is now supported in parallel, the fundamental design decisions on how to integrate and extend AAA signaling for WiMAX needs were taken based on the RADIUS protocol. These can be analyzed from several perspectives, as follows.

3.8.2.1 The Reference Point View

The main reference point where RADIUS is used in the WiMAX network is R3, connecting ASN and CSN, i.e. the ASN-GW and the AAA server in the home CSN. Intermediate proxies are assumed in any visited CSN that the AAA signaling needs to pass through. However, the specifications do not elaborate on their specific role, besides supporting the case that for a specific subscriber the mobility anchor and the 'breakout' of the subscriber's IP traffic to the Internet is assigned to the visited CSN instead of being handled by the home operator.

As part of this standard AAA signaling path, in addition to R3, R5 carries all AAA-related roaming traffic between CSN operators. So an intermediate proxy is logically connected with another proxy, or the home AAA server, through R5.

The second AAA-based interface in the WiMAX network reference architecture, which received considerable effort, is the one deployed between the AAA server and a HA. Although this has not yet been assigned a dedicated reference point, protocol specifications have been created in the WiMAX Forum to detail the messages and attributes exchanged across this interface, which can be deployed in both the home CSN and visited CSN, but is not allowed to cross interoperator boundaries. As a matter of fact, in cases where a HA assigned to the WiMAX session of a roaming MS is located in the visited CSN, the AAA signaling between HA and home CSN AAA server will be routed through the AAA proxy of the visited CSN, so the HA can receive all AAA data through its 'own' AAA entity.

Further applications of AAA protocols in a WiMAX network and between network operators are shown in Table 3.6 below.

3.8.2.2 The Functional View

As discussed in Section 4.2, there are different applications that are supported by AAA protocols in WiMAX. The approach in the WiMAX network architecture is to 'overload' AAA messages between the ASN and CSN that are exchanged in any case when setting up a subscriber session. This includes the RADIUS Access/Accept and Access/Response messages. By default, these messages include attributes for carrying the EAP-based network entry authentication and subsequent authorization sent by the AAA server to the local ASN. In addition, they are used in WiMAX to, for example, carry QoS-related signaling for setting the appropriate QoS profiles of the subscription in the local access and setting up the correct service flows; negotiate basic capabilities between the ASN and CSN,

Table 3.5 Examples of WiMAX-specific RADIUS VSAs

WiMAX RADIUS VSA	Attribute type	Description
WiMAX capabilities	26/1	Allows ASN and CSN to negotiate basic capabilities like the supported WiMAX Release, accounting related parameters, whether hotlining is supported, or the supported IP type like CMIP, PMIP or simple IP. This negotiation takes place within the Access/Request and Access/Accept messages for each specific ASN and CSN, i.e. it is used to negotiate parameters per network pair instead of per MS
WiMAX-Session-ID	26/4	Central identifier to bind AAA control signaling in the WiMAX network related to the same subscriber's session with the network, see Section 3.7.4 for detailed considerations
IP technology	26/23	Describes the IP technology used for the WiMAX session, like PMIPv4, MIPv4 or simple IP
BS-ID	26/46	Identity of the BS that the MS was attached to at the time of AAA message delivery. The BS-ID includes the identity of the NAP (NAP-ID), although this is additionally carried in the WiMAX VSA 26/45
NSP-IP	26/57	The identity of the NSP operator, see also Section 3.7

like supported accounting methods; or distribute IP and security configuration information related to the subscriber's session between the involved network entities in the ASN and CSN.

3.8.2.3 The Attributes View

For WiMAX, in addition to the standard functionality available for RADIUS from the base protocol specifications and additional RFC specifications that can be directly used like those for carrying EAP packets or for accounting, a list of VSAs is defined for specific use within WiMAX systems. The full set of WiMAX RADIUS VSAs can be found in Section 5.4 of [13] and detailed discussion would clearly exceed this book's limitations. However, we categorize a few of the more interesting attributes in Table 3.5 to provide insight about the purpose they were created for. Since VSAs are 'vendor specific', they are all listed as RADIUS type 26, followed by the WiMAX specific attribute type.

Diameter usage in WiMAX, similar to RADIUS, is also composed of the base protocol functionality with some additional RFC specifications (i.e. Diameter 'applications'). To add the specific functionality, WiMAX defined its own set of Diameter applications that extend the basic and generic functionality of the protocol as shown in Table 3.6.

These WiMAX-specific applications introduce the WiMAX-specific AVPs required for Diameter operation. The majority of AVPs for Diameter WiMAX operation are modeled to directly map to the corresponding Diameter AVPs.

Table 3.6 WiMAX Diameter applications

Application	Purpose	Abbreviation
WiMAX network access authentication and authorization	Diameter application between ASN and CSN across R3/5 for authentication, authorization and configuring all information for the subscriber session in the ASN	WNAAADA
WiMAX network accounting	Diameter application for accounting either between ASN and AAA server or between the HA anchoring the mobility session of a subscriber and the AAA server	WNADA
WiMAX MIP4 Diameter	Diameter application for supporting MIP4 operation between HA and AAA server	WM4DA
WiMAX MIP6 Diameter	Diameter application for supporting MIP6 operation between HA and AAA server	WM6DA
WiMAX DHCP Diameter	Diameter application for supporting DHCP operation between DHCP server and AAA server	WDDA
WiMAX Policy and Charging Control R3 Policies Diameter application	Diameter application for the WiMAX-specific parts of the PCC-based WiMAX PCC-R3-P interface[a]	WiMAX-PCC-R3-P
WiMAX Policy and Charging Control R3 Offline Charging Diameter application	Diameter application for the WiMAX-specific parts of the PCC-based WiMAX PCC-R3-OFC interface[a]	WiMAX- PCC-R3-OFC
WiMAX Policy and Charging Control R3 Offline Charging Prime D Application	Diameter application for the WiMAX-specific parts of the PCC-based WiMAX PCC-R3-OFC interface[a]	WiMAX- PCC-R3-OFC
WiMAX Policy and Charging Control R3 Online Charging Diameter Application	Diameter application for the WiMAX-specific parts of the PCC-based WiMAX PCC-R3-OC interface[a]	WiMAX- PCC-R3-OC

[a]See Sections 4.3 and 4.4.5 for more information about the PCC framework in WiMAX.

Here, we can see a major reason why, after RADIUS VSAs had been designed for RADIUS-only operation in Release 1 of the WiMAX Forum network specifications, the later modeling of WiMAX-specific AAA functionality for Diameter has led to a WiMAX-specific set of Diameter applications. Instead of making stronger use of existing IETF RFC specifications for Diameter that in some cases already define similar AVPs, simple translation was chosen as the major design criterion.

Actually, the expectation is that, due to the possibly parallel use of RADIUS and Diameter in WiMAX networking, translation aspects will occur in reality, and therefore WiMAX-specific AAA functionality is clearly designed to make such translation as simple and straightforward as possible without posing any requirements for complex mapping and conversion operations on intermediate translation nodes.

To provide one example, WiMAX could have made more intensive use of AVPs defined in the RFC Diameter NAS application [12]. However, this would likely have resulted in a significantly more complicated translation between WiMAX RADIUS and WiMAX Diameter deployments compared to just defining the same set of WiMAX-specific AVPs that were already in use as RADIUS VSAs.

3.8.3 Routing of AAA Messages and Roaming

Both RADIUS and Diameter operate in a hop-by-hop fashion as explained in Section 3.8.1. In a simple deployment where there are multiple AAA clients that are all directly connected to the AAA server, this is irrelevant. However, when taking into account more complex deployments with intermediate proxies or agents, several AAA servers or several operators involved in roaming, the assumption that the AAA client always knows the correct AAA server address to route a network entry request to becomes quite unrealistic.

Routing of AAA messages is typically based on two possible methods:

1. Preconfiguration, where an AAA entity is statically configured. For WiMAX, this might be the case where a NAP operator is only serving one NSP and routes all AAA messages from the ASN-GW directly to the same NSP's AAA proxy or server.
2. Dynamic routing based on the subscriber's NAI. This NAI is the one received from the MS in the EAP identity response message during network entry and is copied by the AAA client in the ASN to the AAA message header, e.g. when the AAA client forwards EAP packets received from the MS to the AAA server.

Routing based on the NAI is determined by the sending AAA entity looking at the realm part of the NAI. This allows it to forward the AAA message to the default AAA proxy or server of the operator that is identified by this realm.

An additional feature that is supported by WiMAX and that is especially important for roaming is the so-called NAI routing 'decoration' for network selection purpose. This mechanism is defined as part of [5] and is independent of the WiMAX NAI decoration for indicating, for example, that OTA provisioning is requested by the device, see Section 3.6.3. The basic idea is to allow the WiMAX MS to construct the NAI it uses toward the WiMAX network with additional information for routing the EAP messages sent by the MS through the WiMAX network AAA infrastructure.

To understand exactly how the mechanism works, let us consider the following example. A subscriber called Bob of the WiMAX operator 'wo_bestefort.de' uses a pseudonym-based NAI. However, for the sake of avoiding a 128-bit random value here, let us just use 'babel@wo_bestefort.de' as the example NAI for Bob. Bob is not in his home country, but is actually roaming abroad. He switches on his device and discovers a local NAP providing WiMAX service. Through the NAP, three local NSPs are advertised as being available: 'wo_mtcook.nz', 'wo_bestvalue.nz' and 'wo_bestefort.nz' (Figure 3.18).

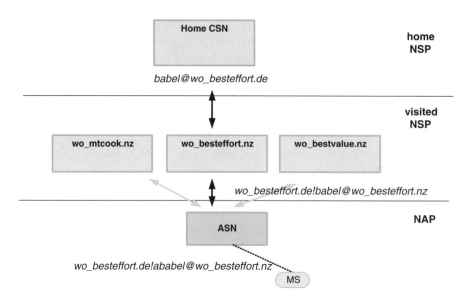

Figure 3.18   NAI-based AAA routing with vCSN selection

Based on the configuration provisioned to the MS, the local preferred roaming partner of Bob's home operator is selected from the set of available visited NSPs and the MS modifies the NAI to the following prior to sending this to the network: 'wo_besteffort. de!babel@wo_besteffort.nz'.

Looking at this example, the decoration actually replaces the original realm part of the NAI that carries the home operator realm by the chosen visited operator realm. In addition, the home operator realm is put at the beginning of the NAI and is separated by '!' from the username part.

When receiving the decorated realm, the NAP will route the resulting AAA message to the AAA server of 'wo_besteffort.nz', which automatically selects the visited operator. Note that the decoration does not put any requirements on the actual routing logic of the NAP; it just sends AAA messages to the AAA server of the NSP indicated in the NAI's realm part. On receiving the AAA message with the decorated NAI, the AAA server of the selected visited NSP will remove the NAI realm part and will move the decorated realm at the beginning of the NAI to the NAI realm part instead, thereby changing our example NAI into the original one that would be used by the MS without any decoration being applied: 'babel@wo_besteffort.de'.

Based on the realm that now points to Bob's home operator, the AAA message is forwarded to Bob's home operator.

This whole process can also be used recursively. Extending our above case of roaming Bob, an additional AAA roaming exchange network used by Bob's home operator could result in a decorated NAI like this one: 'wo_defaultexchange.com!wo_besteffort.de!babel@wo_besteffort.nz', where each AAA server being part of this AAA routing chain would remove the leftmost decorated realm.

MS-based NAI decoration is one way to influence the routing of AAA messages within the WiMAX network. In addition to this, although neither explicitly covered nor prevented by the WiMAX network specifications, any AAA client or proxy in the path to the AAA server can decorate a subscriber's NAI based on deployment-specific needs. This could for instance be a requirement to route the AAA messages through a specific roaming exchange provider's network for interoperator settlement. Of course, to realize deployment-specific use cases, decoration may also be overruled by statically configured routing paths between AAA network entities.

4

Service Provisioning

4.1 Enablers for WiMAX-Based Services

This chapter will provide insights on how WiMAX networks have been designed to support application services across wireless IP broadband access. To align with the main focus of this book and at the same time with the approach taken by the WiMAX Forum standardization effort up to Release 1.5, we will focus on the WiMAX network aspects and especially on the required architecture considerations. We will not elaborate on specific Internet applications except for the IP multimedia subsystem (IMS). In other words, the focus is in fact on the service enablers being available as parts of WiMAX networks, in contrast to the service itself. Such enablers will help operators to deploy applications and services that are integrated into their own network or to provide support for collaborating third-party ASPs.

4.1.1 Motivation and Available Service Enablers

Application services provided to the end user across WiMAX broadband access can be manifold and the most important use case for WiMAX is to provide generic access to the full range of available Internet applications. Hence, the WiMAX network architecture has been designed to help support such applications with generic service enablers rather than adding support for numerous individual applications. Typically there will not be any specific relationship between such application and the WiMAX network anyway, either regarding technical aspects or when considering business-related topics. Some exceptions include the IMS developed by 3GPP as a default system for providing VoIP (but not precluding other VoIP technologies across WiMAX access) or specific considerations of emergency services support. The latter is required by operators to match regulatory requirements while the former can be seen as a simple effort of making an existing service enabling platform easily available for the WiMAX domain.

Deploying Mobile WiMAX   Max Riegel, Dirk Kroeselberg, Aik Chindapol and Domagoj Premec
© 2009 John Wiley & Sons, Ltd

Among the generic service enablers that are available as standardized functions of the WiMAX network, there are:

- Built-in support in the AAA backbone signaling between the ASN and CSN for network services like IP mobility as a network service and generic application service enablers like QoS. This includes for example the negotiation of a set of network features that can subsequently be applied to individual subscriber sessions, or the distribution of user profile settings from the home CSN to the ASN entities. Details are provided in Section 4.2.
- Accounting and charging that is able to efficiently support both simple tariff models and more advanced ones that charge for individual service flows instead of just the whole IP traffic. An overview of possible accounting schemes and deployment choices can be found in Section 4.3.
- Support for the dynamic control of the QoS of IP packet flows and service flows across the wireless interface that is specifically required to enable multimedia applications. Moreover, this makes the 3GPP-specified policy and charging control (PCC) framework available for WiMAX networks. The concepts and available options like different traffic flow types are explained in Section 4.4.
- Location determination of a MS based on both the network and MS measurements, collection of the measured location information in the CSN and means to support applications with such location information. Section 4.5 provides an overview of what is available from the standards perspective and of the use cases that are covered.

There are additional considerations regarding the interfaces to application servers that a WiMAX network can expose for communication with ASPs. However, it is difficult to create meaningful generalized solutions in this area as there are already a lot of existing deployments and standardized formats for the individual service enablers. Furthermore, an effort to replace those technologies by a generic solution would face major challenges to gain acceptance in the market. Hence, the focus of this chapter is on WiMAX network-internal operation instead of elaborating on the wide area of interconnecting network and application service operators.

4.2 AAA Support for Services and Applications

An important part of the control messages flowing through the WiMAX network and between WiMAX operators is covered by the AAA protocols that we looked at in detail in Section 3.8. There, the focus is on security functionality, especially for authentication and authorization in the network entry procedures. Any WiMAX device performs exchanges with the network to get access and IP connectivity, like the exchange of EAP messages that are carried by the AAA protocols between the authenticator in the NAP and the responsible AAA server in the home network.

AAA protocols in WiMAX networks, however, have a much broader scope than just this single 'application'. When taking a closer look at the network functions that are available to enable application services based on WiMAX broadband access, we can see that AAA control signaling plays an important role. It either directly acts as part of a specific enabling service like location, or indirectly provides important pieces like feature negotiation and identity or security information to application servers.

To reflect the importance of AAA protocols for enabling applications across WiMAX, we will provide in the following sections an overview of the areas where RADIUS or Diameter can be applied in WiMAX networks.

4.2.1 Overview

AAA protocols in WiMAX are used for a number of different purposes, or 'applications', and as discussed in detail in Section 3.8, authentication and authorization which make up the first two 'As' in AAA are paramount to WiMAX network operation. However, there are many more elements that are covered by AAA control signaling, where the main areas are:

- AAA support for IP mobility services and DHCP, like negotiating the correct mobility service for a subscriber (from the set of MIP, PMIP or simple IP) and centrally providing mobility session parameters like IP addresses or security keys to the involved network entities in the ASN and CSN.
- Accounting (the third 'A' in AAA) to set up a subscription-specific accounting context and as part of this context report actual usage of the network services from the ASN or possibly a CSN router or MIP home agent to the home CSN AAA server.
- QoS setup for the subscriber's sessions by either static provisioning of subscription-specific profile information or dynamic provisioning and modification of QoS profiles.
- Exchange of a device's location information for network-based location determination between the ASN and CSN.

There are other protocol choices for AAA signaling – or better signaling with the AAA server – than RADIUS or Diameter. For example, within the CSN internal RADIUS/Diameter-based interfaces are only partially specified for WiMAX, like the one between the AAA server and HA or location server as shown in Figure 4.1, but others are left as a deployment decision. Some

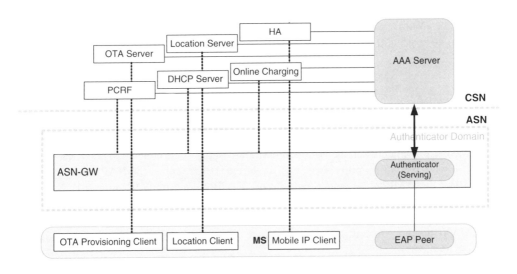

Figure 4.1 AAA protocol usage in WiMAX networks

interfaces like the OTA device provisioning-related one between the AAA server and OTA server have up to now not been formalized. Only the required data elements to be exchanged are identified as part of Release 1.5. However, the development of an AAA-based protocol specification is ongoing for the subsequent release of WiMAX network specifications.

So in many cases it will be up to the deployment and the application server and AAA server implementations to decide which protocol to use for exchanging the required data. This will also depend very much on the specific use case that – in case RADIUS or Diameter is used – either is supported by in-band AAA signaling which just extends the standard AAA communication for network entry of a subscriber, or requires individual AAA message flows.

4.2.2 Mobility Support from AAA

Mobility support of the device's IP connectivity and related IP sessions is an integral service offered by a WiMAX network. Such mobility, as explained in further detail in Section 5.2, can be offered either within the same NAP and ASN where it is handled by the controlling ASN-GW, or between different ASN-GWs and ASNs that hand over sessions without breaking the MS's IP sessions. Mobility that requires handover across the R3 reference point is not limited to control messaging across the R6 or R4 reference points within the same ASN, but requires communication between the ASN and CSN network entities.

4.2.2.1 General Support for IP Mobility Setup

WiMAX offers a set of deployment options for how such R3 mobility is covered. This includes Proxy Mobile IP (PMIP) that relies on the standard MIP infrastructure and control messages between ASN and CSN, but allows for devices that do not implement MIP support by adding a proxy client function for these MSs in the ASN, and also standard MIP, called Client MIP (CMIP), in the WiMAX context.

IP mobility across R3 may be considered a 'network-level' application and it receives some level of support from the AAA framework because the entities providing mobility support, like a WiMAX HA, do not operate completely independently of the WiMAX network. There are three general stages where AAA support for mobility services is required:

1. Negotiation of (R3) mobility services.
2. Setup of a mobility session and related parameters.
3. Modification or termination of a mobility session.

Not only does the first step of negotiating mobility services affect subscriber sessions, but it also has to be performed between the NAP and NSP in roaming environments to allow the NAP to advertise what type of mobility service is supported. For example, the NAP could advertise that MIPv4 and PMIPv4 are supported, but no (P)MIPv6 or simple IP support is available. The NAP and NSP can dynamically negotiate these capabilities per subscriber as part of the network entry-related AAA messages exchanged between the NAP and NSP by exchanging a WiMAX-specific AAA attribute or VSA (see Table 3.5 for further details regarding the protocol elements).

With this, setting up MIP or PMIP mobility support for a specific subscriber is essentially provided as part of the initial AAA messages exchanged as part of the subscriber's network entry procedure. The mobility parameters that are required by the subsequently created mobility session and that are set up via the AAA signaling channel include for example the IP address to be assigned to the subscriber's device, the address of the HA assigned for the device, and security parameters for the MIP itself.

So for MIP sessions, the HA function can actually be assigned via the AAA path by the home AAA server and this includes capabilities for choosing whether the HA as the mobility anchor and as the 'breakout' for the subscriber's IP packets lies in the home operator's CSN or in the visited CSN if the subscriber is roaming. Clearly, MIP support with the help of AAA signaling is closely integrated into the WiMAX network and especially covers the RADIUS- or Diameter-based CSN interface from the MIP HA to the WiMAX AAA server. Interestingly, this interface is not part of the R5 reference point in cases where the HA is assigned for a subscriber in a visited CSN instead of the subscriber's home operator's network. In such cases, the AAA signaling will be routed via the AAA server in the visited CSN, so the HA and the home CSN's AAA server do not deploy this interface directly. Instead, the HA will always contact its local AAA server that will in turn use the already-established AAA interface across R5 to proxy the HA-related signaling.

4.2.2.2 Securing IP Mobility Signaling

Regarding the more complex AAA support for the setup and distribution of MIP-related security parameters, let us first take a brief look at MIP security itself to better understand what needs to be provided, and where the required security information typically comes from. This will allow us to appreciate why WiMAX took its own approach in 'enabling' MIP services security.

As shown in Figure 4.2 for the example of MIPv4 with the MIP client in the mobile device, the FA in the NAP and the HA in the CSN involve three parties. Each pair of these parties requires a common shared secret as part of the security association to protect the messages exchanged. For detailed security considerations the interested reader is referred to [10]. However, it would be impractical and most likely insecure to set up these security associations statically and use them for securing many mobile IP sessions. This holds especially for

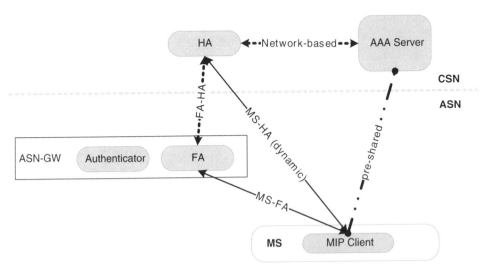

Figure 4.2 Mobile IP security associations

environments like WiMAX, where the HA is typically assigned dynamically, so for a MS it would just not be possible to set up pre-shared security associations with every HA in advance. In practice they have to be dynamically established for every new mobility session, at least for the MS–FA and MS–HA pairing. The security association between FA and HA is not necessarily bound to a specific MIP client and can therefore be based on preconfigured information between the entities, or networks.

MIP itself just assumes these security associations to be available if security is used and mandates at least the availability and use of MN–HA as shown in Figure 4.2 as the security association between the MS and the HA, so all MIP messages are protected end to end. In WiMAX, the MIPv4 mobility service additionally requires that FA–HA signaling is secured and a FA–HA security association is applied. MS–FA security in contrast is possible but only considered as an optional step.

The fundamental question is how the security associations and especially MN–HA, which needs to be established dynamically per mobility session, are created, and here AAA comes into play. With AAA support for MIP-based mobility services, instead of assuming the MN–HA security association to be statically available prior to establishment of the mobility session, it is dynamically created by involving the AAA server that maintains a permanent key with the MIP client.

There are several approaches of how to make use of such a permanent security association between AAA and mobile device or subscription, and the standard IETF one described in [68] requires the exchange of additional signaling extensions between the MIP client and AAA server via the HA to guarantee freshness of the derived keys.

In WiMAX the MIP client and AAA server are not required to share a MIP-specific security association and permanent key in advance. In contrast, a session key for securing MIP operation is derived from the EAP-based network access authentication from which in turn the MN–HA security association is derived. This occurs by making use of the EMSK and further derived keys that are specific to the MIP session. The EMSK is provided in parallel to the MSK for securing the wireless interface, by the EAP method (see also Section 3.3.3).

In fact, this WiMAX-specific method couples the mobility service security with the network access authentication, which in general comes with both advantages and disadvantages. On the positive side, clearly a reduced configuration effort due to the lack of requirements for mobility-specific permanent security parameters as well as reduced signaling effort for setting up MIP security can be identified. This is due to the fact that key freshness is already provided by the EAP method derived keys and does not require additional communication between the MIP client and AAA server as part of the actual MIP signaling. Typical negative impacts might be that it becomes harder to continue the same mobility session and keep the same HA across different wireless access technologies, as for example in WiMAX and WLAN or 3GPP2 access, because the MIP session keys are bound to the access authentication of one specific wireless access.

Any decision about whether to base security bootstrapping of network services like MIP on access security is one that needs to take into account the overall system architecture. Without such architecture being available, as for a general-purpose IETF RFC that is provided rather as a building block with broad applicability, there is a tendency to avoid such binding as it introduces limitations without clear knowledge of the benefits. For WiMAX, the system architecture is known and during the development of the specifications the general opinion was that the benefits of binding the security of network functions like IP mobility to EAP-based authentication prevail over possible disadvantages.

4.2.2.3 Key Distribution

One advantage of integrating mobility service support in the overall network design is that dynamically derived parameters like the security associations for MN–HA and FA–HA protection of MIP signaling can be automatically distributed across the network by piggy-backing them on top of the protected AAA communication that takes place in any case for providing network service to the subscriber. So this comes without any additional cost in terms of message exchanges in WiMAX network-internal control traffic.

Specifically, the security association parameters that need to be shared by the entities applying and verifying MIP security include a shared secret and a SPI value acting as key identifier. This key identifier for WiMAX is made unique per realm controlled by an AAA server. Hence, the responsible AAA server issues the SPI values for all MIP mobility sessions of the subscribers it is responsible for and prevents collisions of SPI values between subscribers under its control. So the SPI serves as a unique handle for identifying the actual session and related parameters like MIP keys and for providing them to other network entities like the HA on request.

The mobility keys and SPI values (as the security association parameters) are transferred by the AAA server to the network entities controlling MIP service in the network as shown in Figure 4.3. This covers the ASN-GW that provides FA functionality for MIPv4 services, as well as the proxy mobile node (PMN) as the MIP client for PMIP. Also, the WiMAX HA implements

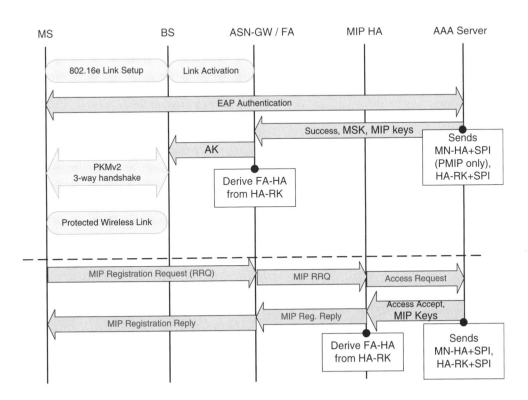

Figure 4.3   AAA-based distribution of MIP security associations

an interface to the AAA server that mainly serves the purpose of AAA support for subscriber IP mobility in addition to accounting that might be performed by a HA.

4.2.2.4 General Design Considerations

One design decision regarding the MIP security associations in WiMAX is that in contrast to MN–HA, i.e. per MS for both the MIP and PMIP cases, MIP security between FA and HA is only applied on a per node basis, so each pair of ASN-GW and HA has a dedicated MIP security association that is not related to any subscription. For the secret key that is part of this security association, the major difference to any MN–HA key is that this is never derived from access authentication but is based on a HA–RK (Root Key) for a specific pair of nodes. Derivation of any HA–RK is internal to the AAA server provisioning the FA and HA. However, distribution of the HA–RK is still performed by adding the HA–RK and assigned SPI value to the existing AAA messages that are exchanged whenever a subscriber receiving mobility service by this pair of nodes enters or re-enters the network.

If a specific deployment were to experience problems related to the binding of network access and MIP services, it would technically be possible to decouple this and make use of independent MIP keys provisioned to the AAA server, at least for PMIP services. The PMIP client in the ASN does not care where the MIP keys are derived from. For MIP, this would require deployment-specific behavior in the MS, so in practice this has to be considered unrealistic.

While looking at PMIP, a valid question is why security associations for this mobility option are per MS. PMIP involves the PMN in the ASN-GW instead of a MIP client in the MS itself, as well as the AAA server in the home network. So the basic security requirements are those for the general ASN–CSN communication where a global security association would be sufficient instead of the many individual per MS ones.

The answer to why this additional complexity has been chosen does not primarily lie in additional security requirements, but rather in reduced complexity of network signaling. AAA support for MIP is designed to support all mobility variants as much as possible through the same mechanisms. Any major distinction between MIP and PMIP support is avoided because it would have meant designing two different schemes where one can solve both cases. Additional complexity is always less favorable for interoperability between different vendor's equipment and different operator's networks. Regarding MIP security, the actual mobility mode – whether PMIP or CMIP is used – for a specific MS remains fully transparent to the HA.

4.3 Accounting and Charging

An important functionality of AAA signaling that is central to the operator's business and key to all possible service offerings is accounting and charging. The network clearly needs to be able to measure the service usage of an individual subscriber and finally 'charge' to the subscriber's account. After the collected usage information is passed through a rating and billing engine in the backend of the network, the resulting bill for the subscriber will be created. The whole functionality described in this section has no impact on the MS, with the potential exception of hotlining where the browser application in the MS can be redirected to a portal page, for instance to recharge a prepaid account. Another possible exception would be an 'advice-of-charge' functionality providing information to the end user about the cost of using a specific service, see e.g. [70]. However, no such functionality has been specified for WiMAX networks as part of Release 1.5 (besides the fact that the availability of such functionality to the end user is in general rather scarce in today's established telecommunication networks).

Otherwise, the functionality described within this section is purely internal to the ASN and CSN.

With our focus on the required network signaling and by imagining different business entities working together for accounting and charging, the importance of well-defined interfaces and data formats between the networks becomes clear. This is especially true as soon as it comes to roaming and also intermediaries that provide charging services to operators as part of a roaming exchange network.

4.3.1 Architecture and WiMAX-Specific Aspects

For WiMAX, the specifics related to accounting and charging include the RADIUS- or Diameter-based signaling and the data exchanged across the reference points. It is therefore important to point out that the focus of this section is on ASN and CSN functionality up to the AAA server that at least for RADIUS-based accounting is the entity collecting the usage data from the network entities performing the actual metering, and correlating it to the related subscription. Both offline and online accounting are available for WiMAX, and this section will give recommendations about valid combinations for deployment. Chargeable events trigger the creation of accounting data that is in the former case finally billed to the subscriber's account, e.g. by sending a monthly bill (called 'postpaid' account). In the latter case, the subscriber's account balance is debited in real time (called 'prepaid' account).

What is not considered is all the functionality 'behind' the accounting server. Hence, the following discussion will not cover functions like combining accounting data collected for different subscriptions that belong to the same human user or billable account, rating the collected accounting data to produce actual charges based on the tariff chosen by the subscriber or based on the type of service, or putting everything together to create the final bill. All such functionality lies in the backend of an operator's infrastructure, but is not specific to WiMAX itself. Greenfield operators starting with WiMAX can likely choose from the generally available equipment in this area, whereas established operators with existing infrastructure will have a strong focus on integrating WiMAX-specific accounting and data formats into their network, avoiding changes to the existing systems and processes as much as possible.

Other functionality that itself is not part of the WiMAX network because it can rather be considered a service from a subscriber's perspective, is hotlining. A typical scenario would be where a subscriber with a prepaid subscription that ran out of credits is temporarily not granted the standard service like best-effort Internet connectivity, but is redirected or 'hotlined' to a special server of the CSN operator that communicates with the device's web browser and allows the subscriber to recharge the account. WiMAX offers support for such hotlining only to the extent that a basic procedure can be triggered whenever required in the network. As a result of this triggering, appropriate filters are installed in the ASN-GW or HA to limit the device that temporarily will only be able to access a specific IP address, for example, but all other services or connectivity are halted at the time of the hotlining. After the actual hotlining-related interaction is out of scope from the WiMAX network's perspective, either the hotlining mode is released and normal connectivity restored for the MS, or the device's session is terminated as a result of the interaction with the server.

Additional examples of where the hotlining capability can be used within the WiMAX network are:

1. Emergency services, where the device's connectivity can be limited to the entities required to place the emergency call, like a specific VoIP server.
2. OTA provisioning of a subscription, where the device's connectivity is restricted to a provisioning server as long as no valid subscription is installed in the device.

To deal briefly with the related terminology aspects and for the sake of simplicity we will focus on the term 'accounting' in this section. This is the most common term for the AAA-related protocol aspects (being the third 'A'). The terminology for accounting and charging-related functionality across the different telecommunication network technologies and organizations creating the related standards does not follow a common set of definitions. So we will keep it simple and not try to end up creating just another set of definitions. Note that the terminology used in 3GPP network standards follows a different approach. Instead, it makes use of the term 'charging', which describes the network functionality 'whereby information related to a chargeable event is collected, formatted, transferred and evaluated in order to make it possible to determine usage for which the charged party may be billed (offline charging) or the subscriber's account balance may be debited (online charging)', see [67].

Let us now look at the high-level accounting architecture of WiMAX networks shown in Figure 4.4. There are a number of specifics motivated by the overall WiMAX network design, such as splitting access and core into ASN and CSN domains to allow different operators independent ownership of these network parts.

A central role in accounting falls upon the accounting server in the home network. It is in charge of collecting data produced by the different metering points across the network and of

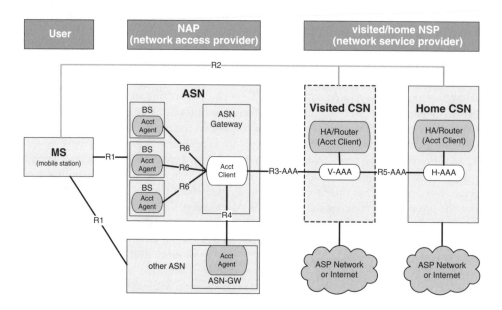

Figure 4.4 Architecture for accounting and charging

relating the collected accounting data that comes in the form of user data records (UDRs) to the subscription generating the related traffic. A UDR contains information about the amount of data exchanged or the length of time for which a session has been active.

As counterpart to the CSN accounting server, the ASN operates the default accounting client that is part of the ASN-GW functionality. In cases where the accounting client does not perform the actual metering itself, an accounting agent functionality is defined. The agent either can move with the MS and the anchor or serving data path function (DPF) like the FA to a different ASN-GW, or can optionally be located directly in the BS whenever the BS itself is required to perform measurements. In the former case accounting data is sent from the accounting agent to the accounting client across R4, while in the latter case it is passed across the R6 reference point. With this in mind, let us once more emphasize that the default accounting client resides in the access part of the network and not in the CSN where accounting is possible as an option.

Within the ASN, the most obvious deployment choice would be where the accounting client also performs the accounting agent functionality in the ASN-GW and the accounting agent may only move to another ASN-GW if the subscriber moves to a new ASN-GW without the anchor authenticator being relocated. However, the WiMAX standards allow for a range of deployment choices that also allow for an optional additional accounting client/agent in the CSN that is located in the HA or CSN router.

Performing accounting in both ASN and CSN in parallel for the same subscriber's session may not seem to be a reasonable deployment choice for an operator running both ASN and CSN. However, it becomes a valid option in all cases where ASN and CSN belong to different business entities. Furthermore, an operator owning both ASN and CSN may choose to only perform accounting in the CSN to avoid the additional complexity of ASN handover support in the accounting procedures. This may make sense if certain options such as per service flow accounting that is limited only to the ASN and cannot be done in the CSN are not required.

Interestingly, in contrast to the default accounting client being located in the access, other network designs like the one used by 3GPP networks define similar functionality as part of the core network that rather maps to the CSN case in WiMAX. This difference is similar to the case of PCC as detailed in Section 4.4.5 where the policy control enforcement function (PCEF) for WiMAX is located in the access domain instead of the network core for the 3GPP design. Subsequently, in WiMAX, specific considerations are necessary to handle the handover between different ASN-GWs and the collected data is more likely to traverse an interoperator interface. At the same time, this allows deployments to keep the actual core (CSN) functionality as small as possible and provides additional flexibility to accommodate the expected broad range of different deployment models.

The basic accounting model that is already available with the initial Release 1 WiMAX network architecture makes use of the same R3/R5 AAA communication path for accounting that is established during network entry of a subscriber. All accounting-related signaling is based on RADIUS [71], where support for offline accounting is the default and can therefore be expected always to be available in products. Support for online accounting is an optional feature, although one of considerable operator interest.

WiMAX accounting can alternatively be based on the Diameter protocol as well as on the PCC framework charging support [56] defined by 3GPP (see Section 4.4.5 for further details and references). The required specifications between the accounting client and the accounting server were introduced as part of the WiMAX network Release 1.5 standards. Also, both options of offline and online accounting support are available for Diameter-based deployments.

One of the basic design criteria behind the accounting architecture is that accounting-related control signaling can be kept logically separate from the AAA signaling for authentication and authorization. The justification for why network standards typically follow this approach is that network access requires control messages to be exchanged in real time, whereas offline accounting is at least not considered to be time critical. The accounting client does not need to send any accounting data to the server at the time of metering user traffic.

The actual data measured in the accounting agents and sent by the accounting client to the accounting server is put together as UDRs. These UDRs follow specifications that are based on the existing IETF formats, but also include a number of WiMAX-specific information elements and identities like the WiMAX-Session-ID (see also Section 4.3.4). This additional information allows the AAA server to relate a UDR including the metered usage data to the correct WiMAX subscription. Several common ways are supported for metering, which allow realization of all common models that the operators offer to their subscribers. For example, it is possible that the accounting client reports the number of exchanged bytes for the MS of the subscriber in the upload and download direction, the number of IP packets, or the duration of active service usage. The UDRs can store such data for the whole IP session of the MS and subscriber (being the default), or optionally for an individual service flow, which allows for finer granularity and would for instance allow for a separation of charging for a VoIP call while causing slightly higher signaling load in the network. Furthermore, it is possible to separate out all the control packets that are exchanged across the radio link with the MS. This can for instance include DHCP packets that are not part of the user data and hence should not be charged to the subscriber.

The WiMAX network specifications create their own extensions to both RADIUS and Diameter UDRs to cover WiMAX-specific session information and identifiers. However, this is very much based on existing formats from other standards organizations that are widely used today. For example, in Diameter-based accounting, the WiMAX UDRs are based on the formats defined by the IETF Diameter specifications, with extensions made by the 3GPP PCC framework and some additional WiMAX-specific data.

There are a number of options for how to perform accounting for a specific subscriber entering the network and establishing a session. The selection of which accounting mode to apply can be based on preconfiguration, which in turn automatically applies the same method to all subscribers entering and using the network. In addition, it is possible to dynamically negotiate the accounting method for a specific subscriber based on the AAA signaling during initial network entry. The appropriate choice taking into account the involved network's capabilities and the subscription profile parameters is negotiated with the help of the WiMAX capabilities attribute, which is explained in more detail in Section 3.8.2.

Regarding the actual protocols being used, it can be seen that as a general rule the accounting protocol follows the AAA protocol chosen for access authentication and authorization. That is, if the AAA protocol for authentication and authorization (A&A) is RADIUS then RADIUS offline accounting will be performed, and in the Diameter case offline accounting will be based on Diameter. Decoupling A&A signaling from offline accounting in the RADIUS case is not a feasible option. On the contrary, theoretically it would be possible to combine Diameter-based A&A with RADIUS offline accounting based on the fact that in Diameter A&A and accounting are realized as logically separate applications (see also Section 3.8). However, this is certainly not considered a relevant combination in practice.

Table 4.1 Accounting protocol deployment options

	RADIUS A&A	Diameter A&A	RADIUS offline accounting	Diameter offline	RADIUS online accounting	Diameter online/PCC
RADIUS A&A		✗	✓	✗	✓	✓
Diameter A&A			✗	✓	(✓)	✓
RADIUS offline accounting				✗	✓	✓
Diameter offline					(✓)	✓
RADIUS online accounting						✗

If online accounting is desired, a mix of different AAA protocols is feasible and the overall options for the concrete protocol choices increase. For example, a network that is already deployed based on a RADIUS AAA backbone for network entry including offline accounting can add Diameter-based online accounting without changing the existing AAA deployment to Diameter. This option is viable as long as the involved equipment like ASN-GW, AAA server and intermediate AAA proxies support such choices.

An overview of the meaningful combinations for the AAA protocols and features is given in Table 4.1. The trivial combinations that are not considered are those realizing the same feature in both protocols, which would overlap and might only occur in theory in a handover case with ASN-GWs supporting different AAA protocols. Assuming that the RADIUS A&A is deployed in a network, basically all the choices for realizing online accounting are available, in addition to optional PCC-based accounting. Any Diameter choice would result in a mixed-mode deployment where, for instance, the negotiation of using Diameter online accounting would be based on RADIUS and the actual accounting would then be based on Diameter.

With the network access already being based on Diameter, it would not make a lot of sense to consider RADIUS-based online accounting further unless, of course, there were requirements to support existing backend infrastructure. Technically this would be possible with statically preconfigured systems and parallel Diameter and RADIUS signaling between ASN and CSN. Static configuration of the network entities interoperating with each other is required due to the fact that with a Diameter-based AAA backbone, the dynamic negotiation of RADIUS-based accounting per subscriber is not explicitly supported by the standards. The same is true for RADIUS-based access with Diameter accounting that is not based on the PCC framework (WiMAX native).

4.3.2 Offline Accounting and Session Concept

The fundamental signaling message flows for offline accounting are relatively simple:

- With the occurrence of certain trigger events such as, in the most obvious case, the assignment of a new IP address to the subscriber's MS, the accounting client in the ASN updates the AAA server and starts an accounting session. This happens by sending an Accounting-Request (ACR)/Start message in the RADIUS case or an ACR/Start message for Diameter-based offline accounting.

- Whenever a trigger event to tear down the accounting session occurs, like termination of the IP session, an Accounting-Request/Stop or ACR/Stop message is sent by the accounting client to the AAA server. This stops the accounting session for the terminated IP session and reports the final resource consumption to be accounted for.
- As an additional option, the accounting client may be able to send intermediate reports on the status of current resource consumption. This is for the ongoing session and takes place by sending Accounting-Request/Interim-Update or ACR/Interim messages to the AAA server. The major motivation here is to allow partial intermediate reports and avoid revenue loss for the whole IP session of an MS in case of ASN failure or loss of a final Accounting-Stop message.

Receipt of any such message is confirmed by the AAA server by sending a RADIUS Accounting-Response or Diameter Accounting-Answer message via the AAA protocol path.

There are two fundamental differences between an accounting session and a WiMAX session that cover the full period of MS attachment from initial network entry to the point of network exit. The first important difference is that accounting sessions can be split. There can be several subsequent accounting sessions for the same IP session of an MS. The main case where such a split can occur is when the MS hands over from one ASN-GW to another. Figure 4.5 depicts the overall accounting message exchanges for a scenario where the MS hands over from ASN-GW A to ASN-GW B based on the RADIUS protocol on R3.

This now allows us to take a quick look at the mobility support for offline accounting. It is in line with the general mobility handling in case of handover actions between different

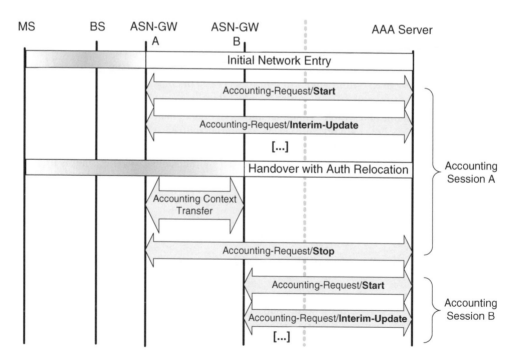

Figure 4.5   RADIUS-based offline accounting with multiple sessions

ASN-GWs that are interconnected via the R4 reference point. In such handover cases where the anchor authenticator remains stable and only the anchor DPF moves alongside the MS to another ASN-GW, the accounting client functionality is also not relocated. It always stays collocated with the anchor authenticator, and only the accounting agent function moves over to the target ASN-GW. However, as soon as the authenticator function is relocated to the target ASN-GW later on, the accounting client is moved. For the protocols, this requires the AAA server to be updated across R3, which takes place with the 'new' ASN-GW sending an Accounting-Request/Start to start a new accounting session. In addition, the 'old' ASN-GW terminates the initial accounting session and informs the AAA server by sending an Accounting-Request/Stop message as soon as the handover has succeeded. It is worth mentioning that these different accounting sessions need to be mapped to the same MS later on by the AAA server. This leads on to the role of the different identities used for WiMAX accounting, which we will look at in Section 4.3.4.

The second difference of an accounting session compared to the overall WiMAX session is related to the accounting granularity. This leads to the question of which type of 'service' can be charged individually. By default, an accounting session in WiMAX covers a MS IP session and runs – possibly being split into several sequential sessions as discussed above – from the IP address assignment to the IP session release or network exit. On assuming that a MS has only one IP session at any given time, this is the most straightforward model and the accounting data basically includes the measured volume or duration of the session.

It is possible to have more than one IP session at the same time with the same MS, e.g. one session for IPv4 and one for IPv6. These would map to two parallel accounting sessions and would result in parallel accounting streams for the same WiMAX session if several IP addresses were assigned at the same time.

More important, however, is the fact that network equipment vendors may choose to implement accounting sessions to cover individual packet data flows within the IP session. Further details about the WiMAX session concept and flows are provided in Section 4.4.3. Therefore, we limit considerations here to the example of a simple video call. The overall video call that may make up only parts of the IP traffic across the IP session is logically summarized as a service data flow. Individual parts within this flow, e.g. the voice data part, are called packet data flows.

In theory, there may be many parallel active accounting sessions within a single WiMAX session. This depends on the decision in the network about how to perform accounting, and on the actual QoS and IP session settings of the MS for the specific WiMAX session.

Any decision about whether to actually deploy such fine-grained accounting models in practice is up to the operator's needs and business models. However, there is certainly an impact on the overall network performance and especially on the load on the operator's AAA backbone. This is due to the fact that any granularity finer than the IP session multiplies the sum of accounting-related control messaging and processing across the network. So the question to be answered is whether the increase in overall network cost can be justified by the benefits of the resulting tariff models.

As already discussed in the introductory part of this section, offline accounting in WiMAX can be based either on RADIUS or Diameter. For Diameter-based offline accounting, there are two realization alternatives: one option is to perform the accounting based on the WiMAX-native parameter set, which can be considered a rather direct translation of the WiMAX-native RADIUS formats for offline accounting into Diameter AVPs; and the other option is to make use of the functionality defined by 3GPP PCC. For the latter option, the architecture shown in Figure 4.4 does not change at a general level. When using PCC, the accounting client maps to the PCEF (see also Section 4.4). The architecture is extended by an offline charging system (OFCS) function. This may be a standalone CSN entity or may be collocated with the CSN AAA server. Due to the fact that all offline signaling in the WiMAX CSN has to go to the AAA server, or at least has to pass through it, the AAA server would proxy all offline accounting to and from the OFCS in the standalone case. The actual interface between AAA and OFCS is not covered by WiMAX specifications as per the Release 1.5 network architecture, so it is a deployment-specific choice about how to realize communication between these CSN entities. However, one benefit of this is that an already existing OFCS deployment following the 3GPP specifications could be reused without modifications. This would require the WiMAX AAA server to perform translation of the actual accounting messages and UDRs between the WiMAX R3 side and the OFCS, such as stripping any WiMAX-specific extension.

4.3.3 Online Accounting

Sometimes subscribers are willing to pay regular fees to the operator, e.g. on a monthly basis. Instead, charging network services on an ad hoc basis is of interest to a large number of users and allows operators to significantly increase their customer base.

Such ad hoc usage, where charging just takes place per use and without any monthly fee, is typically handled via prepaid accounts. Every time the operator's service is used, it is accounted for and directly charged to the prepaid account. Whatever amount of money is loaded to the account, the account is directly charged per actual usage, which is more or less in real time. If the amount of transferred data is the information metered by the network, such accounting is per volume. Otherwise, the time of service usage is metered and accounting takes place per duration.

When looking at the whole concept from a more technical perspective, the important difference to offline charging that motivates the term 'online' is that charging of service usage to the user's account takes place directly. This happens instead of the operator's backend collection of metered data first and then putting together the user's bill later on. As a result, the network and especially the network entities actively providing service, like the ASN-GW, always have to be aware of how many credits are still available. Of course, the WiMAX network is designed to be able to apply either offline or online charging on a per subscriber basis to individually support the different types of subscriptions. Both types of accounting can be performed per IP session, or per service flow.

The fact that network entities like the ASN-GW or the HA (with the former being the default network entity for performing online accounting in WiMAX networks) need to be aware of the remaining credits of a subscriber, and need to stop service in case these are depleted, leads to partially reversed roles in the accounting model. It is no longer the AAA server or prepaid server (PPS) doing most of the active work by collecting accounted data and combining this to charge the subscriber. In contrast, it is now the enforcement point for the IP traffic that is in charge. The AAA client that in offline accounting just sends the UDRs to the AAA or accounting server is now called the prepaid client (PPC). It has to pull a small amount of credits – called a 'quota' - from the subscriber's account in the PPS. The basic concept is very simple: the pulled quota allows the PPC to grant a specific 'amount of service' where the actual service may be subject to a specific tariff group. It is consumed based on the actual service usage in real time. The metering actually takes place in the prepaid agent (PPA). That is, similar to offline accounting, the actual metering can take place in a different entity than the AAA client or PPC. As soon as a defined threshold is reached and the quota is close to being used up, the quota can be replenished by an interaction with the PPS. As soon as the subscriber's account runs out of credits, the PPC cannot get any more quotas and the ASN-GW or HA will stop service. The subscriber has to recharge the account.

Although this model sounds simple, the complexity of handling online accounting is technically higher than what is done for offline accounting due to the real-time nature. This is because a number of special situations need to be supported in a standardized way across the network. Examples of such cases are the following:

- Clearly, a subscriber will not always use up the full quota that has been pulled by the accounting client from the subscriber's account. For example, if such a quota enables a download of 20MB of IP data, the user may consume 7MB and then stop using the service for a while. As a result, the remaining part of the quota, 13MB in our example, will be returned to the PPS and will be recharged to the account. If not returned, the user could connect to a part of the network later on that is controlled by a different PPC, and the remaining part of the quota would be lost.
- Let us assume that the quota enables the PPA to allow service usage for 2 minutes. The quota itself does not contain a specific monetary value due to the fact that the PPC would lack information about how much the usage of the actual service would cost, e.g. in terms of euros. Such a 'rating' of the service and the related knowledge typically lies within the operator's backbone and is done by the PPS. The resulting quota granted to the PPC is already customized to match the metered data, like time or volume. It can also be a token that allows the download of a specific piece of data, like a ring-tone. So different services can be classified into different 'rating groups' and the quota value would typically vary between services belonging to different rating groups. For the same amount of money pulled from the subscriber's account by the PPS, the PPC could receive a quota of 15 minutes of service usage for standard Internet access, or of just 3 minutes for a different and more expensive service. Based on the fact that the PPC has no knowledge about the actual cost of a service, it needs to request a new quota in case a new service is used by the MS that falls into a different rating group.
- Handover between ASN-GWs can result, as shown in Figure 4.6, in a relocation of the PPC to a new ASN-GW. As the PPC is collocated with the (anchor) authenticator, such a relocation will take place together with authenticator relocation for the respective MS. This requires the related MS accounting context to be transferred. Independent of a relocation of the PPC, the PPA as the actual metering point in the anchor DPF can be separate from the PPC and can be relocated to the PPA of another ASN-GW when the anchor DPF relocates. The current quota information for the MS is stored in the PPA, so in contrast to a PPC relocation, a PPA relocation requires the remaining quotas to be transferred to the new PPA across the R4 reference point.

In WiMAX, the initial quota for a session can already be embedded in the AAA message carrying the EAP-Success and confirming network entry to the ASN-GW/PPC. This provides an optimization of the PPC requesting the initial quota after successful network entry with a separate message exchange.

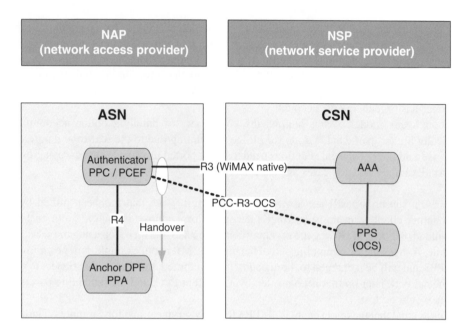

Figure 4.6 Online accounting functional entities with split client and agent

Online accounting for AAA-based network infrastructure can be based on both RADIUS and Diameter. In WiMAX, this is of course not a new invention. For RADIUS, efforts have been made to create an online accounting solution that is based on specifications aligned with IETF-defined protocols and formats. As a result, online accounting messages and formats like the format and encoding of the actual quota data are aligned with [73]. The main difference is that in the WiMAX network specification, only parts of the functionality offered by IETF are actually supported. Some of the more advanced mechanisms like resource pools, where several independent service flows can be accounted at the same time by pulling credits from a single pool of quotas in a coordinated way, are not considered. However, these are not actually prevented, so any vendor could choose to implement them as optional functionality. Such an implementation would be vendor-specific functionality from a WiMAX specification point of view, but would still be based on the standardized IETF procedures.

For Diameter, online accounting is realized from the definitions already used in the 3GPP PCC framework and the interface denoted Gy between the PPC. The PPC functionality maps to the PCEF in 3GPP PCC terminology and the PPS to the online charging server (OCS) for PCC as shown in Figure 4.6. This in itself is based on the IETF Diameter credit control application (DCCA) [72]. WiMAX specifics, however, had to be added to these definitions and the WiMAX network specifications define the resulting PCC-R3-OCS reference point for online charging in [55] that is largely based on Gy.

One architectural aspect that is specific to WiMAX is the way how the online accounting interfaces are routed between the involved CSN entities:

• For RADIUS-based online accounting, the WiMAX native signaling across R3 will always run to and from the AAA server. The AAA server can either include the PPS module itself, or proxy the online accounting-related signaling to an external PPS. However, the actual interface between the AAA server and PPS is not specifically considered by the WiMAX network specifications, so its realization is left to the specific deployment.

- For the alternative Diameter/PCC-based online accounting, the R3 AAA signaling (PCC-R3-OCS interface) can directly terminate in the PPS/OCS of the CSN without involving the AAA server. However, it is still an implementation choice about whether to operate a separate OCS, or to collocate this functionality with the AAA server.

4.3.4 Accounting-Related Identifiers

When looking at the different session concepts related to accounting in a WiMAX network, one may automatically raise the question about identifiers and identities related to such sessions. It may also ask for clarification about how the overall accounting data that is collected for a subscriber's activities in the network can be related to other data, and finally to the subscription that needs to be billed accordingly. The visible subscription identity itself, being the NAI used in AAA signaling during network access, can be a temporary identity as soon as identity hiding is used by an operator. It may change several times within a WiMAX session (see Section 3.5.3), so the NAI is not a suitable identifier for binding accounting-related data within the ASN or any intermediate CSN in roaming. Also, there is no guarantee that the home CSN's AAA server will keep track of all such temporary NAIs for a subscriber. So in summary, the (outer) NAI, although being added to all accounting messages, does not play an important role in accounting.

Every accounting message exchanged between the accounting client and server in WiMAX contains the 'Acct-Session-Id' value. This is the main identifier for correlating all Start/Stop/Interim-Update messages belonging to the same accounting session. It serves well within the same session, but is different between such sessions and will change, for example, whenever authenticator relocation takes place between two ASNs after the MS has handed over to a BS controlled by a different ASN-GW. As a result, an additional identifier is required to correlate individual accounting sessions to each other that take place within the same WiMAX session, or at least are bound to the same IP session of a MS. This second mandatory value for all accounting messages is the 'Acct-Multi-Session-Id'. The value's name hints that it stays the same across multiple accounting sessions of the same WiMAX session. In fact, the WiMAX CSN network that assigns this identity chooses the values of 'Acct-Multi-Session-Id' and the WiMAX-Session-ID to be the same, so all accounting messages can easily be related to a unique WiMAX session across the network.

Regarding the different session types that need to be identified in accounting data, central identifiers for the IP session and the more fine-grained packet data flows are the IP address assigned to the MS for the length of the IP session and the PDFID, respectively. Also, the SDFID, which can bundle a set of PDFIDs, may be used. See Section 4.4.4 for a more detailed summary of the different QoS-related session identifiers that match those used for accounting.

One value that can optionally be used for a number of network-internal features including accounting is the CUI (see Section 3.7.4) value. Its purpose is actually left to the concrete implementation as the available standards do not mandate or describe any specific usage for the CUI. However, one way to utilize it as an additional identifier within the accounting signaling between the WiMAX accounting server and client would be to bind together several WiMAX sessions of the same subscriber. This requires the AAA server to add the CUI value to accounting-related messages and to keep it stable over a period of time to overcome any correlation issues related to identity hiding or changing identifiers like the WiMAX-Session-ID (see also Section 3.7.4).

4.4 Network QoS Architecture

This section will deal with the WiMAX-specific mechanisms that are available to support applications across WiMAX with a specific QoS. Applying a specific QoS means that the service flows over the radio link and the IP packets in the network belonging to the communication generated by a specific application like VoIP can receive individual treatment across the WiMAX network. This helps to match specific application needs where the typical best-effort characteristics may not be sufficient for a good user experience. The section focuses mainly on the network aspects of QoS support across the ASN and CSN. Further details regarding the radio aspects are given in Section 6.2.4. Also, this section is limited to the WiMAX infrastructure aspects of QoS control. In general, good QoS support is a joint effort between multiple layers of the network and especially between the infrastructure and the application layer. For example, the application can contribute to overall user experience by selecting the right codecs, or by being aware of the wireless nature of the communication.

4.4.1 Motivation and Overview

Applications running across IP-based transport can be of quite different nature and can serve extremely different needs. To give just two examples highlighting the broad range, let us consider on the one hand users accessing a map service in the Internet and on the other hand members of a rescue team establishing a voice(-over-IP) call between each other.

With the above example applications it becomes clear at first glance that the requirements for the underlying network and access can be very different and depend on the underlying use case. Access to the map service mainly requires the WiMAX ASN and CSN to offer a reasonable download rate to support fast transfer of map or satellite image data to the MS when the user scrolls or zooms in the map view. In addition, the ability to provide location information to the MS or support the MS with its own measurements is of interest. However, there is no need for any specific service quality like a very short round-trip delay between the MS and application server as this would only increase the overall cost for providing the application to one or more of the involved parties.

Most likely at the other extreme there are voice calls between the parties of a closed group of users that is granted special priority over standard subscribers. The goal is to allow service for this group even in extreme network overload situations. This requires a special treatment, or QoS, applied to the voice data to enable high-quality communication with minimal delays. In addition, the use case requires special priority when entering the network and setting up the service flows required for the voice application – which might also require an overloaded network to temporarily shut down or degrade service for regular users based on the special authorization for the group members.

Even with the above motivation, there is a general discussion about whether operators will really move toward making substantial use of QoS frameworks. This is especially applicable to heterogeneous environments with different interconnected business entities, like several WiMAX operators and third-party ASPs. With a diverse infrastructure across the different involved business partner's networks that certainly introduces an additional layer of network complexity, any end-to-end QoS approach becomes quite difficult. However, although tariffs offered to end users today seem to indicate increasing adoption of flat-rate subscription models, it also shows that best-effort Internet is not always the perfect solution, as we can see today for VoIP calls or mobile IP-TV.

Any answer as to how much differentiation should be applied by the network to different types of application-related data and to how coarse grained certain application classes should be defined can just be considered a rough estimate today. These will probably be given by the market in the long term, based on whether a critical mass of customers will accept higher prices for a perceived quality improvement in the consumed service.

Independent of the answer, which will also vary between different deployments and ecosystems, it is the responsibility of the respective network technology to offer a reasonable set of building blocks to support QoS. These can subsequently be implemented by equipment vendors and enabled by the network operators based on their specific customer and deployment needs.

The focus of the WiMAX QoS architecture is to enable improved control of all connections and service flows created over the air interface and to integrate this with common IP QoS control mechanisms. The related standards are not designed to directly provide end-to-end QoS support throughout the whole WiMAX network, or between two WiMAX devices. As a result, end-to-end QoS has to be realized in deployment-specific ways. The main reason for this is a decoupling of WiMAX-specific QoS mechanisms that are mainly required to control connections across the radio link and simple integration with IP-based QoS control, so WiMAX deployments can be engineered as per IP QoS practices. This approach is conceptually quite similar to what is done in 3GPP networks and it reflects the fact that WiMAX access in many cases needs to be integrated into existing network infrastructure. For such deployments, any technical specification of a WiMAX-specific end-to-end QoS either would provide no real value or would conflict with already existing solutions.

4.4.2 The Basic WiMAX QoS Concepts

As already highlighted in the above introduction, the QoS architecture developed for the WiMAX network intends to support the setting up and control of service flow bearers that allow for fine-grained QoS control of all traffic running across the wireless interface.

Rather than drawing a generic view as shown in Figure 4.7, where the reference architecture for QoS support in WiMAX is shown with the defined logical functions, we will start with a 'top-down' example that begins with the actual application - say an MPEG video stream – which requires a specific QoS level to be delivered with the best possible end user experience.

The example will be an exercise assuming for simplicity that we are in a WiMAX network implementing the feature set of the initial Release 1 network specifications. Additional functionality of WiMAX networks implementing the Release 1.5 feature set will also be considered, but later in this section.

Let us also assume that the preprovisioned QoS profiles for a specific subscription include one profile that especially supports video streaming. Translated into more technical terms, this would occur by a service flow across the wireless link with the 802.16e scheduling type rtPS (real-time Polling Service). An appropriate service flow (SF) will already be prepared during the initial network entry of the subscriber's device. The QoS profile sets all the information

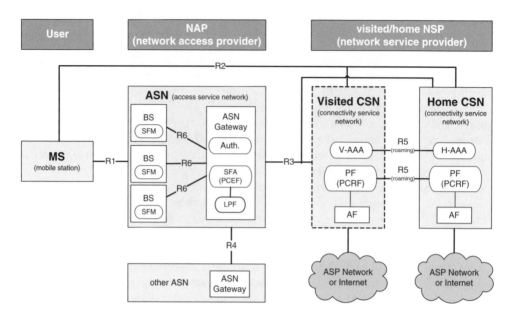

Figure 4.7   Reference architecture for WiMAX QoS support

required to create appropriate service flows matching specific needs of applications. For video streaming, the QoS profile can set a guaranteed minimum traffic rate in bits per second that is sufficient for the expected video data rate, or a reasonably low maximum latency for the IP packets carrying the video stream. More details on QoS provisioning OTA can be found in Section 6.2.4.

In the preprovisioned case the AAA server has sent information describing all QoS profiles for the subscription to the ASN during network entry, and the ASN has created the appropriate service flows – however, these might still be inactive. In more detail, there is always a special initial service flow (ISF) that is created for carrying the initial IP-based communication required as part of the network entry procedure, like DHCP messages or MIP signaling. It can be used later as the standard service flow, e.g. for best-effort Internet data after network entry. The additional rtPS service flow for better video support would likely be prepared but marked inactive until it is activated and used by the MS.

The MPEG video streaming client on the MS will actually trigger the service flow to be switched from inactive to active via appropriate signaling across the wireless MAC layer to the BS. Alternatively, the application server – called an application function (AF) - can also trigger service flow establishment in the WiMAX network. This of course requires that related signaling can be exchanged across an interface between the AF and the WiMAX CSN network.

To apply the settings as described by the QoS profile to the service flows OTA, the WiMAX BS implements a service flow management (SFM) component that is responsible for establishing service flows and for controlling their activation and deactivation. It also applies the required scheduling to the uplink and downlink traffic that actually realizes the different QoS applied to different packet data streams across the wireless link. A BS is only applying QoS to the traffic and is not in control of the QoS policy for a subscriber and the MS. It is also not communicating with the CSN that owns the subscriber-specific QoS profile information.

The latter is centrally performed by the ASN service flow authorization (SFA) component that is realized as part of the ASN-GW. Furthermore, the SFA is the default entity responsible for classifying all traffic according to the active service flows. The classification data itself is simply a description of a flow of IP packets based on common IP packet parameters like:

- source and destination IP address or address range
- source and destination port or port range
- protocol (transport) type
- direction (uplink, downlink or bidirectional).

Additional parameters like an assigned priority or the device's MAC address can also be used. Another important building block in WiMAX that especially addresses fixed deployments is the support for Ethernet services. This is available as part of the QoS framework that allows Ethernet-based service flows carrying Ethernet frames instead of IP packets to be described.

Another main role of the SFA function is to authorize all requests coming from the MS against the QoS profile information of the subscriber. In addition to what the SFA receives from the subscriber's home CSN operator during network entry, the SFA can also make use of a local policy function (LPF) and database to implement ASN-specific QoS policies that might even overrule any of the subscription-specific settings. This in fact leaves several degrees of freedom for the involved operators when agreeing on a common supported QoS policy as part of a roaming agreement.

Looking at the actual user data in our example, the video stream will be carried within the service flow that is active and appropriately prioritized by the scheduler of the BS and MS for the wireless link.

Within the ASN, the BS will map this to and from a specific generic routing encapsulation (GRE) tunnel [54] between the BS and ASN-GW across the R6 reference point. R6 granularity is a per SF one, so each individual service flow will be carried through an individual GRE tunnel between the BS and ASN-GW. Scheduling and prioritization within the ASN can be based on these GRE tunnels. However, across R6 and further on between the WiMAX ASN and CSN network, any prioritization of IP packets carried by a specific service flow is a deployment-specific choice and may leverage common IP-based QoS mechanisms like differentiated services (DiffServ) [53].

Here, it becomes clear that the WiMAX network QoS specifications really focus on the network control of service flows across the wireless interface. Concrete deployments need to consider numerous additional aspects for a consistent end-to-end approach. These include:

- The tradeoff between supporting diverse service flows for the different types of applications (802.16e allows for five different scheduling types that are optimized, for example, for best-effort Internet, VoIP or video). One choice is to make substantial use of them. Another choice is to limit the use of specific QoS measures to the absolute minimum, e.g. to just offer a special service flow type for video (which indeed does not work well without dedicated scheduling). All the other parts of the user traffic would then go through the initial service flow, to keep the overall number of service flows in the ASN low for better performance.
- DiffServ can be used to prioritize different IP packet flows within the same WiMAX service flow, e.g. in the case where the MS just uses one single best-effort service flow for all its traffic. In contrast, different DiffServ classes can each be mapped to an individual service

flow of the same MS, which would of course result in a higher complexity for handling traffic in the ASN and in the MS due to the larger number of parallel active service flows. Note here that a conceptual difference to DiffServ is that WiMAX operates on individual flows that are created, for example, for individual applications, whereas DiffServ makes use of more generic classes of traffic.

- Typically, control signaling like DHCP or Mobile IP packets will be carried across a best-effort service flow only, and better QoS is only relevant for specific user data. However, there may be specific requirements to apply special scheduling with higher priority to control signaling. As an example, SIP/IMS control messages can be given special treatment as soon as they are related to an emergency call.

4.4.3 Flow Management and Dynamic QoS Support

After the rough overview provided by the above example, let us now take a more detailed look at the QoS mechanisms for WiMAX.

We will begin with the service flow concept itself. It is important to note that service flows in the ASN are unidirectional, so there are separate flows for uplink and downlink communication. For example, a video call may consist of:

- a pair of service flows (uplink and downlink) for the video data with a scheduling type rtPS that is optimized for video packets; and
- another pair of service flows for the voice part with a scheduling type UGS (Unsolicited Grant Service) that is optimized for VoIP packets.

It is important to note that the concept of a service flow is internal to the ASN. From the CSN's perspective (and therefore across the R3 reference point) there are service data flows and packet data flows that can be bidirectional and that enable bundling of the different ASN flows related to the same application, as we will see later in this section.

In general, WiMAX service flows can be classified as either preprovisioned or policy framework controlled. Preprovisioned service flows are established at the time of initial network entry as authorized by the home AAA server of the subscriber. They are comparable to 'default' flows in the 3GPP-defined QoS architecture. In contrast, policy framework-controlled service flows are established, modified or terminated as the result of a real-time trigger from the application layer. They are comparable to 'dedicated' flows in the 3GPP world.

A special concept in WiMAX is the initial service flow (ISF). This is actually the first pair of service flows created upon network entry at a time when no IP connectivity is available to the MS. It is bound to the CS type and is essentially needed to carry the delay-tolerant control traffic required as part of the network entry procedure, like DHCP or MIP signaling. The main reason for making the ISF a special pair of service flows is that it initially needs to carry IP control information when the MS does not yet have an IP address assigned. So the ASN needs to create a special binding for routing the ISF. This is based on identity bindings that are initially available, like the MS-ID or the subscriber's NAI, mapping to the service flow identifiers (SFIDs) of the ISF. Prior to IP address assignment across the ISF, no other preprovisioned service flow carries IP packets.

It is essential that the ISF can be activated, because otherwise network entry would not be possible. Also, the ISF always stays operational during the time the MS is attached to the network. Besides the special function during network entry, the ISF can work just like any other preprovisioned service flow, so, for example, it can be used for best-effort IP traffic. The ASN can also update the classifier and QoS policy of the ISF after successful IP address assignment.

The initial Release 1 QoS model used in the above example provided AAA-based support only for preprovisioned service flows with a static nature. QoS profiles are stored in the AAA server and are sent to the ASN-GW during initial network entry. However, they cannot be changed or modified, and new ones cannot be created in mid session. Profile information is carried within the RADIUS Access/Accept message sent by the AAA server if the requested network entry is successfully authorized against a subscription. This is described by (1) in Figure 4.8 showing a high-level message flow for a MS entering a WiMAX network. It also highlights the signaling required to establish service flows for the MS as configured in the subscription profile.

Besides sending the QoS profiles to the ASN in (1), further steps to support service flow establishment within the initial network entry procedure include:

- The exchange of QoS-related capabilities between the MS and the SFA. This is realized within Registration Request/Response messages across the wireless MAC layer together with an exchange of MS_Attachment messages relaying this information between the BS and ASN-GW across R6 in the ASN, see (2) in Figure 4.8.

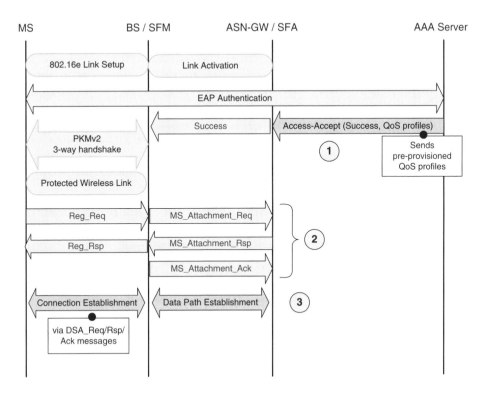

Figure 4.8   QoS setup during initial network entry of a MS

Table 4.2   QoS features per WiMAX release

QoS model	Preprovisioned service flows	Dynamic service flow control (network initiated)	MS-initiated SF establishment
AAA (Rel. 1)	X	—	—
AAA (Rel. 1.5)	X	X	X (no network control)
PCC (Rel. 1.5)	X	X	X (full network control)

- The actual establishment of the service flows, which is subsequently triggered by the SFA as shown in (3) of Figure 4.8. It is based on dynamic service addition (DSA) messages across the R1 reference point.

At least the initial service flow will always be established. Others can be created in addition based on the subscriber's QoS profile information that the SFA received from the AAA server at the time of initial network entry.

The WiMAX Release 1.5 network specifications are based on the same model of QoS profiles provisioned by the AAA server but are extended by additional capabilities like the dynamic creation, modification and termination of service flows in mid session. This is controlled by a policy framework that is either part of the AAA server or a separate standalone function. The functionality available with the WiMAX Release 1.5 network specifications is summarized in Table 4.2, which:

- covers dynamic service flow creation, modification and deletion, so a service flow can also be created or modified in mid session;
- allows both the network and the MS to trigger service flow creation or modification; and
- specifies, in addition to the standard QoS control based on the common R3/R5 AAA signaling path, a PCC framework for WiMAX that is largely based on the 3GPP PCC architecture, see Section 4.4.5.

With the Release 1.5 WiMAX network specifications, it is possible to dynamically create or update a service flow, e.g. by increasing the maximum bandwidth in deployments where the entity in control of subscriber QoS remains the AAA server. The AAA server acting as the QoS policy controller for a MS in the network can trigger such SF creation or update of a preprovisioned service flow by sending a change-of-authorization (CoA) message [8] across R3. This allows the AAA server to initiate a message exchange with the AAA client in the ASN-GW and to attach updated QoS profile information. The SFA in the ASN will subsequently trigger the required interaction with the MS to apply the update received from the AAA server.

The emphasis in WiMAX networks certainly lies on network-initiated service flow control. As an option, the device can also perform MS-initiated service flow creation and modification. However, such MS-initiated action would require tighter integration of the application, operating system and WiMAX driver in the MS than the network-initiated model. This tighter integration does not map well to the 'open device' approach that is so important to the WiMAX ecosystem. For example, it applies to laptops with WiMAX radio adapters and installed applications, where all these components are in general independent of each other. Also, as we describe in more detail in Section 4.4.5, network (CSN) QoS policy control for the MS-initiated model is not provided in a standardized way in the case of AAA-based QoS control. It is only available when using the PCC framework.

With the PCC framework, a different network entity and signaling path from the ASN-GW to a PCRF is introduced. The PCRF acts as the central entity controlling and authorizing the QoS of the individual service flows. With this, it becomes possible to decouple QoS control from access authentication and authorization, and to dynamically provide new QoS rules and establish service flows in mid session. One main architectural difference between the WiMAX-native (AAA server-based) and PCC-based dynamic QoS control is that the PCC framework provides a standardized interface from the PCRF to an AF that allows for interaction between the network QoS control and the actual application server making use of it. WiMAX does not standardize such interaction, leaving it to deployment-specific solutions. One exception to this is the WiMAX Universal Service Interface (USI) that is part of the Release 1.5 network specification. This is intended to cover QoS signaling between the CSN and external applications but does not specifically target interaction with the AAA server in the CSN. In addition, the specification is only available in an initial version that does not reflect any experience from deployments.

Similar to the mobility model that we already described for other control signaling like accounting, also WiMAX QoS control is designed to cover inter-ASN mobility without requiring the ASN to switch R3 signaling over to the serving ASN-GW after a handover took place. Instead, QoS control can be relocated by the original (anchor) SFA to the (serving) SFA the MS is currently connected to. In detail and as shown in Figure 4.9, this means that after handover to a new ASN-GW and relocation of the anchor DPF, the 'old' ASN-GW can still be in charge of QoS control for this MS and the anchor SFA stays collocated with the authenticator

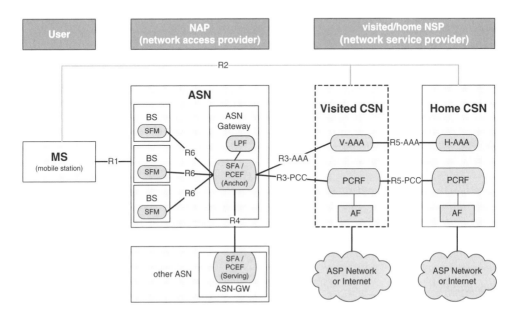

Figure 4.9   WiMAX QoS architecture with alternative AAA or PCC control options

function. However, the anchor SFA only covers the control plane signaling part. The actual QoS enforcement is performed in the serving SFA that is collocated with the anchor DPF or FA and is in the actual data path. Using the built-in capability of splitting different ASN functions across several ASN-GWs can help deployments to avoid time-consuming relocation of control functions to the new ASN-GW at the time of handover. Such relocation might cause a break in service delivery when performed at the time of handover, although the split approach clearly increases complexity in network signaling (see also the additional considerations in Section 8.2). Regarding the split responsibilities in such a setup, the anchor SFA remains the function interfacing across R3 with the AAA server or PCRF, while communicating with the serving SFA across the R4 interface. Similar to the role of the anchor authenticator, this will only change on a relocation of the AAA client from the anchor to the serving ASN-GW that requires an EAP reauthentication.

4.4.4 Identifiers

Controlling a QoS flow based on the WiMAX-native methods with support from the WiMAX ASN and CSN requires the handling and mapping of a set of identifiers. These identifiers describe flows related to a specific MS and subscriber across the different reference points of the WiMAX network and stand for the different granularities they cover.

Starting with CSN involvement in the actual QoS control, the AAA server chooses all QoS profile data based on the subscription and subscription identity (NAI). This profile data is sent down to the ASN during initial network entry, where each service profile is specific to one CS type and is identified by a service profile ID (SPFID). The CSN controls the actually established flows for a MS using the packet data flow ID (PDFID). It may also use a service data flow ID (SDFID) where a service data flow may consist of several individual packed data flows when related packet data flows should be grouped together. Table 4.3 below provides an

Table 4.3 QoS identifiers

Flow type	Identifier	Relevant interface	Example	Described by	3GPP PCC
Service data flow	SDFID	R3	Video call		AF session with associated service data flows
Packet data flow	PDFID	R3	Voice part of a video call	PCC: PCC rule WiMAX: preprovisioned QoS profile	Service data flow
Service flow	SFID	R4, R6 and R1	Upstream flow for voice data		
Connection (air interface)	CID	R1	Upstream flow for voice data		

example and overview of how these identifiers relate to each other and for what parts of the network and reference points they are relevant.

Within the ASN, service flows are controlled by the SFA and are identified by a SFID. Any SFID uniquely identifies a service flow for one MS and is assigned by the SFA. The SFID is sent down to the BS and this puts it into the wireless MAC layer DSx messages to the MS. PDFID and SDFID are assigned by the CSN and are the QoS identifiers used in the accounting data reported by the ASN to the CSN.

Taking a closer look at the wireless interface and at the QoS-related identifiers shared between the MS and BS, we encounter the connection identifier (CID) that is relevant only across the wireless link. There, however, it is the central identity for describing a connection between the MS and BS. It identifies an active service flow, so the difference when compared to the SFID is that for each activation of a preprovisioned service flow, a new CID can be assigned, whereas the SFID remains unchanged over the lifetime of the preprovisioned service flow in the SFA within the subscriber's WiMAX session. In combination with the MS ID (MS MAC address), the SFID can actually be seen as a globally unique identifier across the WiMAX network that can support service continuity across handovers. On the contrary, the CID has only local significance at a single BS.

All of the identifiers discussed above are important within the control signaling of a WiMAX network and the MS itself. Considering data transport inside the ASN, individual service flows are carried by a GRE tunnel between the BS and ASN. Any specific service flow is identified by a GRE key (Figure 4.10) that is mapped within the BS into the SFID and corresponding MS-ID. An overview summarizing the main service flow granularities and related identifiers and also their different scope and relevance in relation to the different parts of the network is given in Table 4.3, but let us also summarize:

- **Service data flow identified by SDFID**: Identifies a service data flow on R3 for the subscriber that comprises one or several individual packet data flows of the full service. One example would be a service data flow describing a video call that is composed of one packet data flow for the video part and another one carrying the voice part.
- **Packed data flow identified by PDFID**: Corresponds to an individual media component of the overall service data flow. Within a video call an example of a packet data flow is the voice part. In case of a pure voice call, the packet data flow actually maps to the service data

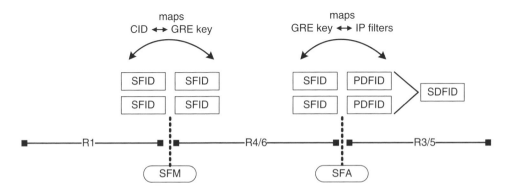

Figure 4.10   Mapping of service flow identifiers

flow, i.e. the service data flow only consists of a single packet data flow. When related to the SFID, the PDFID basically maps to one or a pair of unidirectional service flows.

- **Service flow identified by SFID**: Whereas the above service and packet data flows are concepts relevant to the CSN and AF, service flows are focused on L2 and are therefore limited to the ASN. Also the SFID is only relevant within the ASN. The SFA is responsible for mapping PCC rules describing packet data flows to the individual service flows, and for mapping between the different identifiers.

While considering the adoption of the 3GPP PCC framework for WiMAX as detailed in Section 4.4.5, a mapping between the service flow concept developed by 3GPP for PCC and the WiMAX service flow model as developed for the WiMAX Release 1 network specifications had to be described. For a better understanding, Table 4.3 also shows the comparable terminology of the 3GPP PCC for the non-ASN specific flow types in the rightmost column.

4.4.5 The PCC Framework for WiMAX

For providing advanced mechanisms to dynamically control QoS in the mobile network, the 3GPP specifies the PCC framework. As part of the WiMAX Release 1.5 network specifications this has been adopted for WiMAX needs [55] but with the goal of staying aligned as much as possible with the original framework. For WiMAX the main purpose of adding PCC support as an optional building block is to allow applications like the IMS to dynamically control the service flows related to the application sessions. It is also possible to use specific charging parameters in the access network for the application sessions and relate them to the WiMAX service flows. However, we will focus on the QoS-related functions of the PCC framework in this section.

The 3GPP is producing annual releases of its specifications, and the WiMAX adoption of PCC as per [55] is based on the 3GPP Release 7 solution [56].

In the future, it is planned to further align this with successor releases of the 3GPP PCC solution, such as [57] for Release 8. One of the motivations for this is interworking between WiMAX and the 3GPP core network based on the 3GPP Release 8 evolved packet core (EPC) which in turn requires Release 8 of the PCC framework to enable dynamic QoS control.

The fundamental difference between the standard WiMAX QoS control based on the WiMAX-native AAA signaling path and PCC-based QoS control lies in the way that service flows can be controlled. In the PCC case covered by Figure 4.9, a central entity in the CSN, called the PCRF, controls every creation or modification of a service flow. In contrast to the WiMAX-native mechanisms, this allows for an application-based policy control model especially for MS-initiated service flow creation or modification. It becomes possible to apply a specific QoS only to a specific application based on a standardized interface between the PCRF and the AF.

For the WiMAX-native dynamic QoS control, a MS triggering service flow creation cannot be controlled by the CSN in a dynamic or real-time fashion. This is only possible for the network-initiated service flow creation or modification where the SFA is triggered by the AAA server. In contrast, for the MS-initiated case the SFA has no way to authorize such a request with the AAA server.

As an example, let us assume that a WiMAX operator offers a selected VoIP application in its own network. However, devices may make use of any type of VoIP client that does not have any relation with this operator, but is simply using the broadband IP connectivity. Within a PCC-based deployment, the operator's CSN can control service flow establishment by enforcing a policy where the MS, when making use of the operator's native VoIP application and triggering service flow creation, can receive special QoS. At the same time, any other VoIP application is only allowed best-effort QoS. With the WiMAX-native QoS control, this would not be possible for MS-initiated service flow establishment because the CSN would not be able to control for which type of application a specific QoS level is granted.

When mapping concepts developed for one network architecture to another with a different access technology, typically an important step is to consider the basic definitions and align them to create a sound overall approach for the target network architecture. For WiMAX, one such step is to adopt the PCC architecture included to translate the 3GPP-defined concept of IP connectivity access network (IP-CAN) bearer and session into the WiMAX service flows defined by the WiMAX QoS and data path model. An IP-CAN bearer in WiMAX is simply described by a single service flow or a pair of service flows. However, the exact definition of an IP-CAN session for WiMAX required a detailed analysis and resulted in its unique identification through the IP address assigned to the MS (for cases where Mobile IP provides R3 mobility, this is the HoA address assigned to the MS by the HA) and the subscription identity NAI.

Conceptually one major difference between the 3GPP design of the PCC and WiMAX is that WiMAX distinguishes between the access (ASN) and the core (CSN) network and also allows them to be operated by different business entities. This actually leads to a number of differences in the design. WiMAX, unlike 3GPP, puts the SFA functionality that is mapping to the PCEF in the PCC architecture into the access network instead of the core. As a result, one of the major adjustments when transferring the PCC architecture to WiMAX is to add support for R3 handover or, in more detail, handover from one PCEF to another. Such relocation is not supported by the 3GPP PCC specification as part of 3GPP Release 7. According to the 3GPP design, the corresponding function is already located in the core network and no handover support was required in the 3GPP Release 7 architecture.

WiMAX, however, fundamentally depends on such mobility support for the affected R3 interface that connects the PCEF and PCRF. To keep modifications to the original 3GPP concept as small as possible and make full reuse of the existing interface specifications, a policy distribution function (PDF) was introduced as a new logical entity in the NSP operator's domain. The main role of the PDF is to hide the specifics of the WiMAX access network from a standard PCRF implementation. As shown in Figure 4.11, this keeps the interface towards the PCRF (denoted Gx for 3GPP and Ty for the 3GPP2 derivative) unchanged by hiding WiMAX-specific mobility impacts and the distributed nature of enforcement points like the anchor SFA and serving SFA. The PDF can best be understood as a WiMAX-defined mediation function with a standard PCRF implementation.

Hence, looking at the different PDF realization options, one approach is to integrate the required additional functionality directly into existing PCRF equipment to keep the number of physical entities in the NSP operator's network as low as possible. On the other hand, in cases where PCRF entities are already deployed, or will be used unchanged for other reasons, realization of a PDF as a separate physical PCC interworking unit is possible.

Another major difference of the PCC Release 7 architecture compared to WiMAX that is especially relevant for accounting and charging support is the fact that the PCC is designed to support only a single PCEF at a time for a specific subscriber. In WiMAX, it is in contrast possible to enable accounting support in the CSN operator's HA in addition to the default

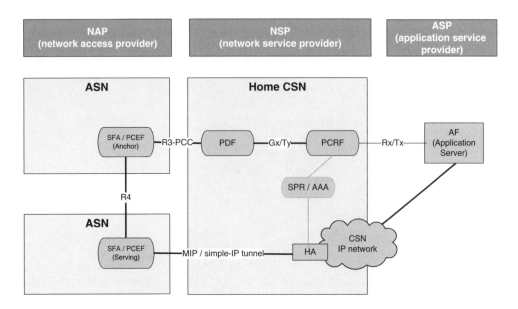

Figure 4.11 PCC-based QoS control architecture

accounting performed in the ASN-GW. With the PCC this would require an additional core PCEF (C-PCEF) function in the CSN that is collocated with the HA.

Within the WiMAX Release 1.5 network specifications this is not fully addressed. The approach followed by the WiMAX PCC architecture is to add the required functionality of distributed policies across several enforcement points to the WiMAX-defined PDF. This is once more to minimize the impact on existing and possibly already deployed components like the PCRF as much as possible. Another approach in cases where an accounting client is desirable in both the ASN and the CSN may be to base accounting on the standard WiMAX AAA communication via RADIUS or Diameter with the home CSN AAA server, while keeping QoS control independent and using the PCC framework for this. Note that for later versions of the 3GPP PCC framework, such a distribution of policies across several enforcement points is possible, which might relax the requirements for keeping the PDF as soon as WiMAX PCC support covers later 3GPP PCC versions in future releases.

When looking at integration of the PCRF into the CSN, any interface to a subscriber profile repository (SPR) that may be collocated with the AAA server, if required data is not configured in the PCRF itself, is left to the specific deployment. No interface specification is available from the WiMAX Forum.

Besides the adoptions described above, a clear design goal for the WiMAX PCC support has been to leave the Rx (Ty for 3GPP2) interface from the PCC infrastructure to the actual AF unchanged. The WiMAX PCC design does not introduce any modifications to Rx and to application functions making use of the PCC framework.

PCC support is not a generic network capability but can be enabled on a per MS basis. As a result, one important piece of information that has to be agreed between the involved networks for a subscriber attaching to a WiMAX network is whether the PCC framework is to be used to apply QoS to the subscriber's sessions. Here, the PCC framework, like other functionalities in WiMAX, depends on the WiMAX capabilities of negotiation as part of the AAA signaling during the network entry authentication and authorization process. Due to the fact that PCC support for a WiMAX operator is only an optional building block, PCC support cannot be assumed to be available in general. The involved ASN and CSN require negotiation capabilities to determine whether PCC support is available. The negotiation mechanism across R3 allows an ASN to advertise PCC support. For any intermediate visited CSN it is possible to remove such an indication for PCC support in case it does not support this itself, or does not want to use PCC. If PCC capabilities are advertised to the home CSN, the AAA server there can choose whether or not to activate PCC QoS control for the resulting subscriber and the related WiMAX session.

4.5 Location Support

One of the more important examples of an enabling service that can be provided by a WiMAX operator is to make the location information of the attached devices available to applications and to the devices themselves. WiMAX provides specifications that in fact cover a variety of different technologies and building blocks, enabling numerous use cases. Within this section, besides giving general information on location technology and the major motivations for operators to deploy location support, these building blocks will be explained and structured to provide guidance about which of the building blocks are applicable to specific use cases.

Location-based service (LBS) can be used to support location-enabled applications in the MS by the network, to enhance application functions within the home CSN or within a roaming partner's CSN, or can be offered to third-party ASPs. They are also required to meet regulatory requirements such as emergency services and law enforcement.

The actual location information for a specific device may be represented in one of two ways:

1. Geospatial coordinates, which describe the position of a device using latitude, longitude and altitude parameters resulting in a point, area or volume in space. The smaller the volume, the higher the accuracy of the location. A common way to measure such coordinates in the device itself is a built-in GPS receiver.
2. Civic location, where instead of representing the location of the target as an area or volume using geodetic coordinates, a set of predefined fields representing an absolute street address are used. Civic parameters vary between different countries and for different uses within countries, resulting in numerous possible formats.

As a result, the effort to provide standardized formats for civic location is harder than the more technical geospatial coordinates. A common format for civic location is available from the IETF GEOPRIV Working Group in [107] and is being widely accepted in many countries including the United States, Canada, the UK and parts of Europe. More general considerations that cover both location formats can be found in [108].

Location information is valuable information that does not offer a service to the subscriber on its own, but enhances the subscriber's user experience with a broad range of applications. Such applications usually have both time and accuracy requirements on the location information and both need to be taken into account when selecting the location determination technique. Network measurements provide a good source of information for determining location. This can be collected from the WiMAX ASN and be used by the CSN or partnering ASPs.

The network operator may also provide services directly in cooperation with the subscriber's device, by collecting location measurements from the MS itself or by providing additional information to the device to allow more efficient location measurements to be made in the MS. An example for the latter case is an assisted GPS (A-GPS) service to speed up satellite discovery for a specific geographical location and improve accuracy when the device intends to use GPS-based location determination.

Although applications that make use of the subscriber's location information are significantly gaining in importance today, from an end user's perspective many of them can work with devices equipped with just a GPS receiver. This provides location information directly to the application in the device. However, autonomous GPS is often not considered acceptable for modern applications as the time to acquire a location is far too long. In practice, such GPS capability can be enhanced by location servers offering A-GPS services. These help to significantly speed up the availability of measurements. A-GPS still has issues in built-up areas and indoors where it may still takes tens of seconds to yield a location. This may be sufficient for numerous applications, but for applications that are time constrained, location that takes longer than a few seconds to produce is not practical. When communicating with a location-aware application server, the MS can alternatively provide the cell or BS ID to the server. No direct involvement of the cellular network operator is required for this but the application server has to be able to resolve the cell ID to the corresponding geographical location.

Applications in the MS may either directly use location data measured by the MS, or communicate with an application server and transmit location measurements to this server. For such application servers the type of wireless or fixed access and IP connectivity used by the device is in general irrelevant, unless information like the cell ID of a mobile cellular network is used.

Taking the above into account, the business cases for operators to support third-party applications directly with network-based location information of their subscribers remain somewhat unclear. For a broadband access technology like WiMAX, most applications will just reside in the Internet. Given this, one of the main incentives for the WiMAX operator deploying location support can be to consistently support devices with location data. The location can in turn be used by subscribers for a better user experience, which as a result can help to make the overall network access offering of the operator more attractive.

Besides improved user experience in applications, one of the main drivers for location determination support by access and ISP networks – and this holds especially for networks providing mobile wireless connectivity – remains the requirement to support emergency calls. Sufficiently detailed information about the actual physical location of the caller is mandatorily required by the public safety answering point (PSAP), such as a police station that is in charge of answering such calls. Requirements in this area are not emerging over time from new application services and the evolving market and customer demand. Instead, the boundaries that have to be met by operators are set by national laws and regulatory bodies.

The above considerations apply to WiMAX networks in general and therefore impact both NAP and NSP operators. Hence, it makes a lot of sense for an NSP to offer location information to authorized parties like VoIP application providers on request and to ensure that there is a standardized location interface to the NAP that is able to perform the actual location measurements. This is of course reasonable for all WiMAX network deployments, but it holds especially in cases where the NSP itself integrates a commercial VoIP offering.

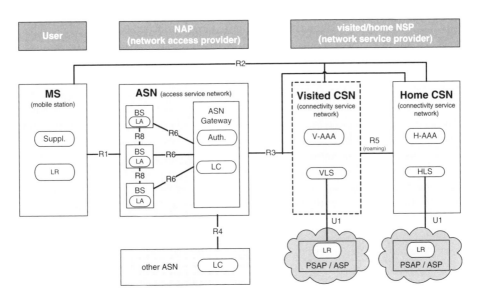

Figure 4.12 Network reference model for WiMAX location support

4.5.1 WiMAX Architecture for Location Support

With the above motivation as to why it makes sense to allow for adding location-based services as an integral part of the WiMAX ASN and CSN, we can now look at the individual building blocks that are added to the standard network reference architecture and that are shown in Figure 4.12:

- Location agent (LA) functionality as the entity measuring the actual location data. LA functionality is located in the BS.
- Location controller (LC) functionality. It resides in the ASN, usually the ASN-GW, and requests measurements from the LAs.
- A location server (LS). This resides in the CSN and is the entity responsible for determining the actual MS location based on measurement data received from the ASN or from the MS itself. WiMAX location support takes roaming into account and as a result a location server can reside either in the home CSN as the home LS (HLS) or in the visited CSN as the visited LS (VLS). Either or both may be involved in providing location information for a single request.
- Location support across the ASN-related R4 and R6 reference points with a set of new control messages and TLVs.
- R3 location support to allow the LS in the CSN to request measurement information related to a subscriber from the serving ASN. This measurement data is used by the LS to determine the location of the subscriber's MS.
- New exchanges across the reference point R2 that cover direct location-related communication between the MS and the location server. The protocol endpoints are the LS in the CSN network and a location requester client within the MS.

In WiMAX the entity requesting location information is referred to as a location requester (LR). Depending on the network operator's policy, external partners like ASPs or responders to emergency calls can operate as LRs by connecting with the operator's LS and requesting location information. These are, however, not considered to be a functional part of the WiMAX network itself. Such LRs logically connect with the WiMAX CSN across a reference point denoted U1 in the reference model as shown by Figure 4.12. WiMAX does not provide a standardized protocol solution for this reference point besides an informative universal service interface (USI) framework [99] that also considers the conveyance of location information. The reason for this is to avoid any overlap with the already existing deployments and protocol choices. Any concrete protocol choice is left to the actual deployment from the WiMAX Forum's perspective. Common candidates, however, would be the HELD Protocol [51] of IETF for the Internet space, or the Mobile Location Protocol (MLP) [100] of the Open Mobile Alliance (OMA) for the mobile cellular world.

Location support in WiMAX covers and allows for combining two different levels of location: (1) control plane location; and (2) user plane location.

For control plane location, extensions to the already existing WiMAX network reference points introduce support for the required communication between LA and LC functions inside the ASN across the R6 reference point. For exchanging location information between the ASN and CSN and between LC and LS, the R3 reference point supports additional functionality.

The term 'user plane location' covers direct communication between the devices and the LS in the CSN. It can rather be seen as the integration of existing protocols in the WiMAX environment that are available from other standards bodies like OMA and IETF. In addition, the subscriber can receive the geographical location of the WiMAX BS through the radio network when offered by the operator. The MS may make use of this information, for instance, to complement location information it measured based on its own capabilities.

Whether user plane or control plane methods are used clearly depends on the individual requirements regarding location-based services and on the capability of the devices attaching to the operator's network. However, a meaningful example is the combination of both, typically denoted 'mixed-plane' location. Here, the LS is collecting MS location measurements from the access network. In addition, location information is provided by the MS itself, and the LS will collect it. The LS can then combine measurement data from both sources to increase the accuracy of the overall location information calculation.

4.5.2 Scenarios and Use Cases

Based on the building blocks that are described in the above section, a number of basic use cases for location determination can be realized. These are classified according to whether the device or the network initiates the communication, whether user plane or control plane procedures are involved, and whether the LS or the device performs the actual calculation of the location from the measurements. The following list can be seen as a rather technical classification of use cases:

- **Device initiated, control plane only**: The MS triggers the LS to initiate control plane procedures, retrieve measurements from the LC in the ASN and calculate the MS location that is provided to the LR.

- **Device initiated, user plane only, device calculates**: This basically covers the case where the MS requests assistance data from the LS across R2.
- **Device initiated, user plane, LS calculates**: The MS sends measurements to the LS to get support for the calculation of the actual location.
- **Device initiated, mixed plane**: The location calculation for the MS-initiated location procedure takes into account both measurements from the MS and measurements received from the LC in the ASN. Calculation can take place either in the MS or in the LS.
- **Network initiated, control plane only**: This case covers a network-triggered location procedure (triggered either by the CSN internally or by an external LR). The LS will retrieve MS measurements from the LS in the ASN and calculate the location. The MS itself is not involved. One example is location retrieval for lawful interception or for network-internal management procedures.
- **Network initiated, user plane, device calculates**: Based on network-triggered location determination, the MS is contacted and provides its actual location information, e.g. based on its GPS capability.
- **Network initiated, user plane, network calculates**: In contrast to the above, the MS only provides measurement data (e.g. from the radio link) to the LS, but calculation of the actual location takes place in the LS.
- **Network initiated, mixed plane**: As for the device-initiated case, both measurements from the MS and from the LC in the ASN are used as inputs to the location calculation.

In contrast to the above classification of location use cases, it is also possible to compile examples that come less from the technical perspective of message flows and signaling procedures in the network. The following list of MS-centric examples collects a number of use cases that can be realized based on the above technical building blocks:

- The MS can collect location information from the BSs in the access network (this is referred to as MS-managed location in the WiMAX context, in contrast to network-managed location, which involves the LS in the CSN). Note that this is not a service dealing with the measurement of MS location; it just conveys the location of the BS that the MS is currently attached to over the radio interface.
- The MS can request GPS assistance data from a location server wherever A-GPS services are used (including for example satellite ephemeris data, GPS almanac data, or lists of 'unhealthy' satellites).
- The MS can measure location information based on MS capabilities like a built-in GPS receiver and can communicate this to a location server in the WiMAX CSN. The location server in turn supports ASPs with location information for the subscriber's active device.
- The MS can add available location information to applications used by the subscriber, including VoIP services. For example, the VoIP client in the device can convey the location information to the VoIP server when placing an emergency call. This location can be measured and calculated in the MS, can be based on location information that the MS requested from the LS in the WiMAX CSN, or can combine both sources for more accurate location information.

Finally, there is the question about why the WiMAX operator should deploy support for MS location. Besides the general considerations at the beginning of this section about location, we can summarize that location support:

- can be mandatory to match regulatory and legal requirements, especially for supporting emergency calls, or lawful interception;
- allows for the better realization of network-internal use cases like traffic management;
- enables services where the operator offers location information collected in the access network to the subscriber as a value-added service, or to make the overall network service offering more attractive to subscribers; and
- allows detailed location information to be offered to ASPs that in turn may be able to improve service quality.

4.5.3 Communication between Device and Network

Messages that are directly exchanged between the MS and LS are logically passing through the R2 reference point as shown in Figure 4.12. Across this reference point, WiMAX is providing support for two application layer location configuration protocols originally developed in OMA and IETF.

Besides technical differences, the two approaches of the OMA secure user plane location (SUPL) protocol and of the WiMAX location protocol (WLP) that is largely based on the work of the IETF Geopriv effort are required by different groups of stakeholders due to their different conceptual approaches. SUPL is driven by the mobile cellular world, whereas WLP is based on IETF location technology and reflects requirements and use cases from the Internet community.

The protocol framework of SUPL is tailored especially to fit into the given environment of mobile cellular networks like those following the 3GPP standards, whereas WLP is based on location technology developed for the Internet – in general, working in any IP-enabled network. As a result, there are a number of conceptual differences between the two protocols. To provide one example, the IETF HELD protocol [51] used for WLP in line with its design for the Internet environment is independent of the underlying access technology, whereas SUPL typically comes with building blocks that are adapted to the access technology, such as the security mechanism. In its basic version HELD provides a simple mechanism to allow a device to request location from a location server in the access network. Allowing a location server to request location information measured by the device is not foreseen in the basic protocol and requires extensions like [69] that are used in the WiMAX domain. The matching communication model is based on the assumption that it is the responsibility of the device itself to provide location information to ASPs as part of the application signaling. So the basic HELD protocol mainly aims at providing measurements gathered by the access network to the device to support location in an Internet environment where the device is not required to be aware of the specific access technology in use, but just locally discovers and contacts a location server. As the actual access technology is regarded as decoupled from the IP connectivity and the applications requiring location, HELD does not consider roaming but assumes the location server to be a local one in the respective access network.

In contrast, SUPL comes with a full-blown protocol architecture to allow network-based location information to be provided, especially to mobile roamers. It specifically takes the access technology into account – leading to the fact that SUPL requires protocol extensions when adding a new access technology like WiMAX. In supporting a roaming environment,

SUPL also requires a roaming interconnect network between operators to be in place on top of the existing relationship in interoperator signaling, to provide its full set of features. In reality, this is not yet available, so existing SUPL deployments are simply based on a global database of cell location information that is consulted by the location server on the request of a MS for assistance data, or of an application server like a map service.

The above is meant to show why WiMAX decided to support two possible solutions for direct (user plane) communication between the device and the location server in the network, and why the choice of which one to support largely depends on the environment and the classes of applications targeted by the deployment. There are many influencing factors in practice. However, in general, if the target is a wireless broadband offer that primarily serves as an Internet service, then HELD/WLP may be more suitable. If a deployment should primarily provide a mobile voice service then either is good, but SUPL may be more compatible with existing comparable network offerings such as 2G and 3G and in a roaming environment.

4.5.3.1 SUPL

Let us take a closer look at the SUPL architecture defined by OMA [48]. This includes a SUPL-enabled terminal (SET) mapping to the WiMAX MS and the SUPL location platform (SLP) mapping to the LS as the main functions (Table 4.4). Application servers making use of SLP-provided location information logically implement a SUPL location agent (SLA) that interfaces with the SLP.

The basic message flow requires that any retrieval of location information from the MS has to be triggered by the LS. Such a trigger is called the SUPL initiation function (SIF). Here, mechanisms including the network sending a simple UDP-based push mechanism (UDP-Push), SIP-based push (SIP-Push) or a mobile-terminated SMS message are supported. An example making use of SIP-Push in the WiMAX environment can be the integration with the IMS, where the LS can trigger the IMS to push a SIP message to the MS requesting the MS-measured location information.

Interestingly, with today's terminal configuration, especially for open platform devices, such a push message to the device may conflict with personal firewall configurations. When applied to WiMAX, device configuration needs explicitly to take such location support into account. One way to bypass such configuration effort that had been considered but was deprecated while developing the WiMAX specifications for location support had been to add a lower layer channel for passing application trigger messages through the wireless MAC layer to the MS. The approach was deprecated because of potential security concerns like the increased vulnerability of devices to denial-of-service attacks. It might allow potential attackers to bypass such firewall functionality by using the lower layer.

Table 4.4 Mapping of SUPL terminology to WiMAX

WiMAX mapping	SUPL terminology
MS	SET
LS	SLP
LR in the application server	SLA

Regarding the actual protocol stack across R2 in WiMAX SUPL-based deployments, the exchanges between the MS and LS are based on the user plane location protocol (ULP) [49]. The communication is TCP/IP based and is secured by the PSK-TLS [50] protocol that is a variant of the commonly used TLS protocol based on pre-shared key authentication. The root session key used by WiMAX systems to enable PSK-TLS is derived from EAP-based access authentication by a mechanism similar to what is done for keying MIP security (see also Section 4.2.2) or OTA provisioning. It is derived from the EMSK.

4.5.3.2 HELD-Based Location Support

R2 communication using WLP is based on IETF standards for formatting and transfer of geographical location information. Location information can be requested by the device from a location server in the access network by using the HELD protocol [51]. Conceptually the LS when mapped to the WiMAX network resides in the CSN instead of the access part of the network, so the terminology used in IETF is different from the more fine-grained WiMAX view where the network side is split between the ASN and CSN. Looking at this from another perspective, the basic HELD protocol does not consider roaming because in the Internet environment a basic assumption is that application protocols are decoupled from the actual access technology, which is just considered an underlying transport. Roaming is part of this transport technology and is therefore often not considered in IP-based application protocols – conceptually the device is always 'at home'.

As mentioned above, the device-centric Internet approach is best reflected by the fact that HELD itself just serves the device requesting location information from the LS. This allows simple scenarios to be enabled (which are, however, also supported by SUPL), like a device requesting location information from the local LS and adding the retrieved location information, for example, to emergency call signaling, so the call can be routed to the appropriate PSAP.

To support the use cases in WiMAX where location information is also provided by the LS to the application service, HELD extensions like [69] are used to support the MS in passing location information measured in the device up to the LS.

Regarding the protocol stack, HELD is carried over HTTP. It is described by an XML document. The location format for both civic [52] and geodetic location information sent via the HELD protocol to the device is contained in a PIDF-LO [52] XML document that complies with the rules defined in [108].

With the HELD approach, location information can be provided in two ways: either directly ('by value') or by returning a URL ('by reference') which when accessed will result in the LS invoking procedures to determine the location of the MS. The latter can be beneficial in scenarios with fully mobile devices, because directly provided location data might become outdated very soon. It also allows different location techniques to be invoked depending on the time and accuracy included in the location request and the identity of the LR.

Considering security, HELD is conceptually different from the access-based keying used to secure SUPL exchanges. It assumes certificate-based authentication of the LS to allow the MS to verify that the LS is not a rogue one. Client verification is simply based on the device's IP address, or another network identifier that is verified by the network, with return-routability checks (meaning that a rogue device would not get the HELD response carrying the actual location information). This design is not based on any cryptographic client authentication, to best match prerequisites and open business models in the open Internet. Rather it is based on the overall network topology being sound. Protection of exchanged data is based on a TLS tunnel with server-only authentication.

So, in fact, keying and authentication would be quite different between plain HELD and SUPL, but the actual protection applied to the exchanged messages is the same, being based on TLS. Within the WLP solution for WiMAX, the approach of aligning the security solutions is recommended, so in WLP security between the MS and LS is also based on PSK-TLS with shared keys based on EAP authentication. This is in fact the same solution as the one described above for SUPL in WiMAX. The IETF HELD security model, however, is not precluded within WiMAX, allowing it to be used in solutions where prearranged keying is less easy, e.g. pre-paid or temporary access situations or wireless broadband scenarios where residential gateways are employed.

4.5.4 AAA Integration of Location Support

Similar to other services that are designed as integral parts of WiMAX network operation, location services are partially integrated with the AAA infrastructure. This helps to cover numerous mechanisms required for LBS, including:

- authorization of location information requests by third parties against the subscriber's profile and settings;
- MS IP address identification;
- MS security key associations;
- LC identification in the ASN; and
- support for accounting to allow network operators to apply proper charging for the use of location information, either to subscribers or to other parties like ASPs.

Regarding authorization, it is the responsibility of the LS to make sure that only authorized requests for the location information of a specific subscriber are served. Such authorized requests clearly include those sent by verified location requesters like PSAPs from public safety organizations in the case of an emergency call. Further, the LS needs to handle authorization of third-party application servers, authorization that the type of requested location information matches the subscriber's profile settings, and authorization of the subscription rights regarding access to the LS in case the LR resides in the subscriber's MS itself.

The AAA must be consulted for every location request, even location requests initiated over the R2 reference point by an MS. The reason for this is that associated session keys are not known by the LS ahead of time. The LS needs to retrieve these from the AAA prior to being able to decrypt an incoming message from the MS. Similarly for the network-initiated case, the LS needs to get the IP address of the MS and the security association keys prior to sending a message to the MS.

For this, the message flows specified in [47] include an AAA-based exchange between the LS and the AAA server, where the AAA server is consulted to decide whether to grant authorization prior to the LS collecting the requested location information from the ASN.

Communication between the LS in the CSN and the LC in the ASN uses the R3/5 reference point and is based on either RADIUS or Diameter, extended by a set of WiMAX-specific attributes or AVPs for exchanging the location-related information.

Conceptually, the AAA communication between the LS and LC is designed to be separate from the basic AAA signaling path for network access between an ASN-GW's AAA client and the AAA server. However, it is important to keep in mind that, at least for the RADIUS-based NWG Release 1 architecture [13], the AAA 'connection' for a specific subscriber and MS between the ASN and CSN is already combining several AAA 'applications'. This includes authentication and authorization, QoS, or mobility support within the same message exchanges (see Section 3.8), and can also include the signaling required for location support.

For the communication between the LC and LS this original approach can be followed as shown in Figure 4.13, where a common mobility pattern is used across the different types of AAA functions, including the QoS SFA that handles mobility as described in Section 4.4.3, or the AAA client for accounting. Figure 4.13 shows the message flow to retrieve measurements and provide location information for the LS when receiving a location request by a LR that is either the MS itself or an external LR. Again, in cases where the MS moves under the 'umbrella' of a new serving ASN-GW and does not perform EAP re-authentication, the original ASN-GW stays in control of the AAA signaling to the CSN. The AAA client does not move and both ASN-GWs relay the control signaling across the R4 reference point. This is also possible for the location-related R3 communication, and is the most obvious approach for any deployment to keep location signaling aligned with the mobility procedures for other network functions that go across R3.

Alternatively, and as shown in Figure 4.14, which displays the same scenario as that in Figure 4.13, it is possible that for the location-related AAA signaling the LS and LC communicate by using location-specific messages that are not merged with other AAA

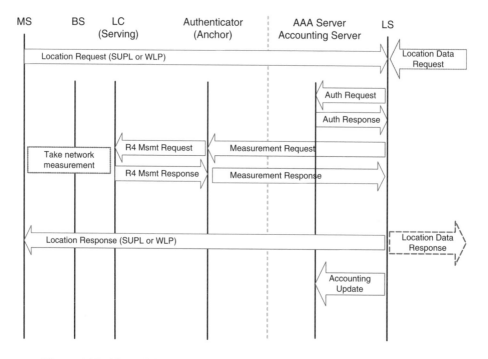

Figure 4.13 Network-based location retrieval after handoff via anchor AAA client

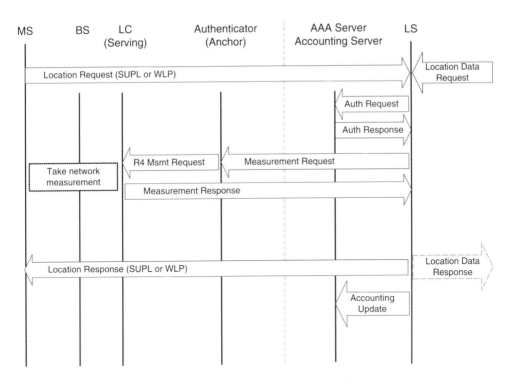

Figure 4.14   Network-based location retrieval after handoff via new AAA client

signaling between the ASN and CSN. After a handover that puts the MS under the control of a new ASN-GW without re-authentication, with this alternative the AAA client in the original ASN-GW would continue to receive location request AAA messages from the LS. However, the new ASN-GW now also runs a location controller, including an AAA client that directly sends AAA location messages to the CSN and the LS. This results in a 'triangular' communication path for the location signaling related to the same MS. One reason to still keep the original AAA client for receiving location requests is that simple requests like those for the BS-ID can directly be answered without consulting the new serving LC.

The reasoning behind this alternative design is to offer a method for reducing the load on the R4 reference point, which may be beneficial in specific deployments. However, at least with deployments that are limited to the functionality supported by the WiMAX location specification of the Release 1.5 network, this will be of minor importance. No periodic location updates from the ASN to the CSN are supported. These would create a higher signaling load when used for a large number of active subscribers in the network.

One aspect that is of possibly less relevance to a single isolated operator but becomes quite important in a dynamic roaming environment with several interconnected ASN and CSN operators is the configuration of the AAA infrastructure across network boundaries. This may be between several interconnected operators or possibly roaming exchange providers. We will highlight this by considering the security configuration between the AAA entities, see also Section 3.8. The AAA concept for RADIUS is a hop-by-hop one, where all neighboring hops must establish – besides further configuration information – a common secret for protecting the RADIUS transactions. With this, an individual AAA architecture for location-based services introduces the need to configure security associations between LS and LC entities using RADIUS, with a potentially large number of home operators or LCs across the different access networks as soon as this takes place within larger roaming exchange networks.

As a result, and depending on the actual deployment, it might well make sense to still 'route' the location communication across the same AAA infrastructure that is used for access authentication and authorization. In more detail, this means that the LS does not establish any communication directly with a LC in the ASN, but still passes the AAA messages to its local AAA server and the AAA server proxies between the LS and LC. Hence, it becomes possible to reuse the existing AAA installations, especially in roaming environments.

4.6 IMS Support

As soon as it comes to deployment models, operator plans, or services to be used with WiMAX access, it is clear that the trivial but still most obvious use case is best-effort Internet. This is probably true both for both operators interested in low deployment costs, especially for the core network, and for end users that are most interested in getting mobile broadband Internet connectivity so they are able to access their individual choice of Internet applications.

In contrast to existing mobile cellular networks, however, WiMAX will typically not inherit a circuit-switched infrastructure for mobile voice calls, unless it is deployed in addition to an existing voice solution. Hence, the question of how to best offer interoperable VoIP support across WiMAX access becomes a very prominent one when considering additional services over the pure broadband data connectivity.

Looking at the current Internet, there are numerous VoIP offerings that can directly be used across WiMAX but that are independent of the WiMAX network, or operator. These are typically non-standardized solutions with most of them being based on the SIP, but each one using an individual set of protocol extensions and supported features – thereby, not considering bilateral agreements between the VoIP service providers, being mostly limited to their own domain.

This is not bad and it integrates best into an open Internet services model. However, a tightly standardized 'default' architecture like the IMS has the potential to achieve a higher level of interoperability. One less obvious but important benefit of such a highly interoperable approach is that it leverages the establishment of so-called 'circles of trust' and more centralized provider ecosystems that are, for example, considered beneficial to counter the expected huge issue of VoIP spam, see [62]. Interoperable and similar systems based on the same common architectural framework can more easily link with each other, can create some trust, and can in turn establish more efficient measures to keep the level of spam acceptable within their interlinked network. On the contrary, a more centralized infrastructure introduces limitations in the flexibility of quickly establishing novel service offerings. It is naturally closer to the traditional PSTN telephony system.

Regarding WiMAX, with a standardized and common set of additional hooks for smooth integration into the underlying network it becomes possible for operators to add significant value to such integrated VoIP service offerings. These enable the VoIP system to make direct use of the service enablers that are covered within the different sections of this chapter, like QoS support, accounting and charging, emergency call support, or location support.

Of course, the overall calculation must still ensure that the related increase in system complexity does not negate the added value.

4.6.1 Architecture Overview and WiMAX Impact

To offer a common option for following this integrated VoIP path, WiMAX has adopted the IMS work of 3GPP [59] that is based on the IETF-defined SIP [58]. The standardized VoIP and application session control offered by the IMS is based on the so-called 'common IMS' approach by 3GPP that tries to ensure interoperability between the different network architectures of the different standards bodies like those developed by 3GPP, 3GPP2, ETSI TISPAN and WiMAX. All of them make use of the common IMS architecture.

Merging IMS with WiMAX access does not require modifications to the existing IMS specifications as developed by 3GPP. Hence, it does not result in an access-specific 'flavor' of the IMS that would of course be undesirable from any IMS vendor's and operator's perspective due to the increased overall cost in maintaining parallel implementations for every individual access network, or in adding additional complexity for translating between the different 'flavors'. An IMS running across WiMAX is fully based on the 3GPP common IMS architecture.

For the general IMS, there is a lot of literature available [60, 61], and it is not the goal of this chapter to elaborate on the global IMS technology. Instead, after providing a brief description of the major components that fall within the focus of WiMAX-specific considerations, this section will limit itself mostly to items that are special for WiMAX and to the discussion why they are so.

The fundamental idea of the IMS is to offer a standardized platform for the provisioning of IP multimedia services like voice, video or instant messaging for both fixed and mobile devices. This is designed in such a way that the resulting system is largely independent of the underlying broadband access and core network.

Let us initially step through a brief description of the major session control functions of an IMS system that are most relevant for the following considerations of WiMAX-specific aspects. All further functional entities of an IMS system can be found in [59] and are not specifically introduced here.

4.6.1.1 The Proxy Call State Control Function (P-CSCF)

From the perspective of an IMS client that runs on the MS, the entry point to any IMS that is the first SIP proxy from a protocol perspective is the P-CSCF. The MS either needs to be preconfigured, or will have to dynamically discover this IMS entity to become aware of an IMS service. The P-CSCF, acting in its role of a proxy, forwards all IMS signaling between the MS and the S-CSCF as the entity in charge of controlling the actual IMS sessions, like ongoing voice calls. Acting as the entity directly interfacing to the MS, the P-CSCF terminates the IMS signaling security with the MS that is based on IPsec, handles the generation of accounting and charging information, and interfaces with the PCRF across the Rx interface [56] in cases where the PCC framework is deployed and used for IMS QoS control.

4.6.1.2 The Serving Call State Control Function (S-CSCF)

The central IMS entity for controlling all sessions of a subscriber is the S-CSCF. It may act in the role of a SIP registrar for registering to the IMS system after the MS is connected to a WiMAX network, holds all session states related to the active subscriber sessions, and applies the session control. Also, the S-CSCF is the central entity interfacing with application functions (AFs) across the 3GPP-specified ISC interface. These are acting as SIP endpoints and provide IMS-based services to the subscribers. As the entity taking authentication decisions and controlling subscription authorization, it is in charge of verifying that the subscription is authenticated and authorized for the requested service.

4.6.1.3 The Home Subscriber Server (HSS)

An IMS subscription is verified at the time of registration with the IMS with the help of the HSS acting as a central subscription repository and authentication server. The HSS is communicating with the S-CSCF across the Cx interface that is based on the Diameter protocol.

Considering authentication and authorization, it is important to note that as well as the IMS being considered a system that extends the WiMAX network but is not part of the WiMAX core network, an IMS subscription is logically separate from the subscription to the WiMAX network. Of course, an operator offering both WiMAX access and IMS-based VoIP can store subscription data for both services against the same subscription profile. Nevertheless, it is important to separate the 'access' from the 'service' subscription. An IMS operator can be a different business entity to the WiMAX network operator. In this case the IMS operator will run a HSS and the WiMAX operator will run an AAA server to control the different subscriptions and both will be independent of each other. The expected deployment, however, is one operator running both systems to allow for a more optimized integration.

4.6.2 Service Discovery and Roaming Aspects

As the default option shown in Figure 4.15, the IMS, when deployed by a WiMAX NSP, is placed in the home NSP domain. For localized support of roaming subscribers it is also possible to offer IMS services within the V-NSP domain where the IMS is reachable directly through the visited CSN. This option is important, for example, when considering emergency services support as explained in detail in Section 4.7. However, it also requires the HA for a WiMAX session of a specific subscriber to be assigned in the visited CSN as a 'local breakout'. The HA in Figure 4.15 is shown as an optional entity. It is not present for a 'MIP-less' deployment based on the simple IP mode, where IP tunneling of subscriber data is subject to the deployment-specific routing configuration.

One aspect that can be seen here and that makes the IMS service (including all other VoIP offerings across WiMAX) depend on the location of the HA is that the IMS control signaling across WiMAX cannot directly be compared to WiMAX network control messages like the AAA-based session control across the R3 or R5 reference point, or EAP. All SIP messages exchanged for IMS session control are part of the subscriber's data traffic from the pure ASN and CSN perspective and, for example, for the ASN-GW or HA all IMS signaling is transparent – these entities are generally not aware of the fact that an IP packet is an actual IMS control message.

The above considerations raise the question of how, after initial network entry, a MS can still be enabled to discover an IMS service in WiMAX with the help of the WiMAX core network. Once more, WiMAX relies on AAA support. Similar to the download of other configuration parameters for the subscriber, like the preprovisioned QoS profiles discussed in Section 4.4, the AAA server in the CSN can add one or more P-CSCF IP addresses or fully qualified domain

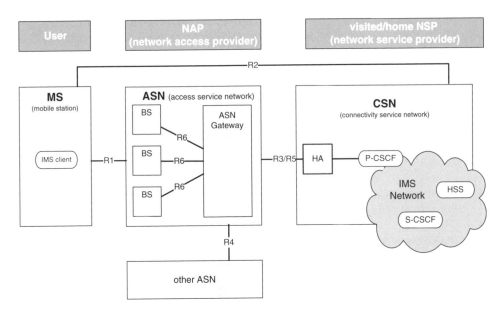

Figure 4.15   IMS integration into the WiMAX CSN

names (FQDNs) to the RADIUS or Diameter messages that confirm successful network entry. To carry such configuration information, WiMAX-specific RADIUS attributes or Diameter AVPs are available.

Through this mechanism the ASN-GW in the access network can be configured dynamically by the CSN with the list of available SIP proxies serving as IMS entry points. The ASN-GW will subsequently act as a DHCP proxy towards the MS and will provide the P-CSCF addresses as part of the DHCP messaging during network entry, if requested, as depicted in Figure 4.16.

Alternatively, the ASN-GW may act as a DHCP relay. For this mode of operation in the network the ASN-GW receives a DHCP server address for the MS instead, and would then relay DHCP messages from the MS to the DHCP server in the CSN that holds the P-CSCF configuration and provides it to the MS, when requested.

To retrieve the P-CSCF information from the network, the MS adds an indication requesting the available P-CSCF addresses or FQDNs to the DHCP Request. This is sent to the ASN as part of the standard network entry procedure for IP configuration. Alternatively, the MS can also send a DHCP Inform message with the same indication. Note that the procedure is not limited to network entry, but can be initiated by the MS at any time during an active WiMAX session.

But why are these alternative solutions required instead of basing IMS discovery on one common procedure? The reason for this is the WiMAX network entry procedure, where the use of the DHCP exchange by the MS shown in Figure 4.16 indicates to the WiMAX network that the MS does not support MIP and hence IP mobility will be handled by PMIP procedures in the network. As a result, any MS that attaches for requesting MIP services will not send a DHCP Discover or Request message and receives IP configuration via MIP procedures, so it is required to use a DHCP Inform message for IMS service discovery.

For any non-roaming case where the subscriber is directly connected to the home operator's CSN, IMS discovery and assignment can be considered a straightforward procedure and

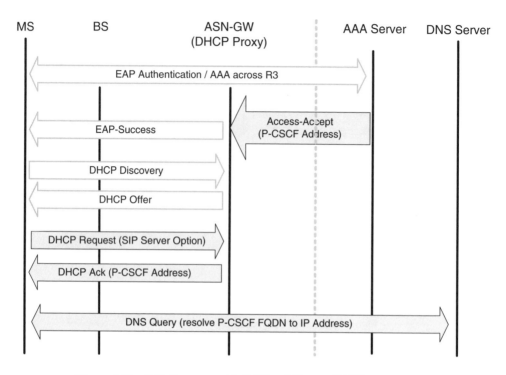

Figure 4.16 IMS discovery with PMIP mobility and DHCP proxy mode

may even be statically configured in the devices. In contrast, such an assignment becomes more complex when looking at a roaming scenario with a visited operator involved that is also capable of offering IMS services to the subscriber.

The main difference to the non-roaming case is that, in case both the home and visited CSN operators want to advertise IMS capabilities and in turn available P-CSCF entities as the entry points, there has to be a mechanism for how to negotiate between the operators on what to finally offer to the subscriber via DHCP. And with such a mechanism, it is important to understand which of the involved stakeholders (i.e. the subscriber's MS, the visited and the home CSN) can really control what IMS service will finally be selected:

- Regarding the negotiation procedure between the visited and the home operator for which the IMS service is offered to a specific subscriber, the solution is to extend the AAA messaging of the network entry procedure by allowing the AAA server of the visited CSN to add a list of P-CSCF addresses that is communicated to the home CSN AAA server.
- It is then up to the home CSN to decide which of the possible P-CSCF addresses are really offered to the MS. The logic behind the home AAA server being in control of this decision is that the choice might depend on specific properties of the subscription that are under the control of the home operator.
- The final choice of P-CSCFs offered to the MS is then sent to the ASN in the same way as for the non-roaming case, see e.g. Figure 4.16.

- In case the home operator does not offer any P-CSCF, the visited operator's AAA server is free to add its own P-CSCF to the Access/Accept message to the ASN.

One important point is that the assignment of an IMS service during network entry has to be in line with the assignment of the HA for IP breakout. In turn, both the HA and the P-CSCF are assigned either in the visited or in the home network and the assignment policies and procedures need to be coupled for generic IP service and IMS service. For example, in roaming cases where the home operator does not support HA assignment in the visited CSN, it would not make sense to advertise a P-CSCF in the visited domain to the MS.

Another result of the above consideration is that if the HA is already assigned after initial network entry, a MS performing P-CSCF discovery later on cannot select local breakout, for example. Hence, the choices for IMS service usage are limited to the domain that the HA is located in. For any other choice, the MS would need to leave the network and perform a new network entry procedure.

But let us also take a look at how much control the MS actually has over which IMS service to choose in case IMS is offered by both the visited and the home WiMAX network.

4.6.3 IMS Identities

We have already mentioned that an IMS subscription and a WiMAX subscription are conceptually different due to the fact that the former is related to the application and the latter to the access and ISP network. Similar to this, the main subscription identities used in IMS are also different from the NAI that is used in the WiMAX network. IMS is based on two identities for the subscription:

1. **Private user identity (IMPI):** This identity is the system-internal identifier for an IMS user. It is the identity that is centrally used in IMS-related signaling procedures like authentication, registration or charging of IMS services. Interestingly, for the representation of the IMPI the NAI format has also been chosen (so it is technically possible to use the same username for both the IMS and the WiMAX subscription). Due to its origin in 3GPP systems, an IMPI carries the IMSI being used as the central subscription identity in classical 3GPP cellular networks, which would also be a binding of the network identity and subscription with the IMS application.

   One important thing to note here is that there can be several IMPIs per subscription.

2. **Public user identity (IMPU):** In addition to the IMPI, one or more public identities are assigned to a user as part of a subscription. The IMPU is the identity that would typically appear on a user's business card or in phone books. It is the one identifying a user to other users or application servers. The idea of allowing more than one IMPU per subscription makes sense, e.g. for splitting business and personal use of the service, or for maintaining sets of IMPUs as aliases. IMPUs are not in NAI format but are identifiers natively used in SIP. Hence they are exchanged in the format of a SIP URI or also a 'tel:' URI [63] that allows phone numbers to be carried within a SIP URI, like 'tel: + 49-89-233-00'.

For a better understanding of the relation of IMPI and IMPU identifiers to each other and for analyzing the mapping of these concepts to WiMAX, let us consider the following examples:

- A user has a single public identity that is known to all potential callers. However, to receive a call, the user needs to be reachable on two separate end devices, like a PC and a mobile phone. This would map to a subscription with a single IMPU but with two separate IMPIs, one per device.
- A user operates a single device, but likes to make use of two separate public identities, one for a group of business contacts and the other for friends. This would result in a subscription with a single IMPI and with two separate IMPUs.

Based on the above examples, the relation between subscription and identity can be characterized as shown in Figure 4.17. An IMS subscription has at least one IMPI and one IMPU assigned. However, it can also have several IMPIs or IMPUs with different relations being possible, as shown by the above examples.

At this point it becomes clear that in general a direct relation between the identities of a WiMAX subscription and a subscription to an application running across WiMAX is hard to create. And this is true also for the IMS. Of course, it would be possible in the simplest case to make use of one single identity in NAI format across both WiMAX and IMS for a specific user. However, this would at least limit several use cases for the IMS subscription as there can be several IMPIs in the IMS but only one central subscription ID in WiMAX. Also, the concept of an IMPU does not translate easily to a WiMAX subscription. This is due to the fact that – leaving aside the different format – an IMPU is a permanent identifier of the IMS subscription that typically does not change between different registrations, or calls, whereas the (outer) NAI in WiMAX access is mainly used for routing purposes and may change on a per session basis.

Another special case for WiMAX related to the handling of identities in the network is created by the identity hiding capability, discussed in Section 3.5.3. The IMS application might

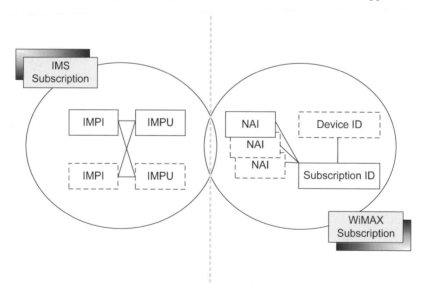

Figure 4.17 Relation of IMS and WiMAX subscription identities

be able to see the NAI as the access identity to support relating WiMAX and IMS sessions which might, for example, be required for accounting and charging of IMS service usage to the WiMAX subscriber's account. However, a pseudonym instead of the real identity may appear in the NAI. In such cases the IMS system cannot learn the real WiMAX identity of the subscriber. This needs to be considered when designing an IMS system for a WiMAX network or when integrating a third-party IMS system to avoid potential issues in the correlation of session information that might be required for proper accounting of both the WiMAX usage and the IMS service.

A solution supporting such correlation might be an interface between IMS entities like the S-CSCF and the WiMAX AAA server that is capable of resolving any pseudonym of the NAI to the real subscription identity. However, at the time of writing no standardized interface has been defined, so the implementation of such an interaction is left for deployment-specific realization.

4.7 Emergency Services in WiMAX

Telecommunication systems provide a large range of services to the end user. Typically, their success is driven by user interest and by what the users or other participants like companies advertising to the end user are in fact willing to pay.

Let us now look at a quite different class of services – at least so far – and at the actual requirements for these: namely, emergency services. Here, most requirements are set by national regulation, which can vary a lot between different countries and is critical to any network operator's business. This stems from the fact that in cases where requirements set by local regulators are not fulfilled by an operator, this operator can in the worst case be forced to go out of business.

4.7.1 Motivating the Problem Space

The most obvious and common example of an emergency service is the possibility for an end user to place an emergency voice call, report an emergency situation to the authorities and call for help. This is done by dialing 112 across the EU or 911 in the United States, as two common examples. Dialing such a number allows the police or fire brigade to be contacted very quickly. It is fully supported in classical voice systems like the PSTN or PLMN.

Interestingly, emergency call support is a major headache for the commercial deployment of VoIP technology nowadays and this is made very explicit by some of the common and widely known Internet VoIP clients when being installed. They simply display a message to the end user that emergency calls are not supported. This is less of a problem for fixed VoIP solutions, like the VoIP offerings bundled together with a DSL broadband package. However, for nomadic usage of VoIP solutions the problem becomes close to impossible to solve with today's deployed technology. Imagine placing an emergency call by trying to dial your well-known home emergency number when staying in a hotel abroad. Either dialing the number will not work at all, or the responding emergency service in your home country will most likely have no idea about your current location and no means to provide timely support for the emergency. In the traditional PSTN, emergency calls are detected by and routed in the local (visited) network. But how could some Internet VoIP service provider located, say, in Luxembourg route an emergency call placed by a user staying in Thailand with no control over the current way that the end user is accessing the Internet and with no interoperability or established roaming agreement with a VoIP service provider in the remote country?

Up to now, we have been talking mainly about a VoIP application's capability or inability to support emergency calls in a meaningful way. However, to make use of any VoIP provider's services, IP connectivity is always required first. Imagine what would happen when switching on an IP phone with WLAN connectivity in a hotel. Leaving aside for the moment the hotel's fixed-line phone, which would of course be the first option in practice, this is certainly not the only example of an area with WLAN coverage: WLAN hotspots are typically not free and nowadays do not provide a free path to the Internet that is especially reserved for emergency calls – they require new users to work through a registration procedure, including payment for access. It is needless to ask how the emergency situation may have developed in the meantime.

This is in fact one of the main differences of WiMAX to mobile cellular networks where circuit-switched voice is tightly integrated. There are two layers involved: the network layer and the VoIP application layer. These might well be totally separate from each other.

4.7.2 WiMAX Architecture for Emergency Services Support

For the above reasons, it becomes clear why WiMAX networks deal with the support of emergency calls and with the regulatory requirements related to them in a different way to existing mobile cellular networks. Logically, this can be highlighted by drafting a network reference architecture as shown in Figure 4.18 for WiMAX emergency services and by identifying the stakeholders involved.

Starting from the general roaming architecture and involving a possible visited NSP, Figure 4.18 shows two additional network domains. These show the potential VoIP service providers (VSPs) and also consider the roaming case where VSP services can be accessed through either the home NSP or the visited NSP, depending on where the 'IP breakout' point (the tunnel endpoint for the MS IP traffic across R3) is assigned for a subscriber's session.

For a feasible emergency call, the business relationship between the NSP and VSP becomes important, and the possible general deployment options are as follows:

- The WiMAX operator and the VSP are separate and independent business entities with no specific relationship, so the WiMAX network serves as a pure broadband access medium in this case.
- The WiMAX operator and the VSP are separate business entities but establish a dedicated business relationship (such as the VSP providing the default VoIP service to the WiMAX operator's subscribers).
- The VSP domain is owned and operated by the WiMAX operator and the WiMAX operator is subject to regulatory requirements for emergency call support.

This relationship is technically established by an interface between the CSN and VSP as shown by the R_v reference point in Figure 4.18. In cases where these are different networks, it is especially interesting that the responsibility for supporting emergency calls falls upon the VSP, not the access provider that maps to the ASN/CSN of WiMAX. As a result, for a WiMAX deployment that only considers broadband data, such regulatory requirements and the resulting responsibilities do not directly apply. It is, however, the clear goal of the WiMAX network

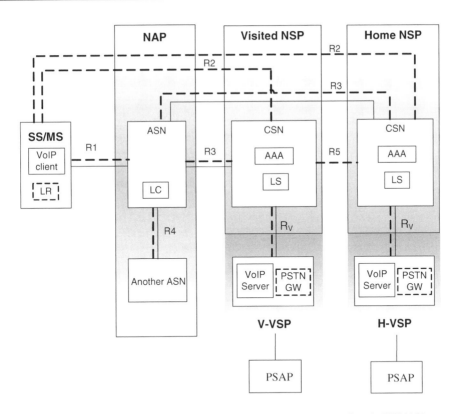

Figure 4.18   Network reference architecture for emergency services in WiMAX

design to offer support to any VSP to enable it to match such regulatory requirements more efficiently. This includes, for example, support by the WiMAX CSN regarding location information or special treatment and an indication of an emergency call in the network layers 'below' what the VoIP application can itself see. Such functionality and interworking are most interesting when partnering CSN and VSP operators. The overall relationship between the involved stakeholders that is of interest for the considerations of this section are shown in Figure 4.19. The different deployment options are shown in Figure 4.18 by the dotted VSP boxes. In cases where the VSP functionality is directly provided by the WiMAX operator, it is the responsibility of the WiMAX operator to meet all regulatory requirements for emergency services support.

The WiMAX Release 1.5 network specifications for emergency services support [146] do not introduce any mandates regarding such an interface between the WiMAX CSN and an ASP domain in the general case or the VSP in those cases relevant to this section. This is because there are already many existing deployments and interface specifications that need to be considered. This also holds for existing deployment-specific or solution-specific choices for conveying location information. Also, such a generic interface would need to cover a lot of quite different scenarios and use cases spanning emergency services and locations as well as identity management and correlation, QoS control by the application, accounting or even payment for a very large range of services. Hence, any effort specifying a dedicated interface

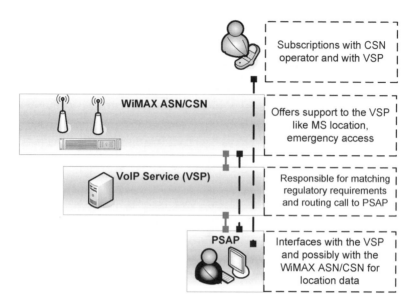

Figure 4.19   Relations between stakeholders in WiMAX emergency calls

and set of protocols for covering all of these use cases serves a good purpose, but is unlikely to be deployed and replace existing, tested and dedicated solutions in practice. Consequently, the approach for WiMAX networks that holds especially for the R_v reference point in an emergency is not to standardize or mandate any specific set of protocols for such interfaces. This is left to deployment-specific choices, or existing frameworks for VoIP emergencies, like the one developed by the National Emergency Number Association (NENA) [106].

One exception here is the effort in the WiMAX Forum networking group to design a USI describing a reference point U1 between the CSN and ASP that is intended to serve as such a generic option. The U1 reference point would map to the R_v reference point for the specific case of WiMAX emergency services. However, only an initial version of the interface specification is available, with no experience from deployments. Also, the USI does not specifically target emergency services, or location. This is contrary to existing deployed solutions, so the above considerations regarding the R_v reference point also hold for the USI.

4.7.3 Design Criteria and Building Blocks

WiMAX network design for emergency services follows the approach of a clear split between the functionality provided by the network and access and the functionality subject to the VoIP application and that may vary significantly between different VoIP services. This is also reflected by Figure 4.19. As a result, there are no elements for emergency services in a WiMAX network that are specific to a certain VoIP system, like the IMS.

Keeping this in mind, it becomes clear that the WiMAX network design without including any specific VoIP functionality as a mandatory part can best offer a set of building blocks for emergency services support. Operators are able to make their deployment-specific selection from a set of available components to best match their local regulatory framework.

The building blocks that are offered by WiMAX and that are relevant for emergency services support include the following:

- A specific network entry indication allows the MS to mark initial WiMAX network entry as an emergency and in turn allows the WiMAX operator to prioritize the network entry attempt or apply specific policies to the emergency network entry.
- Location determination support is provided as a generic functionality, but one of the major reasons for this is of course emergency call support. It is discussed in detail in Section 4.5. WiMAX location support is designed to meet both operators' requirements and regulatory requirements for emergency calls.
- AAA support and hotlining capabilities allow QoS control and restricted emergency access for emergency callers even in those cases where a subscription lacks the authorization to perform normal network entry, as in an empty prepaid account.
- Limited roaming support makes it possible in roaming cases to handle the emergency call locally in the visited CSN in case the VoIP technology offered by the V-CSN matches the one supported by the MS.
- Limited support for WiMAX devices without a valid subscription enables emergency calls even in those cases where a subscriber is roaming but the local WiMAX CSN operator does not have any roaming agreement with the subscriber's home operator, or where simply no subscription has been installed in the terminal.

One specific measure that has been added to the WiMAX network signaling is an indication of the service mode from the MS to the network at an early stage in the initial network entry procedure. This service mode is carried as a WiMAX-specific decoration within the subscriber's NAI that is sent as part of the initial EAP phase. The actual decoration that will be used to indicate an emergency network entry is '{sm=2}'. The mechanism for decorating the NAI is the same as the one used for WiMAX OTA provisioning, discussed in detail in Section 3.6.3.

'Overloading' the NAI string is not in fact the cleanest approach, since the NAI is mainly used for AAA routing purposes in the network and such overloading might conflict with proprietary NAI decorations chosen by other organizations for implementations supporting both technologies at the same time. However, there were clear reasons for choosing this path at the time of the design of emergency services support for WiMAX networks. The air interface communication is handled by chipsets in both devices and BSs, and such chipsets depend on stable specifications that need to be available rather early in the process of developing among the vendor's community the product range required for building a WiMAX network infrastructure. The available wireless interface specification in [1] did not include the means for providing emergency indications from the MS to the network at the time of the design of the Release 1.5 network support for emergency services. With this in mind, it is more logical to put emergency indications into a message that does not impact chipset implementations but still provides such indications to the network at a very early stage in the network entry procedure.

When the MS sends a NAI including the WiMAX decoration and indicating service mode 2 for an emergency, the ASN-GW will be the first entity to interpret the NAI. So, the ASN-GW depending on the locally configured policy can already prioritize the network entry attempt and the resulting AAA message for entry to the AAA server of the home network. The AAA server, on receiving an emergency-decorated NAI, is also able to prioritize the subscriber's network entry attempt and can apply the required policy to the requested WiMAX session that will be used for the emergency call. For example, the AAA server will download a specific emergency QoS profile in the AAA response to the ASN, or put the MS into a hotlining session to grant limited access to the VoIP service for placing the emergency call.

4.7.4 Limited Access to Emergency Services

The functionality that recognizes the network entry of a MS for the purpose of getting access to emergency services also allows the matching of a set of regulatory requirements to grant access to the MS or subscriber when a standard network entry would be denied by the operator.

This is especially useful for handling cases where the subscriber lacks authorization to perform a standard network entry. For example, when a subscriber has an empty prepaid account or would normally be rejected by the AAA server because it is trying to use an ASN that is not allowed by the home operator's policy, the AAA server can decide to still grant access. This would be limited to those unauthorized cases where emergency network entry is required. The MS needs to be limited to use of the emergency service, but should be blocked from regular access.

Interestingly, the above discussion about network entry in cases where normal entry would not be authorized to provide some limited access for emergency cases becomes complicated when looking more deeply into the details. There are numerous cases that need to be distinguished in practice, so let us try the following rather technical classification.

4.7.4.1 Unauthorized Emergency Access

As described above, this covers cases where the normal WiMAX network entry procedure can be performed up to the point of authorization, based on the availability of a subscription to the MS. EAP authentication can be performed, meaning that valid security keys are available as part of the subscription credentials and the home AAA server can be discovered based on the NAI realm identified by the MS. However, the AAA server would deny a normal network entry due to the lack of authorization. Examples of such cases are empty prepaid accounts, timed-out or barred subscriptions, subscriptions not authorized to use a certain NAP to access WiMAX services, etc.

The key differentiator in this scenario compared to unauthenticated access is that the standard network entry procedure can be performed technically. Subsequent decisions, regarding whether the subscription is authorized for specific services, are fully up to the AAA server's policy. Whether to allow emergency network entry and grant limited access by restricting the session just to the VoIP service that allows an emergency call to be placed is in turn also up to the AAA server's policy.

4.7.4.2 Unauthenticated Emergency Access

Allowing emergency calls to be placed with a 'blank' terminal, without a valid subscription being available, is a known feature of 2G/3G mobile cellular networks. Some countries allow so-called 'SIM-less' emergency calls via circuit-switched voice only, others do not. This is subject to regulatory requirements, so as soon as they apply, there is no way for the operator not to support such SIM-less calls. However, as necessary background for this discussion it must be noted that the level of misuse of this feature is dramatic. For the thoughtless, it is of course nice to be able to test whether a used mobile phone works fine by just powering it up and dialing the emergency services number – no need to make use of any valid subscription that can be traced back or charged. Depending on the country and the time of day, hoax emergency calls can make up 80–90% of the overall number of emergency calls received in the PSAPs. This is clearly recognized as a major issue in overloading the public infrastructure for handling emergency calls which involves stakeholders like national regulatory bodies, PSAP operators like the police, and operators.

The overall picture becomes even more difficult with VoIP emergency calls, when no legacy circuit-switched voice service is available. In general, one of the major difficulties for VoIP emergency calls is the split between the VoIP application part and the network access part, with WiMAX falling into the latter category. The WiMAX access network is in general not aware of the fact that a network entry procedure is performed to access emergency services. So this scenario would result in the granting of unauthenticated emergency access, e.g. for blank devices without any subscription it would mean granting open access to the Internet. This would certainly be subject to substantial misuse. WiMAX uses the emergency decoration for emergency network entries, so the network can learn that this entry is meant specifically for an emergency situation. However, this is still a challenge for the network because the WiMAX operator can only control and limit MS access to well-known VSPs. One example of such a configuration is where the operator is running its own IMS VoIP system. Here, IP access can reasonably be limited to the required IP addresses of the IMS system, like the P-CSCF.

Of course, such limited access is required for unauthenticated network entry. Otherwise, fraud would likely become by far the major part of such special network entries, where, after establishing IP connectivity, normal calls could be placed that are beyond the WiMAX operator's control. In fact, this rules out any emergency access to arbitrary VoIP systems in the Internet without any specific agreement between the VSP and the CSN operator.

It is hard to see how regulatory bodies will address such unauthenticated access for VoIP emergency in the future. Pure unauthenticated emergency calls from a blank device might be ruled out in general as unreasonable, based on the known very high level of misuse and the technical challenges of VoIP systems compared to circuit-switched fixed or mobile voice calls. However, there are cases where a valid subscription is available in the MS, yet the subscription does not work for some unknown reason. One class of examples is the roaming cases where the visited WiMAX CSN just does not have any roaming support in place with the home network. As a result, the visited CSN will not be able to discover the home operator's AAA server and authentication will not be performed at all. However, regulation in at least some countries might still require emergency services to be available in the case of missing roaming agreements between operators – assuming of course that the visited network offers a VoIP technology compatible with the VoIP client on the MS.

One interesting thing to mention here is that the architectural considerations for a WiMAX deployment are different from those for plain Internet VoIP. One of the more prominent examples is that, in the Internet, roaming is typically not part of the architectural considerations behind available specifications, as in the IETF ECRIT effort [65], because the assumption is that there is always a 'direct' connection between the end host and the service provider. However, the issues are mostly similar when considering limited access, and similar considerations for Internet VoIP emergency calls can be found in [66].

4.7.4.3 Realizing Unauthenticated Network Access

As briefly considered above, for WiMAX network specifications the goal is to offer appropriate building blocks to operators in this area. Although not made mandatory in the specifications, network access including the WiMAX ASN and CSN is supporting unauthenticated access in a standardized way. This is based on the configuration of network access rather than on additional technology or specifications, or on significant implementation efforts. The solution for any operator to meet local regulation requesting unauthenticated emergency access is based on the EAP-TLS method. EAP-TLS will typically be used to authenticate the device based on its credentials and certificate installed by the device manufacturer.

To realize unauthenticated access, there are two possibilities in general. One is to perform device authentication only. Here, the WiMAX network entry procedure can be performed as usual when just considering the EAP-based exchanges. No changes to the network entry procedure are required besides the fact that the AAA server would require a modified authorization policy and needs to inform the ASN in case limited access is granted to enable appropriate ASN policies for limiting the connectivity. For the devices, support for TLS-based network entry is mandatory, but this also has to be supported for WiMAX OTA provisioning that is essentially based on the same EAP-TLS procedures (see Section 3.6.2). A reasonable step for any such deployment is to install preconfigured QoS profiles for emergency services that are provided to the ASN, and to apply IP filtering with appropriate settings to route and limit all IP communication of the 'unauthenticated' MS to the emergency services domain. Again, appropriate filtering is very important. This is required to prevent any MS from misusing the unauthenticated access to get free access to resources that are otherwise charged for. As an advantage, device authentication can at least provide the authenticated MAC address of the WiMAX device to the network operator. The value of such a MAC address in case of misuse is obviously limited, but it could at least be used to address a number of misuse cases like cloned devices.

Obviously, this solution only addresses the network part, so the remaining issues for a concrete deployment are reduced to the actual application layer technology. This, unfortunately, remains a major one for cases where the actual VoIP systems on the terminal and in the operator's networks are not compatible.

The second possibility as an alternative to the currently specified method for realizing unauthenticated access is to perform a certificate-based EAP method like EAP-TLS without the MS actually providing any credentials to the EAP method exchange. This is technically possible without any change to the WiMAX network entry procedures, other than a modified authorization policy in the AAA server. The EAP method can run successfully, provide appropriate keys for setting up the protected wireless link, and trigger the required AAA exchanges across R3/5 to establish an initial session for the emergency. The difference to the above method supported in the Release 1.5 WiMAX network specifications for emergency support is that no authenticated MAC address would be available to the CSN operator as a result of the EAP exchange. Such a deployment-specific solution would be feasible and again – assuming the availability of support for an unauthenticated EAP-TLS mode in the devices – would mainly depend on appropriate policy configuration in the AAA server and on providing filtering to limit the granted access rights for the session to emergency services access only. The first option of performing device authentication became the preferred choice and the solution covered by the standards, due to the fact that any certified WiMAX device will be shipped with a preinstalled device certificate.

4.7.5 Roaming Considerations

For the remainder of this section we will look at the roaming support for emergency services that is still limited by the Release 1.5 core network functionality of WiMAX. For roaming scenarios the general problem to be solved is that emergency calls must be routed to the PSAP nearest to the emergency caller. Placing an emergency call to the emergency responder in the subscriber's home location clearly is of very limited value when roaming in a foreign country.

Technically, this is either about local breakout for the IP traffic and assigning the VoIP service in the visited CSN, or about the home VoIP provider being able to discover the local PSAP and route the call accordingly. Two general cases can be distinguished:

1. Emergency after performing initial network entry.
2. Emergency with the MS already being connected to the WiMAX network.

The first case handles the requirements rather well when assuming a relatively homogeneous environment and an IMS service available in both the home and the visited domain. This is where the MS connects to the WiMAX network after already indicating the emergency request during initial network entry by sending a decorated NAI. The IMS is required here because this is the only VoIP system directly considered as part of the WiMAX specifications. Hence, for the IMS, the appropriate discovery and service selection procedures are in place.

Typically the steps to establish an emergency call with the visited IMS would include:

1. The MS – assuming it is aware of the fact that the user is requesting an emergency call – performs initial network entry by indicating service mode 2 for an emergency in the NAI.
2. The AAA server recognizes the emergency network entry request based on the NAI decoration and grants access by providing the appropriate QoS profiles to the ASN.
3. The AAA server, being in control of the IMS service assignment (see Section 4.6 for detailed procedures), then selects and assigns the IMS system in the visited CSN and chooses the appropriate P-CSCF address that is in turn advertised to the MS via the ASN (see Section 4.6.2 for further details). The AAA server also needs to assign a HA or router in the visited CSN to set up local breakout, such that the MS is able to reach the selected IMS service (assuming that the IMS is a native service of the visited CSN, otherwise the HA may also be assigned in the home CSN).
4. When IP connectivity is established with the visited CSN and the P-CSCF address of the local IMS is available, the MS connects to the local IMS. It places the emergency call, which in turn can be routed to the local PSAP. The PSAP address has to be known by the local IMS, either by preconfiguration or by making use of a dynamic method like the LOST protocol [156].

Again, this procedure is currently limited to the IMS VoIP system due to the lack of other standardized VoIP systems for WiMAX. In any 'non-IMS' case, the AAA server can still choose to assign a HA or router in the visited CSN and provide local breakout for the IP session. However, compared to the standardized P-CSCF discovery, the WiMAX network would only be able to provide support for the MS regarding VoIP service discovery with deployment-specific solutions. These can be realized in closed deployments but would likely be of limited value in roaming scenarios because of the lack of general support for such solutions in roaming devices.

When looking at the latter case of the MS already being connected to the WiMAX network when initiating an emergency call, current support by the WiMAX network is mostly limited to non-roaming scenarios. In roaming cases where the MS already has a local breakout assigned in the visited CSN, the situation would be rather similar to the above case of initial network entry. The issue with this is that the MS is not necessarily aware of the information on whether the current IP session is terminating in the visited or home CSN, because the WiMAX network does not provide an explicit indication to the MS about home versus visited assignment. If the MS is connected with the IP session terminating in the home CSN, it is not possible to dynamically relocate the IP breakout to the visited CSN, and in turn an IMS in the visited CSN is unlikely to be discovered or assigned. So the two quite limited choices for the emergency caller are either to rely on the emergency call reaching a PSAP in the home location, or to exit the network, perform a new network entry indicating the emergency by using the emergency decoration, and as a result of course delay the emergency call.

In summary, support for emergency calls in WiMAX networks with the Release 1.5 network specifications is covered in a reasonable way for non-roaming cases where the WiMAX operator runs the VoIP service itself, or relies on partners within a controlled environment. It becomes clear that WiMAX offers a number of building blocks to support emergency services at the network and access layers. It also offers the means to support a VSP for handling emergency calls appropriately. However, it also shows that full emergency services support requires a sound integration of the different enablers in the network that are considered throughout this section, like QoS, location support, AAA messaging and the VoIP service. Emergency services are an 'application' that serves as a very good example of how important it is to integrate these diverse enablers across an operator's network and across different networks working together.

It also becomes clear that, due to its complex nature, full roaming support is not yet available in a standardized fashion for WiMAX. And as a general observation, especially in the area of generic VoIP, emergency services support in combination with nomadic or even mobile IP broadband access still face many challenges, although the IETF ECRIT effort has made substantial progress in this area.

4.7.6 Further Aspects of Emergency Support

Up to now WiMAX support for emergency services has focused on emergency calls in combination with the 'default' IMS VoIP system, enabling operators to match their country-specific regulatory requirements. With this, the most important pieces are in place. However, for a broader view let us look briefly at the possible classes of emergency services that can potentially be provided across a telecommunication network, especially a wireless one. In general, three different classes of such services with quite different characteristics are identified:

1. **Citizen-to-authority** services summarize all types of emergency communications where a user of a telecommunication system places a 'call' to authorities like the police or fire brigade to report an emergency and call for help. This covers voice emergency calls as well as SMS, instant messages or other types of rich media communication.
2. **Authority-to-authority** priority services support targets a closed group of rescue forces that make use of a public telecommunication network to allow group members to

communicate and exchange information with each other. The main difference to standard subscriptions that requires additional effort and technology to be in place in the network and in the devices is the fact that the closed group must still be able to use the service even if the network operates in an overload situation, and the network may even pre-empt existing sessions to free up resources. Imagine an earthquake where the local mobile cellular network is subsequently expected to be overloaded due to the too many regular standard and emergency calls. The rescue forces must still be able to make use of the overloaded infrastructure and must get priority over standard subscribers to coordinate the rescue effort.

3. **Authority-to-citizen** emergency services target all scenarios where the public need to be informed very quickly and efficiently. Examples are public alert systems for early warning, for instance the prominent example of 'tsunami warnings', or systems that alert all active subscribers in a certain geographical location, e.g. about approaching severe weather conditions. Technically, these systems may implement a way to broadcast very quickly a relatively short and simple text message to alert all reachable subscribers of a network as fast as possible. Later on, more detailed information about the reason for the alert or actions recommended by the authorities in charge can be delivered. An example of an early warning system for severe weather conditions is the EU 'Meteoalarm' project [64].

As discussed above, WiMAX support for emergency services in the Release 1.5 network specifications covers the citizen-to-authority class with a focus on emergency voice calls. Additional work to extend this scope and to enable authority-to-authority communication within a WiMAX network is ongoing and will be included in future standards releases. Detailed technical requirements and specification work falls within the remit of the Emergency Telecommunications Service (ETS) and is related to the Government Emergency Telecommunications Service (GETS) and Wireless Priority Service (WPS) effort in the United States.

5

Mobility

This chapter provides an overview of the functions in the Mobile WiMAX network supporting mobility with the anchor in the ASN as well as in the CSN. While ASN-anchored mobility mainly addresses fast local handovers for moving terminals, the mobility protocols in place between the ASN and CSN are aimed also at supporting path optimization, wide area nomadic access, roaming and network sharing.

5.1 Mobile Networking

Mobility management is a vital aspect of any wireless network aiming to provide a mobile broadband access service on a large scale, nationwide and internationally. The overwhelming success of the cellular networks that we have witnessed over the past two decades, originally starting with voice services only, already gives a very clear indication as to what the end customer's expectations will be when it comes to mobility. This implicitly sets challenging requirements for any new network architecture in terms of mobility management that is planned to be a success in the market. Subscribers will surely not accept something that is less performing than what they already have today.

Looking back to the time when cellular networks were originally designed, we can consider the creation of their whole network architecture and standards describing a system with worldwide interoperability to be a kind of greenfield project within a walled garden. 'Greenfield' because cellular networks were designed from the ground up and with no external constraints or dependencies, like the need to protect existing deployments or devices. 'Walled garden' because the operator domain of the cellular networks is closed – everything from the physical layer up to the application layer was designed as part of the self-contained cellular network system, resulting in tightly integrated layers that are on the one hand closed to the outside world, but on the other hand simplify interoperability and roaming between operators.

The basic idea behind Mobile WiMAX in terms of user experience was to come up not only with a fully mobile broadband data solution, but also with a 'mobile DSL'. So the design objectives for which Mobile WiMAX was designed are very different from cellular networks. It is focused merely on providing both mobile and fixed IP connectivity through the same radio technology. Hence, WiMAX is meant to be a network where all the services are provided on top

Deploying Mobile WiMAX   Max Riegel, Dirk Kroeselberg, Aik Chindapol and Domagoj Premec
© 2009 John Wiley & Sons, Ltd

of the IP. In this regard Mobile WiMAX is supposed to interoperate out of the box with the applications available on the Internet and with the open Internet in general. Mobile WiMAX should reuse as much as possible the standards upon which the existing IP networks are based, thus being able to interoperate with or, better, to integrate into the existing IP universe as seamlessly as possible. With the above motivation, we will look into the details of WiMAX mobility management in the remainder of this section. After a general overview on IP mobility mechanisms, this section will identify exactly which mechanisms WiMAX is using to deliver on its promise of being a 'mobile DSL' with mobility management performance at the level of the cellular networks or better.

5.1.1 Mobility Mechanisms in Packet-Based Networks

A host in an IP network needs an IP address in order to communicate with other IP hosts. The IP address that is assigned to the mobile host is specific to the IP subnet to which the host is attached and as such the IP address represents a locator. To put it differently, the host is assigned an address based on its current location or point of attachment to the IP network and any packets sent to the host are delivered by the routing infrastructure to the subnet to which the host's IP address belongs. If the host moves to another IP subnet, it is assigned a different IP address belonging to the new subnet.

5.1.1.1 IP Mobility Problem

Looking at Figure 5.1, we can see that the device was at first attached to the router of subnet A where it was assigned the address 'IP A'. When the device subsequently moves to subnet B, it is assigned a different address, 'IP B'. Assuming that the device started a streaming session with some content server while it was at subnet A, the content server, being unaware of the device's movement and its new address, continues to send the packets to the previous address of the device, 'IP A'. Eventually the content server will detect that the device has gone and will terminate the IP session on its side. Under such circumstances in the IP world, there is no attempt at an explicit session teardown between the communicating parties. Each party simply removes any resources associated with the session on its side. A device typically terminates the

Figure 5.1 IP mobility problem

session as soon as its previous IP address becomes invalid, which usually coincides with the movement and acquisition of a new IP address. It is not able simply to continue the existing session with the content server using its new IP address, since the content server would not be able to associate the packets sourced from a different IP address to the existing session. It also does not know that this is the same device from the pure IP perspective. Besides, any attempt to send packets using the source address IP A from subnet B will fail due to the ingress filtering by the router at subnet B. The device is finally no longer able to receive packets sent to its previous address IP A as they would be delivered to subnet A.

At the heart of the problem is the fact that the communication partner is not informed of this IP address change and hence any ongoing application session fails. The original design of the IP networks did not take into account the possibility of a host being able to move around and at the same time maintain session continuity at the IP layer. Even though mechanisms exist today to address this problem, those mechanisms have not seen much deployment in current IP networks. The situation on the global Internet is very much as described in the example above and basically all IP mobility depends on whether each individual application implements specific means to handle such IP address change.

Of course, on completion of the movement the communicating parties are able to start all over again by establishing a brand new communication session, but this would be quite a poor performance compared to the cellular world, not to mention that the device might move again during the second session.

5.1.1.2 Potential Solutions

We can consider an approach where the mobile host, on moving, indicates its new IP address to the parties it is communicating with. Technically, this is a valid approach, but it suffers from the serious drawback that such a solution would require changes to the non-mobile hosts and the IP stacks implemented there. But why should any host or a server somewhere on the Internet, which itself is most probably not mobile, provide modifications to support the mobility of other hosts? There is no incentive for the stationary host for such mobility support. The other issue is backward compatibility and the huge number of legacy hosts on the Internet: there are billions of stationary hosts on the Internet today, and servers providing applications count here as stationary hosts as well. So it is not realistic to expect them to be updated with mobility support – the effort would be simply too high and, moreover, there is no single authority governing all the hosts and equipment vendors which could order such an upgrade; the Internet is just a mesh of uncoordinated, interconnected networks.

Nevertheless, the protocols that allow for a mid session change of the address(es) have been available in the IETF toolbox for some time now and one example of such a protocol is the Stream Control Transmission Protocol (SCTP) [78]. The SCTP has not seen much deployment in the global Internet, and although nowadays it is part of many IP stack implementations, it is still rarely used by applications. This is also probably the main issue with the approach taken by the SCTP: its service is available only to those applications that are being specially (re)written to make use of the SCTP; any legacy application running on the hosts that support SCTP does not profit from the SCTP in any way.

What is needed for a generic approach is a solution that is transparent to the applications so that any random legacy application can automatically and transparently benefit without needing any changes in the application code or the IP stack of the legacy stationary host.

This requirement is easier met if the solution is realized at the network layer (IP) or below, since the transport layer solution (i.e. use of the SCTP instead of UDP/TCP) is going to affect the legacy hosts who are the endpoints for the transport layer connections. There are two basic approaches on how to address this problem at the network level[1] or below: one is Mobile IP, and the other is link layer mobility. Both of them will be presented in the following sections.

5.1.2 Mobile IP

One prominent solution from the IETF community to the IP mobility problem is Mobile IP and its two flavors: MIPv4 and MIPv6. MIPv4 [10] is targeting today's IPv4 Internet while MIPv6 [77] is enabling mobility in upcoming IPv6 environments. There are also various extensions and enhancements to the Mobile IP base protocols provided by numerous add-on RFCs.

5.1.2.1 Protocol Architecture

The basic idea behind Mobile IP is to provision the mobile node (MN) with a stable IP address and make sure that the MN is reachable via that address regardless of its actual location, or subnet. To make this possible a special router has been introduced on the MN's home network – the home agent (HA) – that intercepts all the traffic sent to the MN and tunnels the traffic to the current location of the MN. The MN's stable address on its home network is called the home address, often abbreviated to HoA. The corresponding nodes (CNs) – the communication partners of the MN[2] – are using this stable address when they need to talk to the MN. When the MN is located in its home network, it is able to receive and send packets directly, without any involvement of the MIP processing and the home agent.

When the MN leaves its home link and moves to a foreign network (subnet B in Figure 5.2), it informs the home agent of its new location by sending a MIP registration request[3] message. The address in the foreign network to which the HA should tunnel the MN's traffic is called the care-of address (CoA) and is made available to the HA as part of the registration request message. In the case of MIPv4 the care-of address comes up in two different forms: as a collocated care-of address (CCoA) and the foreign agent care-of address (FA-CoA). MIPv6 only supports the collocated care-of address.

5.1.2.2 Foreign Agent

The FA-CoA mode is typically used when the foreign network offers a foreign agent service. The foreign agent (FA) is a router on the foreign link with additional MIPv4-specific functionality that offers mobility service to the visiting MNs. The service provided by the FA basically consists of providing the tunnel endpoint for the MIPv4 tunnel between the foreign and home networks which is being used to carry the MN's traffic. When the MN

---

[1] The problem could also be addressed by each application individualy at the applicaton layer and a significant number of applicatons today are able to survive the mid session change of the IP address, but what we are looking for here is a mechanism to make mobility service transparent for all applicatons.

[2] Despite its somewhat intimidating name, the CN is nothing more than a regular IP host which happens to be talking to the MN.

[3] This is the message name in the case of MIPv4; if MIPv6 is being used the message is called the binding update.

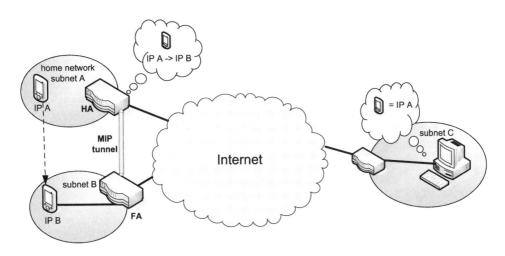

Figure 5.2   Mobile IPv4 architecture in the IETF

registers through the FA in the foreign network, the FA relays the registration request to the HA and the CoA is the address assigned to one of the FA's interfaces. The tunnel from the HA is terminated at the FA. The FA extracts the packets received from the HA via the tunnel and delivers them to the MN based on the destination IP address in the inner header, which is the MN's home address. The MIPv4 protocol stack in the non-collocated CoA mode is shown in Figure 5.3.

The MN located in the foreign network and using the FA service continues to use the HoA from the home network and does not obtain a local IP address that is topologically correct for the foreign network. The use of a HoA that is a topologically incorrect address in the foreign network makes registration with the FA necessary, otherwise any traffic sourced from a topologically incorrect HoA would fall victim to the foreign network's ingress filtering. In the uplink direction, the FA is the default router for the MN and hence receives all the uplink traffic from the MN. The FA has learned the address or the MN's HA from the MN itself as part of the MIP registration process and hence it is able to encapsulate and tunnel the packets from the MN to the MN's HA. The HA removes the outer header from the packets received from the FA and subsequently routes them according to the destination IP address in the inner header.

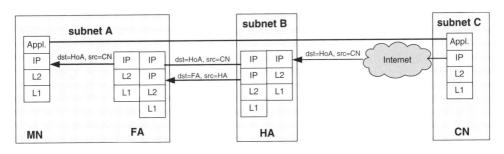

Figure 5.3   MIPv4 protocol stack (FA-CoA mode)

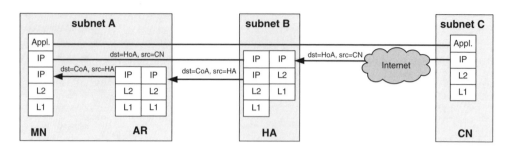

Figure 5.4   MIPv6 protocol stack

In CCoA mode the MN is assigned a local IP address in a foreign network that is topologically correct for the foreign network. The MN assigns the CoA to its interface connected to the foreign network and registers this CoA with the HA; thus in the case of the CCoA mode the MIP tunnel extends between the MN itself and the HA as shown in Figure 5.4 for MIPv6. Since the MN is using the topologically correct CoA to tunnel the packets to the HA, this removes the need for the FA function in the foreign network. And indeed, MIPv6 does not even provide the FA function – the CCoA mode is the only mode supported by MIPv6. But there is a price to be paid for not using the FA and it comes in the form of additional bandwidth consumption. The MIP tunnel is now terminated by the MN, so the outer tunnel header is carried over the air (in the case of the FA-CoA mode the FA takes care of removing the outer tunnel header and only the inner packet is actually carried over the air). The wireless link bandwidth is seen as a precious resource in wireless networks and this is the reason why many wireless operators are reluctant to accept MIPv6-based IP mobility. Although the CCoA mode is supported by MIPv4 as well, the WiMAX specification does not consider it, i.e. in a WiMAX network only the FA-CoA mode is possible with MIPv4.

5.1.2.3 Conclusion

To summarize, the basic idea behind Mobile IP is to provide a device with a stable (home) IP address and an anchor point in the form of the HA. The HA takes care of the delivery of the MN's traffic to/from its current location without impacting IP-based applications.

The MIP requires changes to the MN itself, to the home network which must support the HA function and to the foreign network in the form of the FA for MIPv4. While this may look as a solution requiring bigger changes at different places, it does not impact the CN side at all. With the help of the MIP, the MN is able to freely roam around and change its point of attachment while still maintaining any type of session for IP-based applications. The fact that the MN is moving is completely transparent to the CNs.

Another argument in favor of the MIP is that it is independent of the underlying link layer. Today IP networks run over every conceivable link layer technology, and the fact that the IP is not dependent on any particular link layer makes the MIP a perfect candidate for solving the mobility problem for all kinds of IP networks, irrespective of their underlying link layer technology. This 'implement once – run everywhere' property is a strong argument in favor of the MIP.

On the other side, the MIP was not designed for seamless handover performance. While changing the link, the MN may experience packet loss and the user may notice that there are intermittent connectivity problems. This was felt to be acceptable as IP networks in general and the applications running over them are supposed to be fault tolerant and can cope easily with occasional packet loss. There are numerous enhancements aiming to provide different optimizations to the baseline MIP – as a starting point the interested reader could look at [79], [80] and [81] for MIPv4 – but they all come with the tradeoff of increased system complexity.

While this section focused more on MIPv4, we will look at MIPv6 in greater detail in Section 5.3.2 where we discuss MIPv6 in the context of Mobile WiMAX.

5.1.3 Link Layer Mobility

While tackling the mobility problem at the IP layer may seem an obvious and logical choice to keep it transparent for legacy hosts and applications, it is not the only option. Mobility management can be addressed equally well at the link layer. The main difference is that such a solution is then specific to each particular link layer technology. Resulting mechanisms are potentially more efficient but make any type of interworking across different technologies much harder.

One strong argument for link layer mobility management is the better performance in terms of handover delay and packet loss. This is so because a particular link layer network is a coherent piece of equipment that adheres to a common and well-defined set of standards. The IP, on the other hand, is supposed to work over a wide range of different link layer technologies and as such it cannot make use of specifics available in each of the different link layer technologies. Thus it is inherently less optimized and cannot achieve a performance compa-rable to pure link layer mobility mechanisms.

Another aspect is that the link layer is usually under the closer control of the network operators and equipment vendors. Such control is seen as an important advantage if one has to manage a network offering a service to millions of subscribers, like mobile cellular operators do. In contrast, an operator cannot assume much about a standard IP host like a laptop computer, coming with a standard operating system. The device could be IPv6 enabled but need not be, different DHCP options may be supported or not, the host could be running a local NAT, firewall or antivirus software. All these may impact mobility management at the IP layer. In addition, the user is typically free to upgrade or reconfigure different parts of the host's IP layers and filtering rules, like the DHCP client, NAT or firewall rules. The options for users' intervention at the link layer are usually much more limited.

Providing the mobility service at the link layer results in the movement and reattachment events being hidden from the IP layer, the same IP address will remain valid after the movement, as is apparent from Figure 5.5, and this applies to the whole network no matter what its actual size. For as long as the host does not leave the operator's network, it can continue to use the same IP address and the link layer mechanisms take care of delivering the packets to the current location of the mobile device.

This is actually a common approach as it is the one taken by GPRS and UMTS networks. Link layer mobility is also utilized in the Mobile WiMAX as well, as we will show in the more detailed sections on ASN-anchored mobility.

Figure 5.5 Link layer mobility

5.2 WiMAX Mobility Architecture

Mobile WiMAX networks support a hybrid mobility management scheme comprising two
'layers'. The first layer is the link layer mobility, or ASN-anchored mobility, which is internal
to the NAP's network and does not impact any of the CSN functions. In the event of an ASN-
anchored handover the mobility anchor point where the data paths before and after the
handover are attached is located in the ASN and is not relocated as part of the handover. This
anchor point is called the ASN mobility management (MM) anchor point.

The second layer of mobility management in WiMAX is at the IP layer and is based on the
various variants of Mobile IP with the HA being located in the CSN and the ASN having the role
of the foreign network according to MIP terminology. At this layer of mobility management
(MM), called CSN-anchored mobility, the CSN is actively involved in handling mobility and
the anchor point for this handover is the HA which can therefore be referred to as the CSN MM
anchor point. In the course of the CSN-anchored mobility the CSN MM anchor point remains
unchanged while the ASN MM anchor point in the NAP network is relocated to a different
ASN-GW.

5.2.1 Mobility Anchors

Looking at the WiMAX mobility architecture in Figure 5.6, the ASN-anchored mobility is in
charge of managing mobility across the R6 and R4 interfaces, while the CSN-anchored
mobility handles the relocation of the R3 reference point to a different ASN due to the MS
movement across ASNs.

It must be noted that with the advent of the simple IP feature in WiMAX Release 1.5, the
CSN-anchored mobility became optional and in the context of the Release 1.5 specification it is
possible to deploy networks relying on the ASN-anchored mobility as the only means for MM.
More details on the simple IP feature are provided in Section 5.5.

When the MS hands off to a different BS under the same ASN-GW, the ASN-anchored
mobility is invoked to handle the relocation of the R6 reference point to a new BS. Each BS
hosts the handover control function (HO ctrl) that is in charge of executing the radio link

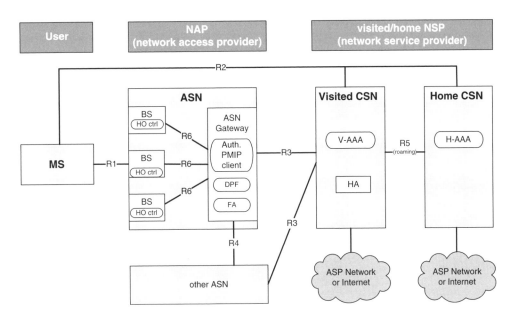

Figure 5.6 WiMAX mobility architecture

handover and coordinating the handover with the ASN-GW via the R6 reference point. The ASN-GW remains the same as both the previous and the new BS are connected to it. If the MS moves to a BS that is under the control of a different ASN-GW, this is still handled by the ASN-anchored MM but now requires interaction involving the R4 reference point. In the course of the handover across R4, the original ASN-GW continues with its role of being the anchor point for the MS session within the NAP (especially from the CSN perspective that does not see any change) and in addition it establishes a tunnel across the R4 interface to the serving ASN-GW that is actually controlling the new serving BS. This configuration is shown in Figure 5.7(a) below. The 'old' ASN gateway acting as the anchor point for the MS traffic is called the anchor data path function (A-DPF) and the 'new' ASN-GW, connected via the R4 tunnel to the A-DPF, is called the serving ASN-GW. Such a configuration results in an extended and hence suboptimal traffic path within the NAP as the MS traffic now traverses two ASN-GWs.

Such a suboptimal configuration, however, is not intended to stay permanently but can be resolved later on by a relocation of the anchor with the help of the CSN-anchored MM. In this case, the CSN-anchored MM will relocate the termination of the R3 reference point within the NAP to the serving ASN-GW, making it the anchor instead and by this removing the old ASN-GW from the signaling and data path for the MS.

From the CSN perspective the relocation of the R3 reference point means that the HA starts delivering the downlink MS traffic to a different ASN-GW due to the MS movement. From the ASN perspective the relocation of the R3 reference point means that the A-DPF is relocated from the previous ASN-GW to the current serving ASN-GW, i.e. to the one which directly controls the BS to which the MS is currently attached. After the CSN-anchored handover is completed, the temporary R4 tunnel between the ASN-GWs is removed. This restores the optimal configuration within the NAP as the MS traffic now once again takes the shortest possible path, R3–R6–R1, without involving R4.

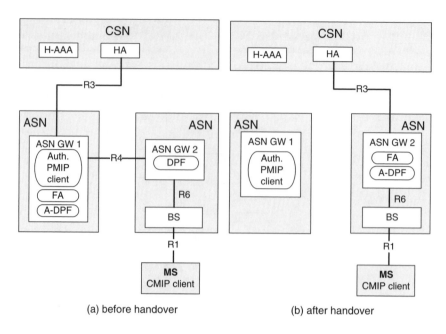

(a) before handover   (b) after handover

Figure 5.7   Configuration before and after the CSN-anchored handover

5.2.2 Hierarchy

But why did WiMAX decide to go with the two-tiered MM scheme? There are several aspects
to this question that need to be taken into account. One of the stated goals of the Mobile
WiMAX design was to minimize the requirements on the CSN. The CSN is seen mostly as a
generic IP network with an infrastructure comparable to an ISP and an ISP was supposed to be
able to easily upgrade its existing network to serve as the WiMAX CSN. Interconnection to the
WiMAX access network – the ASN, which could be owned by a different business entity – is to
be based on the IP standards having wide acceptance among the ISPs and the equipment
vendors. Under such assumptions, the only acceptable way to require some additional
functionality from the CSN was to base any such requirements on the existing IETF standards.
Another point to note is that, according to the architecture of Mobile WiMAX, the CSN is the
entity that 'owns' subscribers and provides them with IP connectivity service and hence the IP
address for the MS must come from the CSN. Taking into account the three important
ingredients that we have so far – the IP-based interconnection for the ASN–CSN leg, the IP
address anchored at the CSN and the need for mobility – MIP-based IP mobility between the
ASN and CSN emerges as the logical choice.

  CSN-anchored mobility support, if deployed by the NAP and NSP, puts the additional
requirement on the CSN that MIP HA functionality needs to be deployed. There are multiple
vendors of standard MIP HAs that behave according to the IETF specifications. WiMAX
requires those vendors to add a number of extensions to their MIP HA product to make it
WiMAX aware, like support for specific information that needs to be exchanged with the AAA
server. However, the extent of these extensions can be considered quite limited.

On the other side, involving the CSN in MM has its disadvantages. For one thing, the MIP was not optimized to minimize packet loss or service interruption periods. It is therefore not a perfect solution, although most applications would not require this anyway. A more serious concern arises from the fact that the ASN and CSN could be different business entities interconnected possibly via intermediate networks and roaming partners. Such a configuration may result in the incurred delay on the ASN–CSN leg that is possibly too high to allow for acceptable handover performance.

Due to rapidly changing radio conditions, providing a good handover performance often means that the handover must be completed within the time frame of a few hundreds of milliseconds. Good handover performance becomes a necessity for VoIP service, and handover performance at the level acceptable for the VoIP service is a requirement for Mobile WiMAX networks.

Here, the ASN-anchored mobility comes into play: the mobility at the link layer is handled internally to the ASN, keeping the handover delay and disruption at the lowest possible level, whereas a relocation using CSN MM can be kept a rare event that can in addition be deferred, only to be triggered in times when the impact on the ongoing sessions is minimal. With this, IP layer mobility is not on the critical path and can be realized without negative effects on end user experience.

On the one hand, without the ASN-anchored MM, WiMAX would not be able to compete in terms of handover performance with the cellular networks where MM is based on the link layer mobility mechanism. On the other hand, adding IP layer MM adds more flexibility in overall network design and the partitioning of the NAP's network into several separate L2 networks.

An important aspect is that MIP as the IP MM protocol needs in principle to be supported by the terminal (also called 'device'). Based on today's devices, however, there is no broad support of the MIP in end user devices. Mandating that every WiMAX device including standard laptops with WiMAX radio support is upgraded with MIP support was seen as unrealistic and such a requirement would seriously limit the number of devices able to connect to the WiMAX network. Hence, another supported variant for IP MM in WiMAX networks is based on the Proxy MIP (PMIP). The basic idea behind the PMIP is to provide a PMIP client in the access network that takes care of the MIP signaling on behalf of the MN. As a result, IP MM is fully transparent to the device. To make PMIP possible, the WiMAX Forum has designed proprietary extensions to MIPv4, leading to PMIPv4, and those extensions are documented for the IETF community in [82]. In the meantime, the IETF designed an additional variant of the PMIP [83], called PMIPv6, which is based on MIPv6, which has also been adopted by WiMAX as part of the Release 1.5 network architecture [116]. We will look at both of these PMIP variants, PMIPv4 and PMIPv6, in the following section.

As an aside regarding terminology, note that the term CMIP, which stands for Client MIP, is often used in the WiMAX context. This simply refers to the standard MIP where the MIP client is implemented in the mobile terminal device itself.

5.3 CSN-Anchored Mobility

Let us take a top-down approach for analyzing the WiMAX mobility mechanisms and begin with a detailed look at the CSN-anchored mobility.

Let us assume in the example shown in Figure 5.7 that an ASN-anchored R4 handover to a new ASN has already taken place for the MS (a). The MS downlink data traffic traverses the R3 reference point to ASN-GW 1 and from there over the R4 tunnel it reaches the serving ASN-GW 2. From there it traverses R6 and finally R1 to the MS. ASN-GW 1 hosts the FA function, which takes care of traffic delivery over the R3 reference point, while the collocated A-DPF takes care of traffic delivery via the intra-ASN R4 tunnel. In ASN-GW 2 we see another instance of a DPF without the anchor attribute. This simply means that it is not receiving traffic directly from the CSN but instead it takes care of inter-ASN traffic delivery.

The temporary tunneling of user data across R4 allows a most efficient handover, but adds a certain overhead. The optimal traffic path can be restored later on by the network relocating the termination of the R3 reference point to the current serving ASN and by the removal of the temporary R4 tunnel. The final configuration after the CSN-anchored handover completes is shown in Figure 5.7(b). Such R3 relocation action in the network will typically not be noticed by the MS in the case of the PMIP. For the CMIP, of course, the MS will have to execute MIP registration

While the traffic path now takes the direct route from the HA to ASN-GW 2, some functional entities can still remain in ASN-GW 1. In particular, the authenticator will only be relocated in case an EAP-based re-authentication is performed, but not as part of the CSN-anchored handover. The reason is that the authenticator holds the ASN security context for the MS and according to good cryptographic practices the secret keys valid for one ASN-GW are not transferred in plain to another ASN-GW. There is also a tight relationship between the authenticator and the PMIP client. The PMIP client is used to execute the MIPv4 signaling on behalf of the MS when the MS itself is not MIPv4 capable. The PMIP client requires security keys which are generated by the authenticator to protect the MIP messages. For simplicity in the network design, the PMIP client is always collocated with the authenticator. Although the PMIP client is a functional entity related to MM, it is relocated only together with the authenticator as part of the re-authentication process, and not as part of the MM procedures.

By looking at the example in Figure 5.7, we can see that the MS did not move at all during the CSN-anchored handover. It is connected to the same BS. The only thing that moves during the CSN-anchored handover is the termination of the R3 reference point in the ASN which is moved to the current serving ASN-GW of the MS. The actual movement of the MS is taken care of by the ASN-anchored mobility procedures.

In theory, ASN-anchored MM alone could be used to take care of the handovers in the complete WiMAX access network and this would leave the CSN-anchored mobility as a superfluous option. In practice, however, there are certain deployment considerations that make CSN-anchored mobility useful. As the MS moves further away from the point where it originally entered the network, the R4 tunnel would need to be extended from the anchor ASN-GW further and further to deliver the traffic to the current location of the MS. The longer the tunnel, the more resources it is using internally in the ASN and the longer is the delay incurred by the packets traversing the tunnel, so the overall performance would suffer. This is where the CSN-anchored mobility comes into play. It moves the traffic anchor point (R3 termination) within the WiMAX access back to the current location of the MS, thus removing the need for the R4 tunnel and restoring the optimal traffic path.

Mobile WiMAX makes use of several different flavors of the MIP and the exact signaling procedure of how the CSN-anchored handover is conducted varies somewhat depending on the

MIP variant being used: CMIPv4, PMIPv4, CMIPv6 or PMIPv6. But the concept is always the same as shown in Figure 5.7, regardless of the actual MIP variant.

5.3.1 CMIPv4

Let us now start exploring the different flavors of MIP that are available in WiMAX by looking at CMIPv4. Of all the different MIP variants defined in WiMAX, this probably is the most appropriate to start with as it represents the straightforward application of the IETF MIPv4 [10] protocol to the WiMAX network architecture, though, in terms of relevance for the deployments, CMIPv4 has low significance so far due to the lack of broad support in typical devices like CPE devices, laptops and smart phones. Nevertheless, looking at CMIPv4 first will give us a good foundation to build upon when looking at the more practical variants such as PMIPv4.

5.3.1.1 CMIPv4 Procedures

We will be looking at CMIPv4 session establishment with the help of Figure 5.8. During the initial network attachment of the MS, the IP layer configuration is started upon completion of the establishment of the initial service flow (ISF). On the network side, the establishment of the ISF triggers the FA in the ASN-GW to send out an ICMPv4 (Internet Control Message Protocol version 4 [84]) router advertisement message to the MS over the ISF. The router advertisement message contains the mobility agent advertisement extension [10] announcing the FA functionality in the ASN to the MS. MIP registration through the FA is obligatory for the MS in this case.

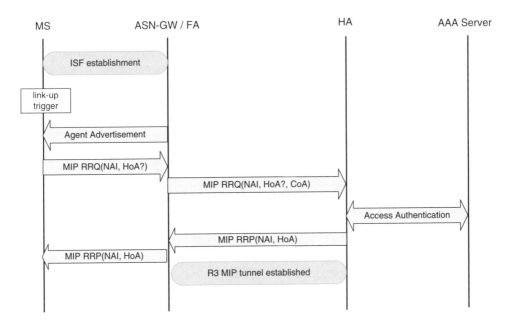

Figure 5.8 Initial client MIPv4 registration

On the MS side, the establishment of the ISF concludes the phase of establishing the link layer connectivity and hence it is typically used as a trigger to deliver the 'link-up' event to the IP stack. The IP stack in turn initiates the IP configuration of the newly available interface. Usually, on the host side, the IP configuration and IP address acquisition are accomplished by means of the DHCPv4 protocol [85]. However, WiMAX disallows any MS attaching to the network with CMIPv4 to use DHCPv4 for IP address configuration. Instead the MS running in the CMIPv4 mode must acquire the IPv4 address via MIPv4 itself. This special rule allows the WiMAX network to determine whether the MS is running MIPv4 or not. As soon as the MS uses DHCPv4-based address configuration, the network will operate in PMIPv4 mode for this MS. In the case when the first IP packet from the MS is not the DHCPv4 message but the MIP registration request (RRQ) message [10], the network will use the CMIPv4 mode.

According to the above rule for the IP address configuration process in WiMAX, the CMIPv4 MS, after receiving the router advertisement (agent advertisement) from the FA, sends a MIPv4 RRQ message to register with the HA and obtain the IP address. The MS must include the NAI as the subscription identity when adding the NAI option (parameter) in the RRQ message, to identify itself to the HA. The RRQ message is sent to the FA and since the MS does not have any IP address at this stage, the source IP address field in the IP header of the RRQ message is set to 0.0.0.0, which represents the unspecified IPv4 address. The RRQ message is protected by an authentication extension, being the default messaging security method for MIPv4. Details regarding the generation of appropriate keys for securing MIPv4 exchanges are discussed as part of Section 4.2.2. The MS is in general not expected to be aware of the address of the HA. Also, the network may dynamically assign a HA to the MS upon network entry. To cover this, the MS will set the HA IP field within the MIP RRQ message also to 0.0.0.0 (or alternatively, 255.255.255.255; both variants are allowed) to indicate dynamic HA assignment by the network.

The FA at this stage is already configured with the HA address that it received as part of the AAA Access Accept message received from the AAA server during the access authentication phase (see also Section 3.8), so upon receiving the RRQ message from the MS, the FA relays the message to the configured HA, including its own FA address as the CoA. On receipt of the RRQ message the HA checks with the H-AAA server whether the MS is authorized for the service. For this the NAI provided as part of the RRQ message is used to identify the MS to the AAA server. If the AAA response is positive, the HA will receive the required security keys to verify the authentication extension of the RRQ message from the AAA. The HA, after successful verification, assigns the HoA to the MS, puts it into the registration response (RRP) message and sends the response back to the FA. The FA in turn relays the RRP message to the MS over the ISF.

The MIPv4 tunnel across the R3 reference point is established and fully operational when the FA received the MIP RRP message [10] from the HA. Typically the encapsulation method for the MIP tunnel is IP-in-IP as defined in [86], but the FA and HA may also negotiate the use of the GRE encapsulation [87] with additional usage of the GRE keys [88] in the MIP RRQ and RRP message as described in [89].

Finally, the MS extracts its home address (HoA) from the RRP message and configures its WiMAX interface with the assigned home address. For verification of the RRP, it will use a MIP-specific security key derived from the EAP-based access authentication to verify the authentication extension that was included by the HA (see also Section 4.2.2). This step marks the completion of the IP address configuration, so the MS has established IP connectivity and is now ready to start IP sessions.

To obtain additional IP configuration parameters, like the DNS, or the address of the P-CSCF for IMS services, the MS is now allowed to use DHCPv4 as the mobility mode is determined and cannot be changed for the duration of the session. In other words, the use of DHCP is only prohibited prior to completion of the MIP registration phase.

5.3.1.2 CMIPv4 in Mobile WiMAX

After considering the registration phase, let us now take a more detailed look at the CSN-anchored MM by looking at the MIPv4 handover with the help of Figure 5.9. The starting network configuration which is a prerequisite to get the CSN-anchored handover triggered is shown in Figure 5.7(a). It is a result of the preceding ASN-anchored handover. The previous ASN-GW is still acting as the FA and is hosting the A-DPF for the MS, and therefore there is a R4 tunnel between the ASN-GWs to carry the MS data traffic.

At any point in time either ASN-GW may request the relocation of the R3 termination point to the current serving ASN-GW. The NWG standard does not define the exact trigger conditions for this event, so this is left to be implementation dependent. For example, if the MS is engaged in a VoIP call, one implementation may choose to postpone any MIP handover until after the VoIP call is finished, to avoid any glitches in the quality of the call as perceived by the user. Another implementation may trigger the MIP handover as soon as the ASN-anchored

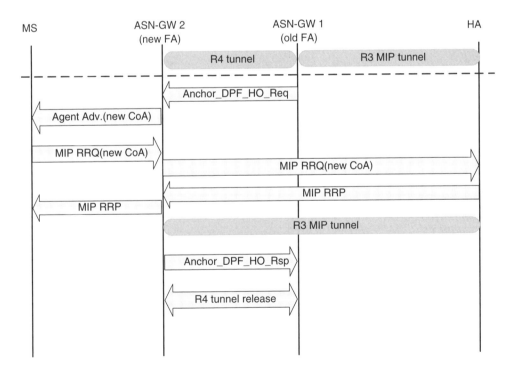

Figure 5.9 CMIPv4 handover

handover is completed, to restore the optimal data paths and free the additional resources consumed by the R4 forwarding as early as possible.

Once the decision is made to execute the MIP handover, the serving ASN-GW sends the router advertisement message to the MS. The mobility agent advertisement option [10] is included in this message. By detecting that the router advertisement message was sent from a new router and that the mobility agent advertisement option is different from the previous one, the MS concludes that it must have moved into a new subnet and therefore executes the MIP registration exchange to inform the HA of its movement so that the HA is able to tunnel the MS traffic to the new location of the MS. For this, the MS sends the RRQ message to the new FA, which relays it to the HA. The new FA acquires the HA address as part of the context transfer from the old FA at the time when the MIP handover was triggered.

As part of the MIP RRQ processing, the HA updates the tunnel endpoint of the MIP tunnel to the address of the new FA and this accomplishes relocation of the R3 reference point to the new FA. Starting at this point, the HA directly tunnels the downlink traffic to the new FA.

On receiving the RRP message indicating success and relaying it to the MS, the new ASN-GW (ASN-GW2 in Figure 5.9) stops tunneling the uplink data traffic to the old FA via the R4 tunnel and instead starts delivering the uplink data traffic directly over the R3 reference point to the HA.

Triggered by the RRP message from the HA, the new FA initiates the teardown of the R4 tunnel to the previous FA. The release of the R4 tunnel is the final step in the course of the CSN-anchored handover and the result is once more the straightforward and optimal configuration shown in Figure 5.7(b).

Note that if the previous ASN-GW was acting as the authenticator for the MS, the authenticator function will still remain with the previous ASN-GW even after the CSN-anchored handover is executed. As we can see, although the MS traffic no longer traverses the previous ASN-GW, this does not necessarily mean that the previous ASN-GW is no longer involved in the control traffic related to the MS session. If the authenticator is to be moved to the new FA, too, an additional authenticator relocation procedure needs to be executed and this involves re-authentication of the MS.

5.3.2 CMIPv6

This section will take an additional brief look at CMIPv6 [77], although the practical relevance of this option for CSN MM seems to be quite limited based on the lack of general support in devices. The introduction of CMIPv6 was mainly motivated to provide WiMAX operators with a way to overcome the depletion of the IPv4 address space. Prior to the availability of simple IPv6 and PMIPv6 in WiMAX Release 1.5, it was the only way to support IPv6 terminals in the Mobile WiMAX architecture.

5.3.2.1 CMIPv6 in Mobile WiMAX

Discussion of MIPv6 in the context of WiMAX mobility is still important because it forms the foundation upon which PMIPv6 is built. PMIPv6 seems to be a more common approach in general and by understanding the basic principles of the regular client-based MIPv6 we should be in a better position later on to understand PMIPv6 in detail and judge its deployment-specific aspects.

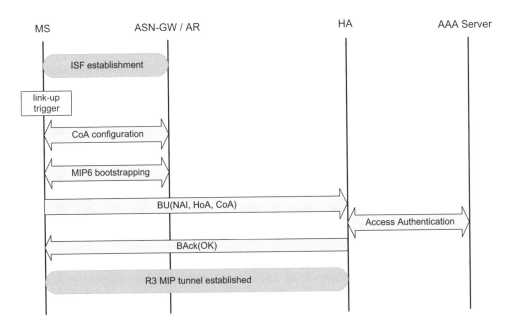

Figure 5.10   CMIPv6 initial registration

The main difference between MIPv4 and MIPv6 in the context of WiMAX is the fact that there is no FA function in the case of MIPv6 and the MIPv6 signaling is exchanged directly between the MS and the HA without being relayed through the FA. This comes with several implications from the perspective of the ASN where the MM signaling is now part of the MS data traffic and as such it is transparent to the ASN.

As shown in Figure 5.10, when the MS attaches to the WiMAX network and the WiMAX interface sends the link-up event to the MS IPv6 stack, the first step for the MS is to obtain the local IP address. While the MS can make use of either the IPv6 address autoconfiguration process [90] for acquiring the IP address, or DHCPv6 [91], the realistic assumption is that most devices will opt for autoconfiguration. The latter does not require a stateful DHCPv6 client on the device. The IPv6 prefix being advertised by the ASN-GW as the on-link prefix is topologically anchored at the CSN. This may seem strange at first glance, but it is necessary to comply with the WiMAX network architecture which prescribes that the point of attachment to the IP network is the CSN. If the ASN were to advertise its own local prefix, this would make the ASN the MS point of attachment to the Internet instead of the CSN. The WiMAX Forum does not specify any means of how the ASN is provided with the right prefix from the CSN; this must be settled in a mutual agreement between the ASN and the CSN operators. This necessity for out-of-band agreements between operators will not be of much practical importance for deployments where the ASN and CSN are operated and owned by a single operator. However, it needs to be considered as soon as the ASN and CSN belong to different business entities.

It must be mentioned here that WiMAX requires that the on-link prefix advertised to the MS is unique to each MS; that is, each MS is given its own IPv6 prefix. The reasoning behind this is that the MSs are not allowed to talk directly to each other, even when attached to the same ASN-GW. Instead, all the traffic must be looped through the HA. Under such circumstances, assigning a different prefix to each MS makes sense because, to the devices, this will look like they are on different IP links even when attached to the same ASN-GW and as a consequence they would as a default not try to talk directly to each other. Any attempt to talk directly to each other may lead to some issues related to the IPv6 Neighbor Discovery protocol (the core part of the IPv6 stack) [93], but the scope of this book does not allows us to detail these concerns here. As a final note on this matter, the 3GPP also adopted the unique prefix per MS approach in its approach of how the IPv6 architecture is applied to 3G networks.

5.3.2.2 CMIPv6 Procedures

Before being able to register with the HA, the CMIPv6 MS needs to obtain the IPv6 address of the HA and it must be able to generate its own home address – in MIPv6 the MN cannot ask the HA to dynamically assign a home address. WiMAX adopted the stateless DHCPv6 [92] approach for bootstrapping the MIPv6 parameters. Basically, this means that the MS is able to utilize DHCPv6 to ask the network for the address of the HA and for the home network prefix that the MS has to use to autoconfigure its IPv6 home address. This information is provided to the MS in form of DHCPv6 options (parameters) by the ASN-GW, which in turn obtains the information from the H-AAA server during the access authentication phase of the MS.

To register with the HA, the MS generates the home address based on the obtained home network prefix and sends the binding update (BU) message to the HA. The MN includes the NAI option [94] in the binding update message as the HA needs this information to identify the MN to the H-AAA server while asking for the service authorization. The binding update message also contains the MN's on-link address as the CoA. As we already discussed, in WiMAX the MN's on-link address is topologically anchored at the CSN, which results in the binding between the CoA and the HoA where both addresses are actually coming from the same CSN. The point is that the MS's CoA changes as it moves around between subnets within the WiMAX network and if it were using the CoA as a source address for application sessions then the sessions would break each time the MS moves into the new subnet. Hence, the HoA can simply be looked at as an enabler for session continuity.

On processing the binding update, the HA creates a binding between the MN's HoA and the CoA and starts delivering the traffic destined to the HoA over the MIPv6 tunnel. Note that the tunnel endpoint in the case of MIPv6 is the MN itself and the encapsulation and decapsulation are performed by the MN. The fact that the outer IP header is transported over the air link is one argument as to why MIPv6 is considered slightly inefficient and not perfectly aligned with the typical requirements of a wireless environment. The capacity of the wireless link is typically considered to be the most expensive resource in the overall system.

When looking at MIPv6-based handover, we see that the absence of the FA function in MIPv6 is causing significant trouble. As shown in Figure 5.11, the starting position is the same as in the case of the CMIPv4 handover and either the serving ASN-GW ('the new one') or the A-DPF ASN-GW ('the old one') can trigger the handover. Once the handover is triggered, the serving ASN-GW sends the router advertisement to the MS advertising the on-link prefix that is different from the prefix that was advertised previously by the A-DPF ASN-GW. The MN deduces that it must have moved into a new subnet, forms the new CoA based on the advertised prefix, and proceeds to register the new CoA with the HA by sending the binding update

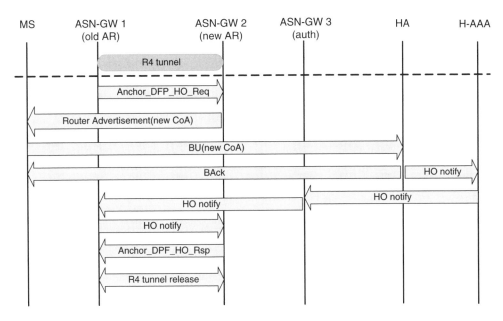

Figure 5.11   CMIPv6 handover[4]

message. The HA processes the binding update and associates the new CoA with the MN's home address. It then sends the binding acknowledgement in response to the MN.

5.3.2.3 R4 Data Path Issues

Everything seems fine so far, though there is a problem that the ASN is not aware that the CMIPv6 handover has been successfully completed and thus is missing the trigger to remove the R4 tunnel between the ASN-GWs. The ASN is not aware because the MIPv6 signaling runs directly between the MS and the HA, without any involvement from the ASN besides transparently delivering those packets like any other MS traffic. In general, there is no elegant way to solve this. In the WiMAX Forum several approaches to solve it were suggested. One idea was that the serving ASN-GW simply starts a timer when it sends out a router advertisement with a new prefix and, after that timer elapses, it should tear down the R4 tunnel. However, as the MS may have valid reasons to delay the handover, this approach was not adopted. Another suggestion was that the serving ASN-GW should inspect the MS packets looking for the binding update and binding acknowledgement messages, so that it would know when it is safe to remove the R4 tunnel. Such a solution would likely come with a serious performance impact, as monitoring and inspecting the user traffic is a very costly operation. Also, a malicious MS could send fake binding updates or any node from the Internet could send

---
[4]'HO notify' is not the name of the real message: different messages are used to carry the handover notification on different legs: RADIUS Access/Accept and Disconnect from the HA to the H-AAA and from the H-AAA to the authenticator, respectively, and within the ASN, the WiMAX proprietary signaling in the form of the Context_Rpt message.

fake binding acknowledgements towards the MS and thus trigger the ASN-GW to remove the R4 tunnel with the result of rendering the MS out of service. The serving ASN-GW would not have any defense against such attacks without being able to prove the authenticity of the CMIPv6 messages. However, it is not in possession of the necessary secret keys – only the MS and HA have access to them.

In the solution that was finally adopted and is now part of the network specifications, the HA informs the serving ASN-GW of the CMIPv6 handover, but the notification message takes quite a detour. As the HA has no direct awareness of the serving ASN-GW to which the MS is attached, it cannot send the notification directly to it. Instead, the HA relies on the AAA signaling channel between the ASN and the CSN to deliver the notification to the ASN and then the internal routing within the ASN takes care of delivery to the serving ASN-GW. On successful CMIPv6 handover, the HA first informs the AAA server of the handover. The AAA server relays the notification to the authenticator in the ASN as a RADIUS message. Within the ASN, the notification is translated into a WiMAX NWG-defined R4 messages and the authenticator sends it to the A-DPF ASN-GW which finally forwards it to the serving ASN-GW (in case these functions do not reside in the same ASN-GW). At this point the serving ASN-GW can complete the handover transaction and initiate the removal of the R4 tunnel. Admittedly, this is not a very elegant solution.

5.3.3 PMIPv4

As we mentioned earlier, MIPv4 did not gain much initial support among the device and mobile computer operating system (OS) vendors, and this is even more true in the case of MIPv6. For WiMAX this is an important aspect to be considered for any deployment, especially when taking into account that WiMAX is not following a walled-garden approach and is targeting generic consumer devices like laptops and smart phones. These devices cannot be assumed to be under strict operator control and are typically not purpose built for WiMAX networks, so they cannot be automatically upgraded by the operator to support MIP. Not being able to provide service to such generic devices was perceived as an unacceptable behavior for Mobile WiMAX.

To counter that problem, another IP mobility variant was introduced in the WiMAX network by adopting the PMIP.[5] The basic idea behind the Proxy MIP is simple: the network takes care of IP MM on behalf of the host, thereby removing the need for the host's involvement in the handling of IP mobility. That is, network-based IP MM works with any standard IP host.

5.3.3.1 The Roots of PMIPv4

At the time when WiMAX was considering the need for PMIP, around the year 2005, there was also an ongoing effort in the IETF related to the network-based IP mobility. This work was being done in the NetLMM (Network Localized Mobility Management) Working Group and the basic idea was the same – providing network-based IP MM to legacy hosts. But the

---

[5]Meanwhile, the PMIP as a concept has gained some traction and there is ongoing work related to the Proxy MIP in other organizations, most notably the IETF and 3GPP.

approach taken in the NetLMM group was a bit different as it was not pursuing the reuse of the MIP. The group was more inclined to design a new protocol for this purpose.

The approach in WiMAX was different because WiMAX wanted to reuse the existing HA products as much as possible. By introducing a new proxy mobile node (PMN) function on the network side which sits in the first hop IP router to which the IP host is connected, the ASN network actually takes care of the MIP signaling on behalf of the host. As soon as the host moves around in the network, the PMN function follows it by being relocated appropriately and executes the MIP signaling on behalf of the host. The MIP signaling would be fully conformant with the MIPv4 standard [10]. In fact, the proclaimed goal in the early days of the PMIPv4 design in the WiMAX Forum was to reuse the regular MIPv4 HA for the PMIPv4 without any modifications, i.e. with the HA not being aware of whether the MIP mode for a specific MS is actually PMIP or CMIP. This approach mostly became true for the MIP signaling itself. However, also note that the interface between the WiMAX HA and the AAA server is quite specific to WiMAX operation, so any HA product to be deployed in a WiMAX network still needs to support a specific set of functions for WiMAX operation.

Interestingly, the adoption of the PMIP concept in WiMAX coupled with the similar thinking of the 3GPP standardization community eventually led the NetLMM group to abandon the concept of a network-based MM protocol that would be independent of MIP and to adopt the PMIP concept as well. While the WiMAX PMIP variant is originally built upon MIPv4, the IETF took a more future-oriented approach and fully based its PMIP variant on MIPv6.

Given that no PMIP specification was available in the IETF at the time of technical completion of the initial WiMAX release of network specifications, WiMAX had to go ahead and designed a WiMAX proprietary PMIPv4. The WiMAX variant is documented in the IETF draft [82] which will finally become an informational RFC document in the IETF. The draft also describes the PMIPv4 dialect used in 3GPP2 networks.

5.3.3.2 PMIPv4 Architecture

A basic idea behind PMIP is that there is a functional entity in the network that is typically located in the access network. This new functional entity is acting as a proxy on behalf of the MS to execute MIPv4 signaling. This functional entity is sometimes called the PMN, but the term commonly used in WiMAX is the PMIP client. Conceptually we think of a PMIP client as a function that takes care of a single MS, so for each MS there is a separate instance of the PMIP client. Figure 5.12 shows how the PMIP model is applied to WiMAX networks. The access network has the task of keeping the PMIP client aware of the actual location of the MS by using lower layer signaling. In WiMAX this signaling occurs across the R6 and R4 reference points in the form of WiMAX-specific messages. When the PMIP client is informed by means of the lower layer signaling that a CSN-anchored handover should take place, this triggers the PMIP client to send the MIP RRQ message to the HA on the MN's behalf, thus enabling the relocation of the R3 reference point to a different ASN-GW.

In WiMAX the PMIP client does not actually follow the MS as it moves within the ASNs. Instead, the PMIP client always stays collocated with the authenticator. The main reason behind this decision is simplicity in network relocation procedures as we will see in the following.

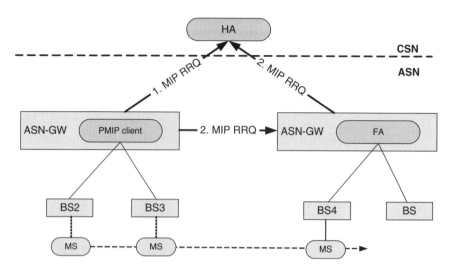

Figure 5.12 WiMAX PMIPv4 architecture

As we know, the MIPv4 signaling messages must be cryptographically protected to provide authenticity and prevent en-route tampering with the messages. So the PMIP client needs some secret keys shared with the HA to be able to provide the required protection. Basically, there were two options on the table related to the protection of the MIP messages in PMIP. One of them was to design so-called 'per node' security associations for protecting the PMIP messaging where the same security association is used by all PMIP clients (i.e. for all active MS sessions) in one ASN-GW. The result would effectively be a single security association per ASN-GW/HA pair. This is a departure from the regular client MIPv4 behavior where obviously each MS has its own private security association with the HA. The arguments in favor of the per node security association were that it is simple and key management becomes much less of a hassle, while there are no drawbacks related to the level of the security protection.

The alternative to this is to use the per MS security associations for PMIP to keep it aligned with CMIPv4 as much as possible. There is no significant difference in the security level here compared to the per node approach. An attacker gaining access to an ASN-GW would have all the keys at its disposal anyway – all of them would become equally compromised. So the main reason that per MS security was adopted in WiMAX is to keep the whole security architecture for the MIP in WiMAX as a single one that is independent of the actual IP mobility mode, thereby reducing the overall network complexity by removing the need to implement different options for deriving and sharing security keys.

Returning to the collocation, the per MS decision further led to a firm marriage of the PMIP client to the authenticator to avoid an unnecessarily complex transfer of state for the PMIP client, including security keys. The authenticator is in possession of the secret keys used to protect MIP messages sent by the PMIP client and as good security practices recommend avoiding key transfer whenever possible, the cleanest way to deal with this was to make the PMIP client and authenticator live together and never part. In fact, the only way to relocate the PMIP client is to relocate the authenticator, and the PMIP client is moved together with it as part of the authenticator relocation procedure.

5.3.3.3 PMIPv4 Initialization

PMIP operation where the MS initially enters the network is quite straightforward. On ISF establishment, the MS, since it is a IP host, starts the DHCPv4 messaging as per [85] to obtain the IPv4 address and other IP configuration parameters. The arrival of the DHCPv4 message at the ASN-GW, as opposed to the alternative of receiving the MIP RRQ message in the CMIP case, triggers the PMIP mode and the PMIP client is invoked to generate the MIP RRQ. In the meantime the DHCP process is gated at the ASN-GW until the MIP registration is completed. The MIP RRQ message includes the NAI option identifying the MS and is protected by the keys obtained from the collocated authenticator. The generated message is then internally relayed to the collocated FA, which in turn sends it to the HA. By looking at the received MIP RRQ message itself, the HA is not able to tell whether this message was originated by the PMIP client or by the MS itself. However, the HA is able to figure it out with the help of the AAA server and the MIP security context that it needs to fetch from it to be able to authenticate the request and sign the MIP RRP message sent in response. The HoA that is delivered to the ASN in the MIP RRQ message is taken over into the DHCP module in the ASN-GW and is assigned to the MS as part of the DHCPv4 configuration process. The described tight coupling between the DHCPv4 signaling and the MIPv4 procedure is also apparent in Figure 5.13.

The model where the ASN-GW acts as the DHCP server for the MS is called the DHCP proxy mode. The name 'DHCP proxy' is something of a misnomer as the DHCP module in the ASN-GW is not actually proxying (i.e. interworking) the DHCP message between the MS and the DHCP server. The DHCP proxy is simply a DHCP server in the ASN-GW that does not have its own address pool and instead obtains the address to be assigned to the host by some external means, like the MIPv4 signaling in cooperation with the PMIP client or via the AAA signaling as part of the MS access authentication. The DHCP proxy mode is the common deployment feature supported on the network side. The WiMAX specification also allows for an alternative

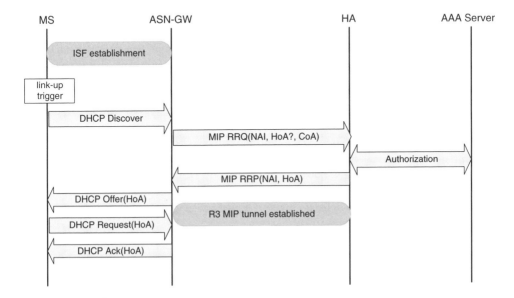

Figure 5.13   Initial PMIPv4 registration in DHCP proxy mode

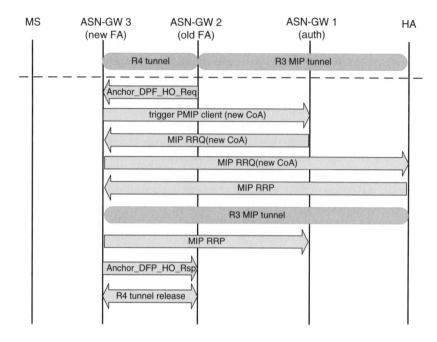

Figure 5.14   PMIPv4 handover[6]

mode where the ASN-GW acts as a DHCP relay and forwards the DHCP messages to the DHCP server located in the CSN. This seems more in line with the generic DHCP model that we know from the Internet. Yet the separate DHCP server in the CSN would need to interface to the HA as it must be assured that the address assigned by the DHCP server is actually the same as the HoA registered with the HA. Another aspect of the DHCP relay mode is the potentially different lifetimes of the home address in the DHCP server and in the HA as each entity assigns its own lifetime and that lifetime gets extended individually. It was generally felt that the DHCP relay solution is burdened by those additional issues that are not present in the case of the DHCP proxy and thus the DHCP proxy is preferred by implementations.

5.3.3.4 PMIPv4 Handover

When we look at the PMIPv4 handover scenario and compare it to the CMIPv4 one, we see that there is now an additional entity involved, namely the PMIP client. As the PMIP client is collocated with the authenticator, due to the previous movements of the MS, the authenticator may be located in an ASN-GW that corresponds to neither the current nor the new FA (i.e. the authenticator was 'left behind'). So in the most generic case the PMIPv4 handover involves three distinct ASN GWs (Figure 5.14).

The difference to the CMIPv4 handover case starts at the point where the new FA is supposed to send the router advertisement to the MS and trigger it to execute the MIPv4 handover. For PMIPv4 the new FA sends a trigger message to the ASN-GW where the authenticator and PMIP

---

[6] For reasons of brevity and clarity, this message flow represents the logical action taken by different entities and in particular the message names shown in the figure do not always correspond to the actual R4/R6 message names.

client live. As part of the trigger message the PMIP client gets all the information, like the new CoA, that is needed to generate the MIP RRQ message. The PMIP client generates the MIP RRQ message and relays it to the new FA, which in turn forwards it to the HA. The processing at the HA side is no different to the processing that takes place in the case of plain CMIPv4.

The reception of the MIPv4 RRP message at the new FA accomplishes the relocation of the R3 tunnel endpoint to the new FA. The new FA relays the MIP RRP to the PMIPv4 client which updates its internal state accordingly. This completes the PMIPv4 handover at the IP level. What remains to be done is release of the temporary R4 tunnel and this step is exactly the same as described earlier when we discussed the CMIPv4 case.

It is interesting to note that not only did the MS not actually move – this is again exactly the same as in the CMIPv4 handover case – but there is no single message directed to or originated by the MS during the PMIPv4 handover. The PMIPv4 handover is transparent to the MS, which is completely unaware of the network-internal operation. This makes it possible to execute the PMIPv4 handover while the MS is in the idle mode and indeed the NWG specification explicitly allows for the PMIPv4 handover during the idle mode.

We can also infer from Figure 5.14 that the MIP lifetime is extended as part of the PMIPv4 handover, but the same is not true for the DHCP lifetime. This reiterates the fact that there is no direct association between the DHCP lifetime and that of the MIP session. There is nothing wrong with this, but it is worth remembering that the DHCP lifetime and the MIP lifetime are not kept in sync.

There is an additional aspect related to the DHCP handling in the case of the DHCP proxy mode. According to the WiMAX network architecture, the DHCP proxy is collocated with the FA function which terminates the MIPv4 tunnel on the ASN side. So as part of the PMIPv4 handover, the new FA automatically takes over the DHCP proxy function for the MS. A problem arises because the MS remembers the address of the DHCP server that originally assigned the IP address to the MS and it will use this cached address to send any new request to the DHCP server to extend the IP address lifetime. As the MS is not made aware of the PMIPv4 handover, it cannot know that this DHCP server function has already moved to a different ASN-GW. Since after handover the previous ASN-GW is no longer serving the MS, it would simply drop such requests and the end result would be that the MS is left without a valid IP address as soon as the current lifetime expires. From the various approaches to solve this issue, the one adopted was where all the ASN-GWs in the operator's WiMAX access network should be configured with the same IP address as the address of the DHCP proxy function. Thus the address of the DHCP server from the MS's perspective will not change as every ASN-GW in the operator's domain will respond to the DHCP messages from the MS at the same IP address. It is worth noting that the DHCP relay mode does not suffer from this limitation as the address of the DHCP server that is located in the CSN remains stable across the PMIPv4 handovers.

5.3.3.5 Ethernet over PMIPv4

Mobile WiMAX supports the transport of Ethernet frames across the radio interface. This capability requires the deployment of the Ethernet convergence sublayer (ETH-CS). This section will look briefly at how the connection between the ASN and CSN is managed for such a MS based on PMIPv4.

Today almost anything can be tunneled over IP and tunneling of Ethernet frames over IP is really nothing new. Applying PMIPv4 mobility to the ETH-CS case simply means that the

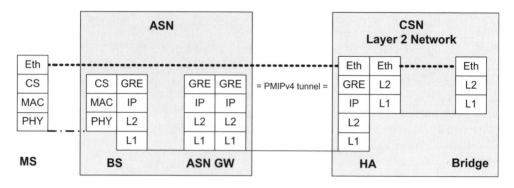

Figure 5.15   Protocol stack for Ethernet CS over PMIPv4

PMIPv4 tunnel carries Ethernet frames instead of encapsulated IP packets as shown by Figure 5.15. The IP tunneling technology already supports tunneling of Ethernet frames, so what remains to be done is to extend the PMIPv4 signaling to be able to negotiate the transport of Ethernet frames instead of regular IP packets that are assumed by default. And since in the Ethernet world the devices only use an IEEE 48-bit MAC address instead of the higher layer IP address, the PMIPv4 extensions must take care of this part as well.

As part of the Release 1.5 specifications, NWG standardized the application of PMIPv4 to the ETH-CS case [116]. Much of the additional pieces needed for this work were already available as existing building blocks, and WiMAX needed to glue those various pieces together and provide normative semantics for the resulting combination of options.

The WiMAX NWG specifies the following differences at the protocol level to the regular PMIPv4 operation that enables PMIPv4 to be used with ETH-CS:

- The MIP RRQ message must carry the MS's MAC address in the PMIPv4 device ID extension that is defined in [82].
- The HoA field should be zeroed out in both the MIP RRQ and the MIP RRP as the IP address is used neither by the ASN-GW nor by the HA when providing the Ethernet service.
- GRE encapsulation and using GRE keys [89] for the PMIPv4 tunnel are mandatory.

When the HA receives the initial PMIPv4 RRQ message, it creates the binding between the MAC address of the MS contained in the PMIP device ID extension and the CoA. It also marks the created binding entry as the one being used for the transport of Ethernet frames. After the PMIP registration is completed, the HA is ready to receive Ethernet frames from the MS through the ASN that are sent via the PMIPv4 tunnel. For any such frame the PMIP header is stripped off and the resulting Ethernet frame is passed on to its destination within the CSN. In the downlink direction, the HA intercepts any Ethernet frames being sent to the registered MAC address of the MS and sends them via the PMIPv4 tunnel to the ASN where they are delivered to the MS.

When supporting ETH-CS operation, the HA in fact acts as a kind of PMIPv4-enabled Ethernet switch where some of the ports are created on a dynamic basis and on top of that they feature the PMIPv4 encapsulation. The HA may even support the learning bridge mode: it may inspect uplink Ethernet frames coming in over the PMIPv4 tunnel and any unknown source

MAC address in such frames could be added to the binding entry related to the tunnel over which the frame was received. The uplink frame with an unknown source MAC address is related to the existing binding entry with the help of the GRE key which is part of the tunnel encapsulation header. Once the new MAC address is added to the binding entry, the HA may start intercepting the Ethernet frames for that MAC address and send them to the ASN via the associated PMIPv4 tunnel.

5.3.4 PMIPv6

Compared to the CMIP variants and to PMIPv4, PMIPv6 [83] is a later development that was designed in the IETF by the NetLMM Working Group. The group originally started to work on a completely new approach for network-based IP mobility independently of the MIP, but as we discussed earlier the changed industry perspective led to the adoption of the PMIP in the NetLMM group. The IETF specification does not exactly match the PMIPv4 approach taken by WiMAX and the IETF designed PMIP in a way that is not transparent to the HA. The latter makes the benefits of basing the solution on the existing MIP somewhat questionable since the HA needs to be modified to support the PMIP, but presumably the necessary code changes will in the end amount to much less new code than what would be needed for a completely new MM protocol.

5.3.4.1 PMIPv6 Architecture

PMIPv6 by default supports mobility for IPv6 hosts, but there is also an accompanying work [95] that enables PMIPv6 to support IPv4 hosts as well. Instead of the PMIP client, the PMIPv6 introduces a mobility access gateway (MAG) that performs the MIP signaling on behalf of the MS. The HA was given a new name and according to the PMIPv6 terminology it is termed the local mobility anchor (LMA). The MAG bears some significant similarities in terms of functionality to the MIPv4 FA as it represents the tunnel endpoint for the PMIP tunnel. When comparing the IETF PMIPv6 design to the WiMAX PMIPv4 architecture, we can consider the MAG as a combined PMIP client and FA functions. The other notable difference is that the PMIPv6 security in the IETF is based on the per node concept, i.e. the MAG and the LMA use a single security association per MAG–LMA pair to protect the PMIPv6 messages sent on behalf of all the MS.

In PMIPv6 the MAG is an entity that handles both the mobility-related signaling on behalf of the MS as well as the tunneling of the data traffic. In contrast to WiMAX PMIPv4 where the PMIPv4 client is 'left behind' in the previous ASN-GW but still continues to take care of PMIPv4 signaling, in PMIPv6 it is the new MAG that originates the proxy binding update (PBU) to the LMA and the previous MAG is not involved in any way. The PBU message is the adapted MIPv6 binding update message serving the same purpose – registering the MS with the HA – the main difference being that it is sent by the MAG on behalf of the MS instead of being sent directly by the MS itself.

While PMIPv6 is based on MIPv6, it does not actually require the transport network between the MAG and the LMA to be IPv6 enabled. The supplement IPv4 support document for PMIPv6 [95] provides the means to transport the PMIPv6 signaling messages as well as the user traffic over the IPv4 network by means of tunneling IPv6 over IPv4.

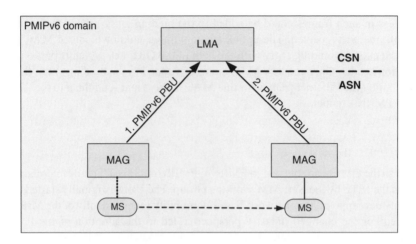

Figure 5.16   Generic PMIPv6 architecture

When the architecture from Figure 5.16 is applied to Mobile WiMAX, the LMA and the MAGs might be located in different administrative domains because the CSN and ASN could be owned by different operators.

5.3.4.2 PMIPv6 Initialization and Security

The scenario for initial attachment in WiMAX represents a straightforward application of the PMIPv6 specification [83]. The ASN-GW hosts the MAG function and during the access authentication the ASN-GW learns from the AAA server the mobility mode that needs to be used for the MS. In the case of PMIPv6, the AAA server also provides the address of the LMA in this step. On completion of the access authentication, the MAG may send a PBU message to the LMA to register the MS with the LMA and establish the PMIPv6 tunnel. In the PBU, the MAG includes the NAI used by the MS during access authentication to identify the MS or subscriber to the LMA and asks the LMA to assign an address for the MS. The MAG can ask for an IPv6 address, for an IPv4 address or for both. Any IPv6 address actually takes the form of an IPv6 prefix that is unique to the MS and is called the home network prefix (HNP). The reasons behind assigning the unique prefix to each MS are basically the same as those that we have already briefly touched upon in the section on CMIPv6. The PMIPv6 standard by itself requires that the hosts that are attached to the same MAG have a point-to-point connection with the MAG and cannot talk directly to each other. WiMAX conforms to this requirement as each MS has its own private radio link connection to the WiMAX access network. The LMA allocates an address to the MS and returns it in the proxy binding acknowledgement to the MAG. The PMIPv6 tunnel used to carry the data traffic is a shared tunnel, i.e. there is a single tunnel between the MAG and the LMA that carries the traffic of all the MSs attached to that MAG.

What remains to be done at this stage is the address configuration on the MS side. For this, the MS, unaware of any PMIPv6 MM at the network side, simply makes use of the standard Internet protocol. This is DHCPv4 [85] or in the case of an IPv6 MS it can be either stateless autoconfiguration [90] or DHCPv6 [91]. Although the PMIPv6 standard [83] allows the MS to obtain both IPv4 and IPv6 connectivity simultaneously, the approach in WiMAX according to the network Release 1.5 specification is that the MS should opt either for IPv4 service or for IPv6 service, but it should not try to use both simultaneously.

The simultaneous IPv4 and IPv6 usage in the current WiMAX network Releases 1 and 1.5 is not very well defined; it is neither explicitly supported nor explicitly disallowed. This is planned to be improved by future releases of the network specifications as part of an effort to properly cover IPv4/IPv6 transition mechanisms as IPv6 is expected to gain in importance.

The PMIPv6 specification recommends use of IPsec to protect the PMIPv6 signaling between the MAG and the LMA where the security association between the two should be bootstrapped with the help of the Internet Key Exchange protocol (IKEv2) [96]. WiMAX, however, decided not to follow this recommendation and instead to recommend relying on the network domain security (NDS) concept as known from the 3GPP [114]. In contrast to mandating the use of a specific solution, this rather puts the concrete decision about deploying the security mechanisms in the hands of each individual operator. If the operators of the ASN and CSN consider their deployments secure, they can decide not to add the overhead of the additional PMIPv6 security to what is already available. In such a case the MAG and the LMA themselves are not involved in providing any security to PMIPv6 messages. However, it must clearly be spelled out that for any deployment-specific decision not making use of the available PMIPv6 security mechanisms, the risk of potentially introducing vulnerabilities fully resides with the operators themselves.

For cases where the network domain security is not available, WiMAX defined a fallback security mechanism based on the authentication protocol for MIPv6 [97]. The authentication protocol for MIPv6 is considered to be a lightweight solution when compared to IPsec. It basically mimics the authentication extensions available in MIPv4 [10]. A WiMAX-specific piece that is added to PMIPv6 is to cover the gap where no particular method is specified for how the shared secret keys needed to compute the message signatures are obtained. WiMAX defined its own way of bootstrapping those keys as part of the MS access authentication process for MIPv4-based IP mobility. The keys are individually derived based on MS authentication and the PMIPv6 messages sent on behalf of different MSs are protected by different keys unique to each MS. Thus not only is a different method used, but it also represents a more serious departure from the PMIPv6 security concept as it is based on the per MS security associations. This was actually done on purpose, and the justification for the per MS security was the compromised MAG threat as well as alignment with the existing MIP security procedures in WiMAX not to end up with a magnitude of different security mechanisms. For the former and when the per node security associations are used, a compromised MAG can cause havoc in the whole PMIPv6 domain by sending PBU messages to the LMA, claiming that other MSs are attached to it while they are in reality attached to some other MAGs. Given that the PBU messages from the compromised MAG contain a valid cryptographic signature, the LMA has no reason not to trust them and refuse the update. By accepting the PBUs and redirecting the MS traffic to the compromised MAG, the MSs are rendered out of service. Such an attack is not possible with the per MS security associations as the compromised MAG is not in possession of the individual keys of the MS that are not attached to it.

5.3.4.3 PMIPv6 Handover

While one might expect a PMIPv6 handover to look significantly different from a PMIPv4 handover because there is no separate PMIP client, this is not really the case. As it turns out,

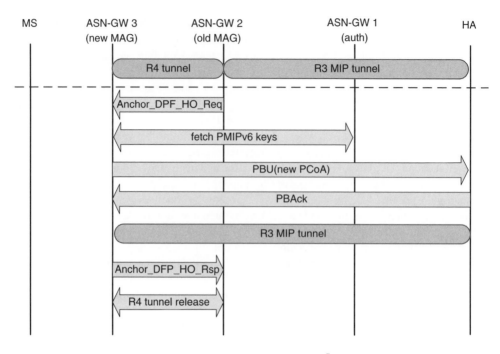

Figure 5.17   PMIPv6 handover[7]

the PMIPv6 handover procedure is very similar to the PMIPv4 one and this is apparent from Figure 5.17. After a PMIPv6 handover is triggered, the new MAG first fetches the keys from the authenticator using the same message that is used in the course of the PMIPv4 handover as well. Based on the cryptographic material that was generated during MS authentication, the authenticator derives the secret key specific to this MAG and the MS undergoing the handover and delivers it to the new MAG. The new MAG sends the PBU message to the LMA containing the MS NAI and its own address as a CoA. The PBU message is signed with the MS specific key obtained from the authenticator in the previous step. The LMA updates the CoA in its binding cache entry to the address of the new MAG and confirms the successful handover in the PBAck message. On receipt of the PBAck message the new MAG completes the handover transaction by informing the previous MAG that the handover is completed and by releasing the R4 tunnel.

Looking at PMIPv6 from a more general perspective and comparing it to PMIPv4, we can take a look at what the actual benefits of PMIPv6 are or identify what the functionality is that cannot be achieved with PMIPv4.

Of course, PMIPv6 is able to support IPv6 hosts. We have already seen how PMIPv4 was extended to provide mobility for Ethernet devices, so it could have been extended along the same lines to provide mobility for IPv6 hosts as well. The ability of PMIPv6 to use IPv6 transport on the ASN–CSN leg does not seem to play a major role in the near-term future,

---

[7] This figure shows the logical message flow and not every message that is exchanged over the wire is shown here as an individual message.

although it is clearly relevant in the long run. It is also true that PMIPv6 is not going to replace PMIPv4 very soon in WiMAX (although it would be technically possible to deprecate PMIPv4 entirely and replace it with PMIPv6). Support for PMIPv4 is widely available in products while the benefit that would justify the costs of adding support for PMIPv6 is not clear.

There is no compelling reason within the WiMAX ecosystem itself to generally deploy PMIPv6, besides the fact that PMIPv6 as an IETF standard quickly became popular with other standardization organizations as the preferred network-based mobility protocol, while PMIPv4 will remain a WiMAX proprietary solution. Indeed, the main use case for PMIPv6 in the context of the NWG Release 1.5 environment seems to be interworking with other technologies. The prominent example is to enable the WiMAX ASN to connect to the evolved 3GPP packet core (EPC) on assuming the role of a trusted non-3GPP access network from the perspective of a 3GPP operator's core network. The MM for this scenario, being in line with the 3GPP-defined architecture, is based on PMIPv6 with MAGs being located in the WiMAX access network, while the LMA is located within the 3GPP evolved core network. For details on the 3GPP architecture related to the interworking with non-3GPP access networks the interested reader is referred to [98].

5.4 ASN-Anchored Mobility

The ASN-anchored mobility in WiMAX is the MM at the link layer, i.e. the layer below the IP data path. This is the MM that takes care of the actual handover between BSs as the MS moves from one cell to another. The IEEE 802.16 standard [1] itself mainly describes how the handover in WiMAX is executed in terms of the messages exchanged over the radio link. It does not specify the architecture and supporting procedures within the backhaul network. Although the 802.16 standard does imply that the BSs talk to each other during the handover, it does not provide any details as to how this actually happens, which messages are used or which data gets exchanged. This is where the ASN-anchored MM procedures specified by the WiMAX Forum NWG step in, by providing the necessary glue and specifying how the different entities must behave and interact with each other to make the radio link handover possible in the first place.

There might be an expectation that the ASN-anchored MM must be more complex than the CSN-anchored mobility given that it has the task of taking care of the actual handover on the radio link. This is not the case. The complexity of the CSN-anchored mobility arises from the need to support various (P)MIP flavors and from the fact that it is based on standards which are not under the control of the WiMAX Forum, but are specified by other organizations like IETF. Luckily, the situation with the ASN-anchored MM in this regard is much better due to the fact that an ASN-internal procedure is being defined and therefore fully controlled by the WiMAX Forum standardization effort. This enabled the ASN-anchored MM to be designed in a much tighter way, as we will see in the rest of this section.

The ASN-anchored mobility is split into two parts: the intra-ASN mobility, also called R6 mobility, and the inter-ASN or R4 mobility, as indicated by Figure 5.18(a) and (b), respectively. (Strictly speaking, there is also the possibility of an intra-ASN R4 mobility since an ASN might include more than one ASN-GW connected via an ASN-internal reference point R4. This detail is omitted here for the sake of clarity.) During the R6 mobility, both the previous and the new BS are controlled by the same ASN-GW and the handover involves moving the R6 tunnel endpoint from the previous BS to the new BS, hence the name R6 mobility. In the case of the R4

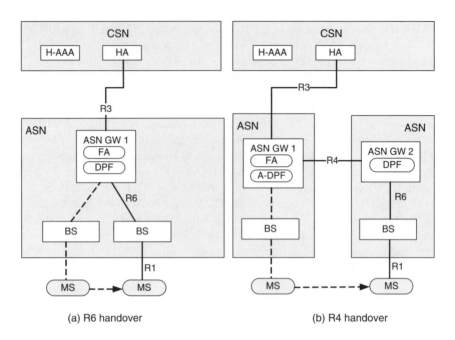

(a) R6 handover   (b) R4 handover

Figure 5.18   R6 and R4 handover configurations

handover, the new BS is connected to a different ASN-GW than the previous BS, which makes it necessary to tunnel user traffic across R4 between the ASN-GWs and also exchange state information for the MS session between the old and the new ASN-GW via R4. Hence the name R4 handover. As far as the handover signaling messages on the radio link are concerned, there are no differences between a R6 and R4 handover since the setup of the R4 tunnel remains internal to the ASN. It is transparent to the MS.

Note that the R4 tunnel scenario as in Figure 5.18(b) can be understood as a temporary situation, to be resolved on the next occasion by a R3 handover described in Section 5.3. However, there may be scenarios where a R3 handover is not recommended or may not be feasible at all; in this case the R4 tunnel will not be interim but become permanent.

5.4.1 R6 Mobility

The radio link handover is typically triggered when the quality of the connection between the MS and the serving BS drops below some threshold and another BS offering better radio link quality is discovered nearby. The handover can be triggered either by the network side or by the MS. In both cases the signaling message flow is almost the same, with some minor differences at the initial stage of the handover signaling.

In order to enable the MS to assess the link quality of the neighboring BSs, the serving BS broadcasts on a regular basis the basic information about its neighbor BSs, including their frequency and other PHY properties. This enables the MS to quickly switch to the channel served by the neighbor BS and perform measurements without having to perform a full scan over the complete frequency band. Without such help from the network side, the MS would

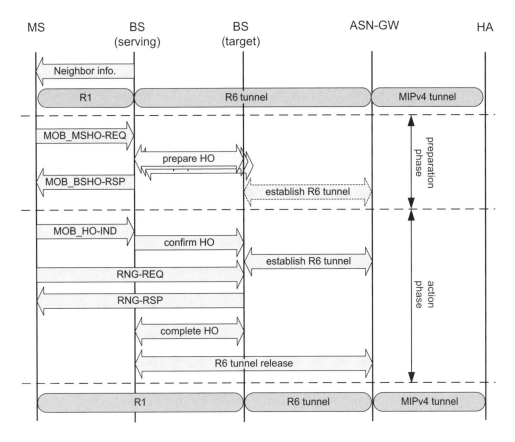

Figure 5.19   R6 handover

have to apply a full frequency scan, which would be associated with a significant amount of time. As the MS is not reachable for the current serving BS while performing the frequency scan with a neighbor BS, thus unable to receive or send any traffic, this would have a major negative impact on the quality of the handover and would make seamless handover impossible.

Generally, the radio link handover, both R6 and R4 handover, is divided into two distinct phases, the handover preparation phase and the handover action phase, that are clearly marked as such in Figure 5.19.

5.4.1.1 The Handover Preparation Phase

The handover preparation phase is initiated when the MS informs its current serving BS that it intends to execute a handover and indicates the likely target BSs for the handover. (This is for MS-initiated handover. In the case of network-initiated handover, this initial message from MS to BS would be absent.) The MOB_MSHO-REQ message that is sent by the MS to initiate the handover may contain a list of several target BSs in decreasing order of preference. The serving BS maps the MAC addresses of the target BS as received from the MS onto their respective IP addresses and sends a message to each of them to announce the possible MS arrival. (These BS to BS messages can be sent directly via the reference point R8, or, in the absence of R8, they may be carried via R6 and relayed by a relay function in the ASN-GW. For simplicity, a direct transfer is assumed here.) The message conveys the identity of the MS and the MS context. This context contains, besides many other pieces, the addresses of the authenticator ASN-GW and the A-DPF ASN-GW and the active service flows of the MS together with their QoS descriptors. The target BSs evaluate the requirements of the MS in terms of the necessary resources and respond to the serving BS informing it whether they are able to accommodate the MS or not. If a candidate target BS can accept the MS, it may already reserve some resources for the MS at this time, in preparation for the MS arrival. In particular, it may already fetch the AK context (necessary for radio link security) from the MS authenticator in order to speed up the later handover phases. It may also pre-establish the R6 tunnel with the serving gateway, though the tunnel is not operational at that time. It is in hot standby state, ready to be activated as soon as the MS attaches to the target BS. For the time being, the serving ASN-GW continues to deliver the traffic via the existing R6 tunnel to the serving BS. Having received the response from the candidate target BSs, the serving BS notifies the MS which target BSs are willing to accept it by including their MAC addresses in the MOB_BSHO_RSP message.

Features like the target BS pre-establishing a R6 data path with the serving ASN-GW, or fetching the AK context from the authenticator or not during the preparation phase, are optimizations that depend on the actual support in implementations and that need to be considered individually for any deployment. If these steps are not performed during the handover preparation phase, they must be accomplished in the later phase of the handover procedure, where they might be on a critical path and add to the total handover delay. However, allocation of the resources during the handover preparation phase will pay off only for one single BS, the one that will be selected by the MS as the final target and to which the MS will actually hand over. The other candidate target BSs will need to release the preallocated resources once the MS completes the handover at the other BS. So there is a tradeoff between reserving too many resources by each candidate target BS during the handover preparation phase versus having to reserve those resources by the selected target BS during the actual handover action phase, where this is on a time-critical path and may negatively impact the overall handover performance.

5.4.1.2 The Handover Action Phase

The handover action phase commences when the MS sends the MOB_HO-IND message to the serving BS, indicating the selected target BS. MOB_HO-IND is the last message that is sent by the MS to the serving BS. On receipt of this message the serving BS immediately stops scheduling the radio link transmission slots for the MS and stops delivering any traffic to it. The serving BS also informs the selected target BS that it was selected as the final target for handover by sending the HO_Confirm message to it. If the target BS did not preallocate the resources for the MS during the handover preparation phase, the establishment of the R6 data path and fetching of the AK context will now take place. If the data path was pre-established in the handover preparation phase, activation of the pre-established data path by switching it into the active state will be performed.

The next step is the actual arrival of the MS at the target BS. The MS announces its presence by sending the RNG-REQ message to the target BS and by identifying itself with its MAC address. The RNG-REQ message includes some special bits to indicate that the MS is executing a handover and not trying to perform an initial network entry. This is the very last moment when the target BS must acquire the AK context and establish the R6 data path to the serving ASN-GW if it has not already done so. The MS is already present and expects to be provided immediately with the connectivity. The AK context is needed as the target BS needs to verify the cryptographic signature provided by the MS as part of the RNG-REQ message, to be sure of the MS identity. The BS will do nothing before the MS identity is positively confirmed, otherwise this would pose a serious security risk. It is now also the time to establish a data path or activate the pre-established one. To accomplish this, a transaction with the serving ASN-GW is necessary, informing it that the MS actually arrived at the target BS and that the previously prepared data path is now to be activated. Note that activation of the R6 data path is typically executed on receipt of the HO_Confirm message. By doing this only when the MS had already announced its presence at the target BS, the handover delay would be unnecessarily prolonged and hence such an implementation would be suboptimal. Once the target BS has successfully verified the RNG-REQ message, it responds with the RNG-RSP message, indicating the successful handover. In the RNG-RSP message the MS is also told the CIDs allocated at the target BS and their mapping to the service flows. Thus the MS is able to resume sending and receiving the user traffic immediately on receipt of the RNG-RSP message; there is no need to execute the separate service flow establishment procedure as this is taken care of automatically as part of the handover. The sending of the RNG-RSP message is the moment when the target BS officially becomes the new serving BS.

Once the target BS sends the RNG-RSP message to the MS, it also indicates to the (now previous) serving BS with the help of the HO_Complete message that the handover transaction has been completed successfully. This triggers the previous serving BS to initiate the release of its now obsolete R6 data path and on completion of the R6 tunnel teardown the previous serving BS releases any internal resources related to the MS. This resource cleanup at the previous serving BS is the final step of the successful handover procedure.

The careful reader may have noted that there were possibly many other candidate target BSs that reserved some resources for the MS and possibly also have pre-established R6 data paths with the serving ASN-GW. As soon as the previous serving BS is notified of completion of the successful handover, it is supposed to send the HO_Confirm message to the unselected target BSs notifying them that the MS elected another BS as the new serving BS and that they can safely release their resources. If for some reason the previous serving BS does not send out this notification, the unselected target BSs will release those resources eventually on their own, based on an internal timeout period.

5.4.2 R4 Mobility

In the case of the R4 handover, the current serving BS and the target BS are associated with different ASN-GWs. The fact that the target BS is under the control of a different ASN-GW is fully transparent to the MS and, as far as the MS is concerned, the handover procedure looks exactly the same as in the case of the R6 handover. While this is true for the radio link, there are differences on the network side as there is a need for the R4 tunnel between the ASN-GWs to carry over the user traffic between the ASNs. We will focus on those differences in this section when discussing the R4 handover and not mention again the parts that are common to R6 and R4 mobility.

5.4.2.1 R4 Handover Procedure

When the MS scans the radio channels to gather the possible target BSs, it cannot tell whether the particular target BS belongs to the same ASN as its current serving BS or to a different one. The MS initiates the handover as usual, indicating all promising BSs in terms of signal quality as possible targets in the MOB_MSHO-REQ message. The serving BS proceeds to notify the target BSs as usual. The first difference in the message flow is notable when one of the target BSs from another ASN attempts to pre-establish a R6 data path with the ASN-GW. It does not establish the R6 data path directly with the serving ASN-GW of the MS. Instead it goes to the default ASN-GW that it was preconfigured with and this is the ASN-GW that is in control of the target ASN and not the ASN-GW controlling the serving BS. As part of the R6 data path setup transaction the target BS signals to its default ASN-GW that the data path is being established due to the handover and provides it with the MS-ID. The ASN-GW looks up the MS in its memory based on the received MS-ID and since it cannot find any context related to the indicated MS, it correctly concludes that this MS must be coming from another ASN, i.e. an R4 handover is taking place. In such cases it helps that the R6 data path transaction also conveys the address of the anchor ASN-GW (A-DPF) of the MS, which is the ASN-GW that terminates R3 for this MS and is the tunnel end point for the MIP. The target BS received the address of the A-DPF ASN-GW as part of the handover notification (HO_REQ) from the serving BS. At this point, the ASN-GW is in possession of the address of the A-DPF ASN-GW and may pre-establish the R4 data path to the A-DPF ASN-GW.

The handover procedure then continues as usual: the target BSs respond to the serving BS indicating whether they are able to accept the MS, the serving BS compiles the answers gathered from the target BSs and notifies the MS of the available target BSs. The MS proceeds to select the target BS and indicates its choice to the serving BS. The serving BS in turn notifies the selected target BS in the other ASN that it has been selected as the target for the handover. At this point the target BS must assure that the data path for carrying the MS traffic is correctly set up and operational. The target BS either initiates the R6 data path setup or activates the pre-established data path. The activation of the R6 data path triggers the ASN-GW in the target ASN to activate the R4 data path to the A-DPF ASN-GW. This completes the data path setup and the data traffic that enters the WiMAX access at the A-DPF ASN-GW now traverses the R4 data path to the new serving ASN-GW and from there it takes the usual route via the R6 tunnel to the new serving BS.

What remains to be done is to notify the previous serving BS about completion of the handover so that it can initiate the teardown of the R6 data path and release any other resources including the unselected target BSs – this is again the same as already discussed in the course of the R4 handover.

The resulting data path configuration is visible at the bottom of Figure 5.20. What is really new here is the R4 data path that was inserted between the R6 data path and the MIP tunnel. We have discussed the pros and cons of such a design earlier in Section 5.2, and we have seen how the CSN-anchored MM may serve to remove the R4 tunnel and restore the optimal traffic path within the ASN.

In Figure 5.20 we also see that at the start of the handover the A-DPF ASN-GW, or in short the anchor ASN-GW, is simultaneously the serving ASN-GW. That is, besides terminating the R3 tunnel, it also terminates the R6 data path to the current serving BS. However, in the most

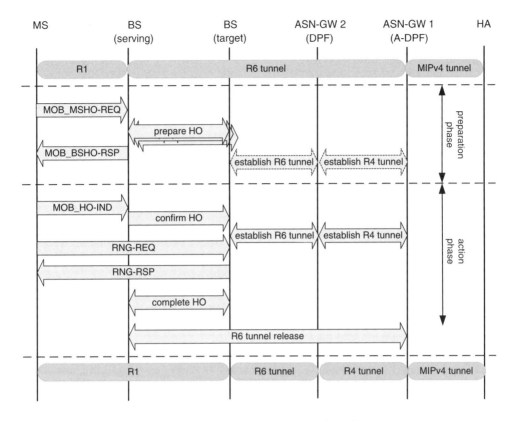

Figure 5.20 R4 handover (MS initiated)

generic case, the serving ASN-GW and the anchor ASN-GW need not be the same at the time when handover is initiated and such configuration is discussed in the next section.

5.4.2.2 Interaction between R4 Mobility and CSN-Anchored Mobility

Before we conclude this section on R4 handover, let us briefly consider the case where the serving ASN-GW (before handover) is not the A-DPF ASN-GW. This is the situation shown at the end of Figure 5.20 where ASN-GW2 is now the serving ASN-GW. It means that another R4 handover is triggered before the CSN-anchored mobility succeeded in removing the R4 tunnel and restoring the optimal data path. Such an event can of course happen and this is a perfectly legal situation. In turns out that ASN-anchored MM can cope with such a situation quite well. The naive way would be to simply add yet another R4 tunnel extending from the previous target ASN to the new target ASN. But such concatenation of the R4 tunnels would have detrimental effects on the overall performance as more hops are involved, which leads to longer delays and increased traffic round-trip time. The trick deployed by the ASN-anchored MM is simple: the new target ASN-GW does not establish the R4 tunnel to the current serving ASN-GW (the one that was the target in the course of the previous R4 handover) but instead it establishes the R4 tunnel directly to the A-DPF ASN-GW. On completion of the handover, the step where the R6 data path in the previous serving ASN is torn down is extended to include release of the R4 tunnel between the previous serving ASN-GW and the A-DPF ASN-GW. So no matter how many consecutive R4 handovers are executed, there is no chaining of the R4 tunnels and the final outcome after each R4 handover is that there is only a single R4 leg with the complete data path being composed exactly as shown at the bottom of Figure 5.20: R1–R6–R4–R3.

5.4.3 Unprepared Handover

Conditions on the radio link, due to its nature, can change abruptly. The MS that may have an acceptable connection quality at one moment may find itself in the next moment in a situation without a usable radio link connection. Once we acknowledge the fact that radio link quality can change drastically within a very short time, it becomes clear that there are cases that need to be supported where the MS and the network lack enough time to prepare the radio link handover as discussed in the previous sections. So let us look at what happens in cases where the MS loses connectivity with the serving BS prior to being able to complete the handover preparation signaling. Clearly, full new network entry would imply that the MS loses its communication sessions and is therefore not an option. Fortunately, even in such cases the MS is still able to execute a controlled handover. The handover may not be as smooth as the one that was prepared in advance via the serving BS, but the performance degradation is expected to be relatively low in comparison to the interruption caused by a new network entry.

In the case of unprepared handover the MS simply appears at the target BS without prior notice to the target BS of its arrival via the serving BS. In such cases the MS will provide the MAC address of its previous serving BS as part of the RNG-REQ message that is sent to the target BS. The target BS identifies the previous serving BS based on this MAC address with the help of its default ASN-GW and retrieves the MS context from the serving BS. The MS context contains all the relevant MS data of the existing session, such as the addresses of the authenticator ASN-GW, the A-DPF ASN-GW, a description of the service flows, etc. With this information, the target BS is able to quickly establish the missing pieces like the R6/R4 tunnels and to update the other network entities on the new MS location. The overall handover delay may be slightly longer than in the case of the prepared handover as the MS context is fetched from the previous serving BS only upon MS arrival at the target BS and the rest of the resources are set up following this step. Nevertheless, this still allows for a fairly acceptable handover performance.

5.5 Simple IP

Mobile WiMAX has been designed from the outset with full mobility support, but clearly there are deployment scenarios where WiMAX technology is used to provide service for mostly fixed or nomadic devices, as in the case of a wireless alternative to DSLs. Deploying WiMAX technology for such scenarios becomes especially attractive if there are variants that allow the overall deployment and maintenance costs to be reduced. WiMAX targets this requirement by allowing network operators to avoid deploying selected equipment or functions that are mostly required to serve mobile devices, like the HA, but still be in line with the WiMAX standards. One of many examples is deployment in sparsely populated rural areas where the costs of laying copper cable far exceed the possible revenue, but also deploying a wireless alternative needs to be carefully calculated due to the limited total number of potential customers. Yet WiMAX may prove to be a viable and cost-effective option for such deployments with such a stripped-down, low-cost core network.

5.5.1 HA-less Architecture

Deploying mobility as part of the system is fine as long as it does not incur any extra costs. This concern does not seem to apply to the ASN part of the system as the BSs and the ASN-GWs are needed anyway, regardless of whether the operator will provide mobility to the customers or not. However, the situation within the CSN is different. In the case where WiMAX is used to provide fixed access service there is no compelling need to deploy and administer a (P)MIP HA in the CSN and a (P)MIP FA in the ASN. (Of course, if the CSN is serving both mobile and non-mobile users, it will need to provide a MIP service on the link to the ASN anyway. If an ASN connects to such a CSN, it may have to include an FA even if the ASN is used for stationary or nomadic users.) In general, it would be unwise to force operators not requiring IP mobility services to buy HAs and install them in their networks without a business need, just because of specification restrictions.

For such deployments, the NWG standard starting with Release 1.5 supports a simple IP mode for R3 operation. Simple IP is an optional feature that – in contrast to the NWG Release 1 specifications that are fully based on the MIP across R3 – in turn makes MIP-based mobility across R3 optional. Simple IP is a WiMAX system that is able to function without the HA in the CSN, so it is sometimes referred to as the HA-less architecture.

In the simple IP architecture shown in Figure 5.21, the endpoint handling the user traffic at the CSN side is no longer an HA. In simple IP the HA is replaced by a core router (CR) that is responsible for delivering the user traffic to and from the ASN. Given that there is no HA, there is also no FA on the ASN side either, so the functional entity terminating the R3 reference point within the ASN-GW is termed the access router (AR).

5.5.2 Unspecified R3 Data Path

As we know, MIP is the protocol to interconnect the ASN and CSN when looking at the user traffic. In the HA-less architecture there is obviously no MIP available and an alternative tunneling method is required. This protocol is left unspecified on purpose. This may seem strange at first glance – after all, standards are about reducing the multiplicity of the options and preferably making a single default protocol available that has support in all affected products and in turn creates good interoperability between different vendors.

So why did WiMAX leave this open and decide not to specify a default tunneling method for the simple IP deployments? The reason is that it would cause undue restriction. There are numerous solutions on the market with more or less significant deployment footprints, and picking any one of them as a default would mean negatively impacting different parts of the WiMAX installations. It showed that operators were not too concerned about the lack of specification on this specific aspect of non-mobile deployments based on simple IP, because in many cases both the ASN and the CSN are expected to be owned by the same operator. While the architecture allows for separation of the ASN operator (NAP) and the CSN operator (NSP), as for the MIP-based architecture, and even allows for roaming scenarios with a visited CSN in

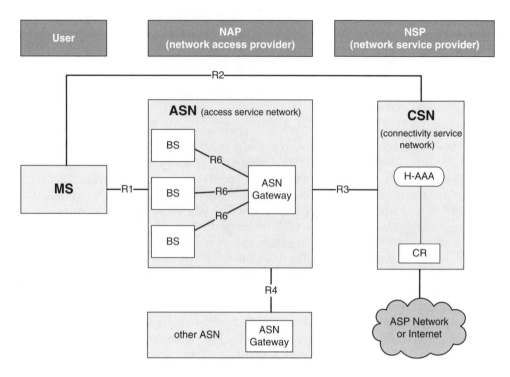

Figure 5.21 Simple IP architecture

between, these are rather theoretical deployment choices inherited from the underlying Mobile WiMAX reference architecture and are not expected to have much practical relevance.

In summary, WiMAX does not provide any specific means or hints (and therefore does not introduce any specific limitations) on how the data path between the ASN and the CSN should be set up in simple IP-based deployments. Even the wording was chosen carefully: the specification does not mention the word 'tunneling' in the context of simple IP so as not to imply any particular solution.

Interestingly, in a simple IP deployment a good degree of mobility is still possible. ASN-anchored MM is still possible, and this includes not only the R6 handover but also the R4 handovers, so it may well cover a large geographical area. What is not possible is to dynamically relocate the R3 termination point at the ASN side as the MS moves and changes its point of attachment to the network. For this the (P)MIP variants with HA functionality in the CSN need to be chosen for any standards-compliant system. Note that with simple IP the only way to remove an R4 tunnel spanning different ASN-GWs that was created after MS movement is to force the MS to perform full initial network entry at the new location, with the consequence that the MS loses its previous sessions.

5.6 Mobility Restriction

The simple IP features enable operators to save some costs by not having to buy and operate a HA as part of their CSN, but, as we can see, the HA-less design does not automatically drop mobility, it simply restricts mobility to the extent of the ASN-anchored MM. Depending on the particular deployment and network configuration, an ASN still enables handovers and session continuity and can potentially cover a large geographical area.

There are deployments where any form of mobility is not desirable, either for cost reasons or because it is explicitly prohibited. Such stringent exclusion of any form of mobility is usually governed by the regulator and the license terms (which again comes down to deployment cost) under which the WiMAX operator obtained the rights to use the spectrum. In such cases the operator is not allowed to offer services in a mobile fashion.

The non-mobile deployments are usually known as stationary networks, but they actually come in two different forms. In the case of *fixed access*, the subscriber is allowed to attach to the network only at a single geographical location, e.g. at home. In the case of *nomadic access*, the subscriber can attach to the network at any place without restriction, but once attached the device will be denied any handover to another BS outside a clearly defined group of BSs (that in the trivial case consists of just a single one). It remains fixed at the place of attachment for the duration of the whole WiMAX session with the network. A nomadic subscriber may terminate the current session, move to another location and reattach to the network from there if the location is allowed by the subscription policy, but at the price of losing any ongoing communication sessions and having to start them anew.

5.6.1 Dynamic Limitation of Attachment and Handover

The WiMAX standard introduces explicit support for such fixed and nomadic deployments as an optional feature of the Release 1.5 network specifications. When looking at the technical solution, this can mostly be considered as a dynamic limitation of mobility support in the network based on either the operator's overall policy or specific limitations applying to individual subscribers, or devices. As a result, the necessary additions to the baseline standard to support fixed and nomadic networks (as well as the resulting implementation effort for vendors) are actually quite small.

5.6.1.1 Subscriber Profiles

In the case of limitations to an individual subscription, the subscriber profile in the home AAA server flags the MS as being authorized either for fixed or for nomadic service (or for full mobility that is also assumed in case there is no such flag, which just maps to normal or, better, unrestricted WiMAX operation). When the subscriber is authorized for fixed service only, the profile also contains an indication of the geographical location where the MS is allowed to obtain service. This information is made available to the ASN as part of the initial subscriber authentication.

For any fixed subscriber, the ASN-GW compares the location given in the subscriber profile with the location of the BS where the MS is trying to enter the network, and if the two do not match, the ASN-GW denies access and instructs the BS to reject the MS's attempt to enter the network. As an alternative, the AAA server can at the time of initial network entry evaluate the BS's identity where the MS is entering the network and deny network entry itself based on limitations of the actual subscription.

Nomadic subscribers may attach to the WiMAX network from any place that is allowed in the subscription profile, but are not allowed to hand over to another BS (leaving aside a potential set of neighboring BSs that the operator may hand over to for load balancing reasons). This is easily implemented by letting the serving BS know that the subscriber is authorized for the nomadic service only. In this case the serving BS rejects any attempt by the MS to execute a handover. Alternatively, the authenticator in the ASN-GW can verify whether the target BS asking for the AK context of the device in the course of a handover is an allowed destination. If not, the authenticator simply prevents the handover by denying the request for the AK context. The standard supports both alternatives, with the authenticator-based enforcement being the default that is mandatorily implemented.

One reason behind the decision to make the authenticator the default entity to enforce such handover decisions in mobility-restricted deployments is that standardized support for this is only available as an optional feature in the Release 1.5 WiMAX network specifications. For existing deployments, it is therefore much easier to roll out this feature by upgrading only the ASN-GWs in a first step but not require modifications to all the BSs. The second step in upgrading BSs to support mobility restrictions in the standardized way can then be considered as an additional performance optimization that might only be required for specific deployment needs. Also, operators might deploy BSs from different vendors in their network, where not all of them support or enforce fixed or nomadic services, and again it would be a lot easier to upgrade the ASN-GWs only.

5.6.1.2 Reattachment Zone

Besides applying any mobility restrictions, operators may want to redirect a MS from one BS to a neighboring one for the purposes of better signal quality, load balancing or due to the need for maintenance of the BS. This actually means that the handovers are allowed as long as they are used exclusively for these purposes and not because the MS has moved. To enable such handovers to a limited set of BSs, the network can keep as part of the MS context a list of BSs to which the MS is allowed to hand over, and any handover request is allowed to proceed as long as the target BS is on this list. The list is called the reattachment zone and is typically initialized at the time of network entry. It typically contains the immediate neighbor BSs of the BS where the MS attached to the network, thus restricting handovers to the immediate vicinity of the location where the MS entered the network. Alternatively, a different list of BSs can be stored against the subscription profile and used to allow specific cases like nomadic services covering, for example, a large hotel complex or campus that is served by a limited set of BSs. Here, the list of allowed BSs would cover exactly those BSs serving the hotel or campus. In this case, the list of allowed BSs can also be communicated by the network to the MS as a guide to allow more accurate handover decisions in the MS.

6

WiMAX Radio Interface

6.1 Physical Layer

The physical (PHY) layer of Mobile WiMAX is designed to support high-speed broadband wireless networks with the inclusion of advanced radio features such as OFDM multiple access, multiple antenna (MIMO), advanced coding and modulation, fast feedback and hybrid-ARQ. It also supports scalable channel bandwidth from 1.25 to 20 MHz in various worldwide spectrum allocations such as 2.3, 2.5 and 3.5 GHz. The mobile system profile, which describes the functional features of Mobile WiMAX certified equipment, is built on the IEEE 802.16 standard. The system profile contains sets of mandatory and optional features and also serves as a toolbox allowing network operators to customize their Mobile WiMAX systems according to individual needs.

6.1.1 Overview of OFDMA

Orthogonal frequency division multiplexing (OFDM) has become a preferred modulation technique for advanced wireless communication systems such as digital audio broadcasting (DAB), IEEE 802.11a/g/n (Wi-Fi), IEEE 802.15.3 (Ultrawideband), 3GPP LTE and IEEE 802.16 (WiMAX). This is mainly due to its bandwidth efficiency and robustness to signal distortions and multipath fading resulting from mobility and wireless channels. The concept of OFDM relies on parallelization of data into a number of orthogonal streams or subchannels. The orthogonality among subchannels is maintained by a simple mathematical transform that can be efficiently implemented using the widely available fast Fourier transform (FFT) algorithm. Since each OFDM subchannel essentially occupies a fraction of the bandwidth and carries data at a rate much lower than that of a single carrier system with the same bandwidth, the frequency-selective channel is converted into a number of frequency non-selective channels. Hence, complex channel equalization is no longer necessary. In addition, a cyclic prefix, which uses a fraction of transmission bandwidth to append a cyclic replica to its own signal, can be used as a guard interval to alleviate the effects of the channel delay spreads and intersymbol interference at a small loss of bandwidth efficiency. Figure 6.1 illustrates the use of a cyclic prefix against the multipath components of the received OFDM symbol. Furthermore,

Deploying Mobile WiMAX   Max Riegel, Dirk Kroeselberg, Aik Chindapol and Domagoj Premec
© 2009 John Wiley & Sons, Ltd

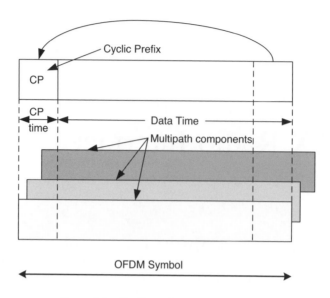

Figure 6.1   Cyclic prefix and OFDM symbol

some of the subcarriers may be nulled in order to provide additional guard bands and reduce interference to and from other systems operating in the adjacent frequency bands. In addition to channel coding, which provides additional protection against transmission errors, interleaving is normally used to spread the signal distortion over frequency and time and further increases the efficiency of channel coding.

Orthogonal frequency division multiple access (OFDMA) is one variant of OFDM, in which multiple users occupy non-overlapping sets of frequencies. In other words, the OFDM subcarriers spanning the allocated spectrum are divided and grouped into a number of subchannels that can be assigned to multiple users, or even one user if necessary. Therefore, unlike OFDM, several users in OFDMA can be scheduled to transmit or receive data simultaneously. Flexibility in allocating resources with fine granularity has been one of the major advantages of OFDMA over other multiple-access technologies. For example, the usage of radio resources as well as the level of frequency and time diversity may be dynamically adjusted to reflect the current demand for bandwidth and channel conditions at any given time. In the Mobile WiMAX System Profile Release 1.0 [146], the essential OFDM parameters such as FFT size, subcarrier spacing, cyclic prefix and nulled symbols were carefully chosen to match the expected operating scenarios such as vehicular mobility, 2.5–3.5 GHz carrier frequency and moderate delay spread [148]. Table 6.1 summarizes the OFDM parameters for WiMAX.

6.1.2 Frame Structure

Figure 6.2 shows the frame structure for Mobile WiMAX with time division duplexing (TDD) allowing the downlink and uplink transmissions to share the same transmission medium. Due to the prohibitively complex filtering necessary to separate the uplink and the downlink, transmission and reception cannot occur simultaneously. In addition, further time gaps such

Table 6.1 Mobile WiMAX physical layer parameters

Parameters	Value			
Channel bandwidth (MHz)	1.25	5	10	20
FFT size	128	512	1024	2048
Sampling frequency (MHz)	1.4	5.6	11.2	22.4
Subcarrier frequency spacing (kHz)		10.94		
Data symbol time (μs)		91.4		
Guard time (μs)		11.4		
Frame size (ms)		5		
Number of OFDM symbols		48		

as the transmit/receive transition gap (TTG) or receive/transmit transition gap (RTG) are needed between the subframes in order to avoid interference between the signals transmitted from the BS and other MSs. The TTG allows sufficient time for the BS to switch from the transmit to receive mode while the RTG provides time for the BS to switch from receiving to transmitting. The size of these time gaps depends on the cell radius, the transceiver turnaround time and other implementation constraints.

The frame is composed of a number of OFDM subcarriers (in frequency) and symbols (in time) forming the downlink and uplink subframes. The number of OFDM subcarriers and symbols in the frame depends on the bandwidth and frame size. At the very beginning of the downlink subframe, the first OFDM symbol contains the preamble with a known and repetitive pattern indicating the start of each frame. Preambles are designed to have good cross-correlation and low peak-to-average power ratio (PAPR) properties, which increases the probability of detection while minimizing interference to other BSs transmitting their own

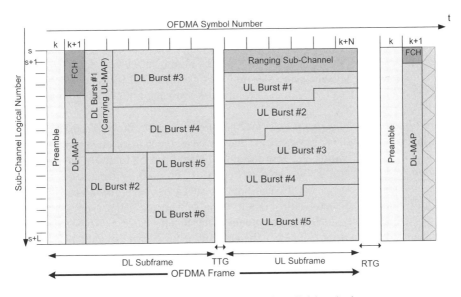

Figure 6.2 Frame structure with time division duplex

preambles in close proximity. Frame-level synchronization is commonly required in the cellular networks in order to avoid interference between the uplink and downlink transmissions. Furthermore, the Mobile System Profile Release 1.0 [146] imposes strict frequency and timing synchronization requirements. Therefore, the Mobile WiMAX preambles can be transmitted simultaneously without interfering with the data portion of the frame. A MS can use the preamble to synchronize the timing and frequency parameters with the selected BS. In addition, since the preamble is unique for each BS within the same geographical area, it can be used to identify the BS during network entry or handover. Immediately following the preamble, the BS broadcasts basic control information that is necessary for network entry and handover. Such information includes frame configurations, modulation and coding schemes, resource mapping, and uplink transmit opportunities for synchronization and network entry. The rest of the downlink subframe is used for other management messages and data bursts. Similarly, the uplink subframe is divided into multiple regions for data bursts and management signaling such as acknowledgements, synchronization and feedback.

6.1.2.1 Subcarrier Allocation

In order to best serve users with different channel conditions and requirements, Mobile WiMAX allows different methods that map the logical resource allocation to the actual physical OFDM symbol allocation. Depending on the mapping, the OFDM subcarriers and symbols are grouped to form a subchannel, which can be assigned to each individual MS. The mapping techniques can be generally categorized into distributed and localized subcarrier allocations:

- **Distributed subcarrier allocation**: With the distributed subcarrier allocation, subcarriers forming a subchannel are randomly distributed throughout the allocated frequency to maximize frequency diversity. In addition, intercell interference is controlled by the careful design of the permutation sequences that minimize the probability of having the same set of subcarriers being used in the nearby cells. This allocation method is suitable for mobile environments where the channel conditions change very quickly. Although the data subcarriers are spread throughout the allocated band, the subcarriers designated as pilots may be common for all subchannels or they may be dedicated for each subchannel. The following are examples of subcarrier permutations available in the Mobile System Profile Release 1.0 [146]:
  - *Partially used subchannelization (PUSC) permutation:* In the downlink, the partially used subchannels are grouped into segments that can be assigned to different sectors within the same cell. In other words, the entire frequency band can be logically and dynamically divided to form a sector. Frequency diversity is achieved by the use of permutation sequences designed to randomly distribute the subcarriers over the allocated bandwidth while minimizing the probability of the same subcarriers being used in the adjacent sectors or cells. Since the IEEE standard mandates that the basic control signaling that the MS needs to obtain prior to entering the network be broadcast using the PUSC permutation, the ability to transmit and receive bursts with the PUSC permutation is mandatory for all Mobile WiMAX equipment.
  - *Fully used subchannelization (FUSC) permutation:* Similar to PUSC, FUSC uses a permutation sequence to achieve full frequency diversity with fixed and variable sets

of pilots that can be adjusted to trade off the allocated power and frequency diversity of pilots. The FUSC permutation applies to the whole allocated frequency without logical segments.

- **Localized subcarrier allocation**: In the scenarios where the need for frequency diversity can be relaxed (e.g. low or fixed mobility users), adjacent subcarriers may be grouped together to form subchannels with their own dedicated pilots. The combination of operating scenarios (i.e. low mobility), subcarrier grouping, pilot allocation and fast channel feedback enables a quick adjustment of modulation and coding since the BS has knowledge of localized channel and interference conditions during the channel coherent time. In addition, the BS scheduler can utilize knowledge of channel conditions from active MSs and dynamically select the best subchannel for each MS. This technique is also known as 'water pouring' or frequency-selective scheduling. The following is an example of localized subcarrier allocation from the Mobile System Profile Release 1.0 [146]:
  - *Advanced modulation and coding (AMC) permutation:* With AMC, each subchannel contains sets of adjacent data subcarriers in frequency with embedded pilots. Different groupings of adjacent subcarrier sets allow flexibility in channel estimation. For example, subchannels may contain adjacent subcarriers in either frequency or time, or both time and frequency.

Since the permutation sequences rearrange the OFDM subcarriers to provide frequency diversity or frequency selectivity, they cannot be used simultaneously over time. In order to switch from one permutation method to another, a number of contiguous OFDM symbols sharing the same permutation are grouped together to create a logical zone. The zone boundaries are indicated in the information elements contained in the resource allocation mapping (DL-MAP and UL-MAP) messages. The downlink and uplink subframes may contain more than one zone allowing different types of MSs with different permutation requirements to coexist within the same frame. Since each zone can be configured with different layer parameters such as permutation or MIMO selection mode, the actual radio conditions may vary from one zone to another. Because of the way pilots are allocated, the AMC zone is normally placed at the end of the frame. Figure 6.3 shows the zone switching procedure as indicated in the resource allocation mapping message (DL-MAP and UL-MAP). The highlighted zones are optional and may not appear in every frame. Note that basic control signaling necessary for network entry and normal operations are always transmitted in the first zone with the PUSC permutation.

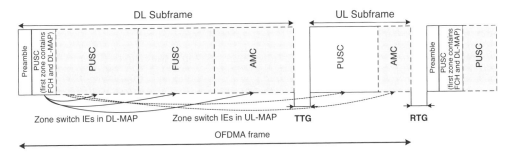

Figure 6.3   Multi-zone frame structure

6.1.2.2 Modulation and Coding

In order to provide the error correction capability, the IEEE 802.16 standard [1] specifies a number of modulation and coding options that can be used in conjunction with subcarrier allocations, link adaptation, retransmissions and power control. The basic block diagram for modulation and coding includes randomization, forward error correction, interleaver and modulation. The randomizer is applied on all PHY bursts, except the frame control header, in both the uplink and downlink directions. Not only does it scramble data from the higher layer, but it also increases the efficiency of channel coding. The tail-biting convolutional code and convolutional turbo code are the only two mandatory coding schemes defined in the Mobile System Profile Release 1.0 [146]. Different code rates are available through puncturing of the mother code. Prior to the modulation block, bit-level interleaving is performed on the stream of coded data to provide much needed diversity for transmissions over fading channels. Lastly, three modulation schemes, quadrature phase shift keying (QPSK), 16-ary and 64-ary quadrature amplitude modulation (QAM), can be dynamically combined with channel coding to create a balance between robustness to channel errors and bandwidth efficiency. For essential control messages such as the frame control header and the downlink and uplink MAP, repetition coding can be used to increase the link margin necessary to reach MSs in the low-signal area or cell edge.

Another important modulation and coding feature is hybrid automatic retransmission request (H-ARQ). It effectively increases time diversity and robustness of channel coding by transmitting incremental redundancy for data bursts requiring retransmissions. The receiver then tries to decode the packet by combining multiple received versions of coded data. Since H-ARQ operates within the link layer, it provides fast retransmissions and powerful error correcting capabilities.

6.1.2.3 Multiple-Antenna Systems

In addition to adaptive modulation and coding and different subcarrier permutations, Mobile WiMAX supports a number of multiple-antenna technologies, which can enable significantly higher throughput than a conventional single-antenna system. Space–time coding is one form of multiple-antenna technologies, in which the correlated signals are transmitted in both the spatial and temporal dimensions. The simplest example is the use of the well-known Alamouti code [151], which achieves better spectral efficiency than other systems using the same bandwidth at no extra cost. Multiple input, multiple output (MIMO) allows the use of multiple antennas at both the receiver and transmitter.

Mobile WiMAX based on the system profile Release 1.0 [146] supports the following multiple-antenna technologies:

- **Space–time block coding (STBC)**: STBC can be efficiently implemented with an OFDM system. The modulated symbols are simply encoded into space–time codes immediately after constellation mapping and prior to the inverse FFT operation. Then subcarrier permutation is performed independently for the signals belonging to each antenna.
- **Cyclic delay diversity (CDD)**: Instead of transmitting space–time-encoded signals from multiple antennas, the transmitter sends the cyclically shifted versions of the same signal over multiple antennas simultaneously. This allows for extra spatial diversity without adding much complexity to the receiver.

- **Spatial multiplexing MIMO**: In addition to diversity achieving techniques such as space–time coding and CDD, spatial multiplexing MIMO encodes multiple data streams and then splits the signal over multiple antennas. This allows significant throughput enhancements at the cost of reduced diversity and receiving complexity, which is similar to a decision feedback equalizer [152]. Collaborative spatial multiplexing is one important application of spatial multiplexing MIMO, which allows two single-transmit antenna users to transmit their own signals over the same set of OFDM subcarriers simultaneously.

6.1.2.4 Frequency Reuse

Traditionally, cellular networks assign different frequency allocations in sectors or cells to separate interference among them. For example, the reuse factor of three is quite common and indicates that three independent frequency allocations are available in the network. The major drawback of such conservative frequency reuse is that the total system capacity is reduced as the whole bandwidth is divided to provide interference protection. In a CDMA system such as UMTS, however, the single reuse factor can be applied since the CDMA spreading codes provide interference separation for both users within and between the cells. In Mobile WiMAX where OFDMA is used, the frequency reuse of one that allocates the same frequency to all sectors within the cell can be easily supported in addition to a conventional frequency reuse of greater than one. This is due to the fine granularity and different permutation sequences that allow flexibility in allocating OFDM data to users. Although the single reuse factor greatly alleviates the burdens in frequency planning, interference from other cells sharing the same subcarriers may be unpredictable and causes unpleasant user experiences.

Mobile WiMAX additionally supports fractional frequency reuse, which combines the benefits of simplicity from the single reuse factor and better interference separation from a less aggressive reuse factor. Specifically, it allows a higher reuse factor for users at the cell edge that normally experience the higher level of interference from neighboring cells while maintaining the single-frequency reuse for users who are well within the cell coverage and have distance as another degree of interference protection. This flexibility can be easily achieved with the use of multiple permutation zones and zone switching. Moreover, coordination with neighboring cells over the backbone is often performed to ensure the minimal level of interference. For example, users suffering from high interference may be switched to a coordinated PUSC permutation zone and switched back to another permutation zone such as FUSC or AMC when the interference level is lower (i.e. users move inside the cell). This switching occurs without the need for handover or network re-entry. Figure 6.4 illustrates the use of different zones to switch between fractional frequency reuse and the reuse factor of one.

6.1.2.5 Summary of Key Profile Features

A summary of the key features is given in Table 6.2.

6.2 MAC Layer

The MAC functionalities of Mobile WiMAX are contained in three essential sublayers: the service-specific convergence sublayer, the MAC common part sublayer and the security sublayer as shown in Figure 6.5. The service-specific convergence sublayer (CS) is responsible

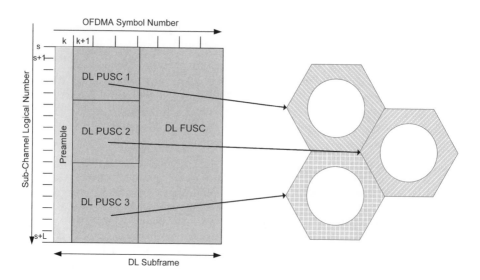

Figure 6.4   Fractional frequency reuse with partially used subcarrier permutation

for classifying a higher layer protocol data unit (PDU) from an external network into an appropriate MAC service data unit (SDU) and assigning the QoS parameters and identifier associated with the SDU. The 802.16 MAC PDU contains the header and payload that are formatted according to the IEEE 802.16 standard and suitable for over the air transmissions. Currently three CSs are specified in the IEEE 802.16-2009 standard [1]: the asynchronous transfer mode (ATM) CS, the packet CS and the generic packet CS (GPCS). Figure 6.6 shows an overview of CS operations in both the uplink and downlink. The MAC common part sublayer (MAC CPS) provides the supporting functionality to handle radio and network resource access, management of bandwidth, QoS, and connection establishment and maintenance. It exchanges classified data with the CS and manages the over the air transmission according to the QoS parameters and identifier provided by the CS. These QoS identifiers such as the service flow identifier (SFID) and connection identifier (CID) are crucial to the QoS support of Mobile WiMAX. Lastly the security sublayer provides subscribers with privacy and authentication by applying cryptographic transforms to MAC PDUs carried across the connections. While the credentials of WiMAX users are authenticated as part of the core network services (see Chapter 3), the focus of the security sublayer is to provide additional protection to link-level messages. Mobile WiMAX supports cipher-based message authentication code (CMAC) and hashed message authentication code (HMAC).

In Mobile WiMAX, the BS is the central entity that controls and schedules access to system resources for all terminals attached to its network. With the point-to-multipoint topology, the BS uses the broadcast control messages to announce the scheduled allocations for both the uplink and downlink. In the downlink, the BS is the only transmitter in this direction to multiple MSs. All MSs scheduled to be active during this portion of the downlink subframe need to listen and process the control messages and 802.16 PDUs addressed to them. On the other hand, the BS is the only receiver for the transmissions of multiple MSs scheduled to transmit during this uplink subframe. To maximize the utilization of uplink resources, the BS allocates transmission opportunities to the MSs according to the agreed level of QoS, requests for transmission opportunities from MSs and

Table 6.2   Physical layer key profile features [147]

Category	Feature	BS support	MS support
Subcarrier allocation	Downlink PUSC	Y	Y
Subcarrier allocation	Downlink PUSC with dedicated pilots	Optional	Y
Subcarrier allocation	Downlink FUSC	Y	Y
Subcarrier allocation	Downlink FUSC with dedicated pilots	N	N
Subcarrier allocation	Downlink AMC with 2×3 allocation	Y	Y
Subcarrier allocation	Downlink AMC with 2×3 allocation, dedicated pilots	Optional	N
Subcarrier allocation	Uplink PUSC	Y	Y
Subcarrier allocation	Uplink AMC with 2×3 allocation	Y	Y
Modulation/coding	Repetition	Y	Y
Modulation/coding	Randomization	Y	Y
Modulation/coding	Tail biting convolutional code (CC)	Y	Y
Modulation/coding	Convolutional turbo code (CTC)	Y	Y
Modulation/coding	Block turbo code (BTC)	N	N
Modulation/coding	Low-density parity check (LDPC) code	N	N
Modulation/coding	Interleaving	Y	Y
Modulation/coding	Downlink QPSK	Y	Y
Modulation/coding	Downlink 16-QAM	Y	Y
Modulation/coding	Downlink 64-QAM	Y	Y
Modulation/coding	Uplink QPSK	Y	Y
Modulation/coding	Uplink 16-QAM	Y	Y
Modulation/coding	Uplink 64-QAM	N	N
Modulation/coding	H-ARQ: Chase combining with CC	N	N
Modulation/coding	H-ARQ: Chase combining with CTC	Y	Y
Modulation/coding	H-ARQ: Chase combining with LDPC code	N	N
Modulation/coding	H-ARQ: Incremental redundancy with CC	N	N
Modulation/coding	H-ARQ: Incremental redundancy with CTC	N	N
Modulation/coding	LDPC code	N	N
Modulation/coding	Interleaving	Y	Y

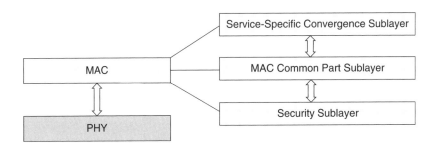

Figure 6.5   MAC layer division

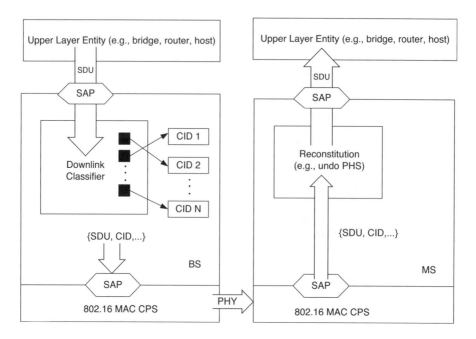

Figure 6.6   Overview of convergence sublayer

scheduling constraints. Only the MSs that were informed of their uplink allocations in the particular uplink subframe are allowed to transmit, as indicated by the relevant control messages from the BS. A direct communication between MSs is not possible.

Given the granularity and flexibility of OFDMA resource allocation in Mobile WiMAX, the radio resources can be allocated in both time and frequency. In addition, all dedicated resource allocations within a frame are considered orthogonal and do not cause intracell interference. For the MSs that do not have dedicated resources, such as those trying to enter the network or requiring additional bandwidth, contention-based uplink transmit opportunities are also provided. Furthermore, advanced antenna and coding techniques provide additional dimensions for resources to also be allocated in space or code.

6.2.1 Protocol Data Unit

The IEEE 802.16 MAC Protocol Data Unit (PDU) is the WiMAX-specific data unit exchanged between the MAC layers of the BS and the MS. It consists of a 6-byte generic MAC header and a variable size payload. Optionally a cyclic redundancy check (CRC) may be added for error detection of the payload when the error detection capability is needed as part of the retransmission techniques such as ARQ, or H-ARQ. The MAC header contains basic information about the PDU such as type, length, connection identifier, header check sequence and, if applicable, encryption information for the payload and extended subheader field. The total PDU size including the header, payload and optionally the CRC is limited by the 11-bit length field, yielding the maximum PDU size of 2048 bytes. Figure 6.7 illustrates the MAC PDU with the generic MAC header, optional subheader, payload and CRC. The CID in the

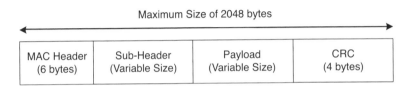

Figure 6.7 MAC PDU with optional subheader field

header field is a 16-bit value that identifies a unique connection between the MAC layer of the BS and the MS. Not only can the CID provide information regarding the source/destination of the PDU, it also classifies the types and the QoS mapping associated with the PDU. Similar to the CRC used to detect errors in the payload, the header check sequence (HCS) is an 8-bit field that can detect errors in the header. Optionally, subheaders and extended subheaders may be included in the payload in order to carry supplement control information.

When the header indicates a non-zero-sized payload, each MAC PDU contains either MAC management data or CS data as indicated by the type of CID (e.g. transport, basic, management) shown in the MAC header. MAC management data is generated by and exchanged between the MAC CPS entities of the MS and BS to handle various management issues such as network access, bandwidth request and grant, connection establishment and QoS. When the length field indicates no payload, the PDU contains only the MAC header that carries very short and time-critical information such as bandwidth request, uplink transmit power report, CINR feedback and feedback header. This special arrangement can only be used in the uplink direction and uses minimal bandwidth to carry essential management data.

Due to the limited amount of space available in the generic MAC header, supplement control information may be transmitted in the subheaders and extended subheaders as indicated in the type and extended subheader field in the header. Most extended data carries short control signaling that helps process the PDU. Some examples of subheaders and extended subheaders are given below:

- The fragmentation subheader indicates whether the MAC SDU is fragmented and transmitted in one or more MAC PDUs as well as the orientation of SDU fragments in the payload. This process allows efficient use of available bandwidth relative to the QoS requirements by fragmenting a large data packet into smaller chunks that are easier to schedule and less prone to transmission errors.
- The packing subheader indicates whether multiple MAC SDUs are packed into one MAC PDU. Unlike fragmentation, packing reduces the inefficiency of transmitting multiple small packets by aggregating them into a single packet that is suitable for the current bandwidth allocation.
- The grant management subheader is used by the MS to indicate its bandwidth need for a particular connection to the BS. Depending on the level of service, the MS may also include additional bandwidth management data such as frame latency, poll request and the amount of uplink bandwidth request in the piggyback mode.
- The fast feedback allocation subheader is used to allocate a small uplink communication channel, over which the MS can use to provide certain feedbacks to the BS.
- The PDU sequence number extended subheader indicates the sequence number of the PDU. Once enabled for a H-ARQ connection, the transmitter should apply the sequence number of every PDU on this connection. The receiver may use this sequence to ensure proper PDU ordering.

- The downlink sleep control extended subheader is sent by the BS to activate or deactivate the power saving operations of the MS. It can also be used to instruct the MS to transmit the channel quality indication during its availability intervals or deactivate the report.

6.2.1.1 Summary of Key Profile Features

A summary of the key features is given in Table 6.3.

6.2.2 Basic Control Signaling

On successful synchronization to the preamble transmitted in the first OFDM symbol of the frame, the MS needs to obtain basic control information such as frame control header, channel description and resource mapping, in order to gain access to the system. Immediately following the preamble, frame control signaling is first transmitted. All active MSs need to decode the frame control header (FCH) and the MAP messages contained in the frame control section of every frame in order to enter the network or continue normal operations. Figure 6.8 shows the WiMAX TDD frame structure illustrating the logical allocations of the preamble and frame control section, as well as data bursts. The FCH and MAP messages are transmitted with one of the most robust mandatory coding and modulation schemes in order to ensure that all MSs, regardless of their location or status (active, idle, sleep or trying to enter the network), are able to obtain essential information contained in these messages.

Table 6.3   PDU format key profile features [147]

Category	Feature	BS support	MS support
PHS	PHS	Y	Y
CS	Packet, IPv4	Y	Y
CS	Packet, IPv6	Y	Y
CS	Packet, 802.3/Ethernet	Optional	N
CS	Packet, 802.1Q VLAN	N	N
CS	Packet, IPv4 over 802.3/Ethernet	Optional	N
CS	Packet, IPv6 over 802.3/Ethernet	Optional	N
CS	Packet, IPv4 over 802.1Q VLAN	N	N
CS	Packet, IPv6 over 802.1Q VLAN	N	N
CS	ATM	N	N
CS	Packet, IPv4 with ROHC	Y	Y
CS	Packet, IPv6 with ROHC	Y	Y
MAC PDU formats	Reassembly at Rx	Y	Y
MAC PDU formats	Fragmentation at Tx	Y	Y
MAC PDU formats	Packing of fixed length MAC SDUs	N	N
MAC PDU formats	Packing of variable length MAC SDUs at MS	N/A	Y
MAC PDU formats	Packing ARQ feedback payload	Y	Y
MAC PDU formats	Extended subheader support	Y	Y
MAC PDU formats	Capability of receiving bandwidth requests using grant management subheader	Y	N/A

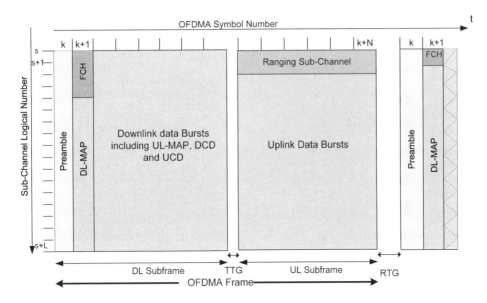

Figure 6.8   WiMAX TDD frame structure with basic control signaling

The FCH is the first control message that every MS needs to decode prior to accessing the resource mapping (MAP) and the actual downlink (DL) and uplink (UL) allocations. The location, size as well as the modulation and coding of the FCH are fixed so that it can be immediately decoded. The FCH carries the downlink frame prefix, which indicates the coding scheme and the variable size of the DL-MAP of the current frame. The MAP is a MAC management message indicating the two-dimensional (time and frequency) allocation of resources for all transport and control data. An information element contained in the MAP messages indicates time–frequency resource mapping for the MS being allocated resource within the MAP relevance interval. As the name indicates, the DL-MAP is specific to the downlink resources and connections while the UL-MAP indicates the allocations in the uplink. The DL-MAP and UL-MAP also contain information regarding the frame duration, the frame number, the BS identifier (BS ID), the DL and UL subframe duration and the UL allocation start time. The UL-MAP also indicates the ranging region(s), as shown in Figure 6.8, that the MS can use to transmit a ranging code and adjust its UL transmit parameters during initial network entry.

A MAP information element (MAP IE) utilizes a CID, which is unique to a connection belonging to the MS, to provide a logical mapping between resource allocations and the relevant MS. Further details of the CID are provided in Section 6.2.4. The MS may receive more than one MAP IE within the same MAP message. For example, a MS may be scheduled to receive a few MAC management messages and some data bursts in the DL and also be granted a UL allocation to transmit a bandwidth request. Therefore, the size of these MAP messages is proportional to the number of DL and UL allocations, the number of users contained in the messages and the level of modulation and coding applied to the MAP. To ensure successful decoding at the cell edge, the MAP is normally transmitted with a robust transmission scheme requiring higher radio bandwidth. The lengthy MAP messages reduce the available space that could be used to transport data and thus negatively impact the system throughput. The BS needs to find a balance in adjusting the transmission scheme and the number of information elements of the MAP to fit the current channel conditions, individual cell configurations and the number of active connections.

It is shown in [149] that the amount of MAP overhead alone can easily rise to 23% of bearer data. In addition, when there are many small-bandwidth allocations such as those used in VoIP, the MAP messages can grow very long and become limiting factors for the overall capacity [150]. In order to reduce the amount of MAP overhead, especially in a short frame, Mobile WiMAX also allows the use of the compressed MAP, whose MAC header portion is substantially suppressed. The compressed MAP can contain regular MAP IEs as in normal MAP messages. In addition to the compressed MAP, the sub-DL–UL MAP can be used for a subset of users with good channel conditions. As a result, the sub-DL–UL MAP may be transmitted using a more efficient coding and modulation scheme that results in smaller overhead. The sub-DL–UL MAP may be placed as the first burst of the relevant zone or immediately after the compressed MAP as shown in Figure 6.9.

In addition to the FCH and MAP messages, the downlink channel descriptor (DCD) and the uplink channel descriptor (UCD) define the characteristics of the DL and UL channels, respectively. They contain essential system operating parameters such as DL and UL burst profiles, handover type and triggers, timers and ranging parameters. The BS broadcasts these messages periodically to allow new mobiles to enter the network and the existing mobiles to receive updates regarding operating parameters. The MS needs to obtain, at the very least, the DL-MAP, DCD, UCD and UL-MAP messages before initiating the network entry procedure and continue to receive the updated messages in order to maintain normal operations in the network.

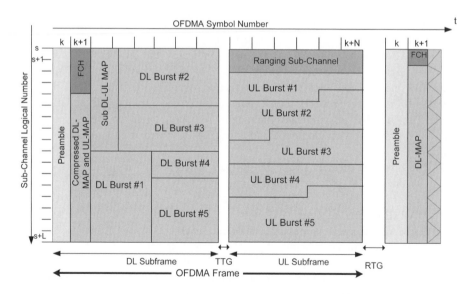

Figure 6.9 WiMAX TDD frame structure with compressed DL–UL MAP and sub-DL–UL MAP

6.2.3 Network Entry and Initialization

Initial network entry is a process initiated by a MS with the intention of gaining access to and receiving services from the network and may start upon powering on, reinitializing of the MAC after losing the connection with the BS or being instructed by the BS. The procedure contains multiple phases involving a number of management message exchanged and negotiated between the intended BS and the MS. Generally, a new MS needs to synchronize with the DL transmissions of the selected BSs, obtain the necessary system parameters and perform basic UL adjustments for the purpose of continuing with registering and obtaining basic network services. Figure 6.10 shows the overall process of network entry and initialization, which includes synchronization and transmission parameter adjustments, capability negotiation, authentication and registration, IP connectivity and connection setup.

6.2.3.1 Synchronization and Transmission Parameter Adjustments

Prior to obtaining services from the network, a MS needs to discover available networks and obtain the basic control parameters that are necessary in deciding the potential target BSs and executing further network entry steps. The process involves acquiring the list of available channels of the DL frequency bands. In some scenarios such as dense and/or multiple-frequency overlaid deployments, the number of candidate channels arising from autonomous network scanning may be prohibitively large and may cause significant delays in the scanning time. Therefore, the MS may use the preferred frequency list provided by the network operator or from the last known operational parameters to speed up the scanning process. The actual algorithm used in obtaining the frequency list and deciding which target BS will be used for further network entry is implementation specific and beyond the scope of this book. Once the potential BS has been identified, the MS may start the initial network entry procedure by synchronizing and adjusting the transmission parameters.

During the course of synchronization, the MS starts its PHY-level adjustment and gathering broadcast control information to determine whether the current channel and the associated network are suitable for its purpose or continues scanning for another channel or network. The MS uses the preamble located at the beginning of the frame to initiate signal acquisition and tuning of its primitive parameters such as frequency and timing. Then the MS decodes the basic control messages such as the FCH, DL-MAP, DCD, UCD and the UL-MAP to learn about the frame structure, DL and UL channel characteristics, resource mapping and UL transmit opportunities. The MS uses the UL transmit opportunities provided by the BS to initiate a closed-loop adjustment of UL transmit parameters. This process is also known as ranging. The MS remains in synchronization with the current BS as long as it is able to decode the DL-MAP and DCD messages and continues to have the most recent DL parameters. Similarly, the MS is not allowed to use the UL channel unless it continues to receive valid UL parameters from the UL-MAP and UCD messages.

Ranging is an interactive process in which, based on feedback from the BS, the MS adjusts its UL timing, frequency and power to align with the reception threshold of the BS. The adjustments may take a number of iterations until all transmitted parameters are within the acceptable range of the BS. In particular, the MS first contends for the ranging opportunity by transmitting a CDMA code that is randomly selected from a preallocated subset in one of the designated UL ranging slots. The BS is required to respond to decodable ranging attempts by

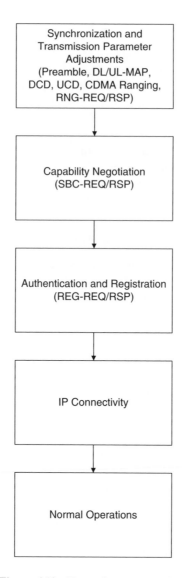

Figure 6.10 Network entry procedure

indicating whether ranging was acceptable (*Successful*), required further adjustment (*Continue*) or cannot continue (*Abort*). Contrary to the popular carrier-sensed multiple access and collision avoidance (CSMA/CA) scheme used in IEEE 802.11 (Wi-Fi), ranging in Mobile WiMAX uses a contention mechanism without carrier sensing. Since the same ranging code and ranging slot may be selected by multiple MSs, collisions may occur and result in failure to decode the ranging code at the BS. Since the MS is not required to sense the medium, it may not even be aware of the collision. However, the lack of any ranging response implies that the ranging attempt was not successful due to collisions or insufficient UL transmit power. A truncated binary exponential backoff scheme is used by all MSs that failed to receive any ranging responses in order to avoid

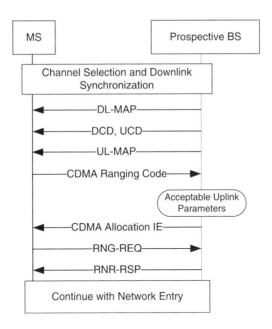

Figure 6.11   Synchronization and transmit parameter adjustment procedure

future collisions due to repeated ranging attempts. The MS can also increase transmit power in the subsequent ranging attempts. The various sets of ranging codes provided in the UCD message serve different purposes such as initial ranging, periodic ranging, handover ranging and bandwidth request ranging. Figure 6.11 outlines the procedure involved in synchronization and parameter adjustments.

During contention-based ranging, the BS only responds to the ranging code without the ability to identify the MS performing ranging. The ranging response (RNG-RSP) message is simply broadcast with the received ranging code and transmitted slot information to inform the MS that transmitted the particular ranging code. If the received ranging code is within the acceptable range of the BS, the BS will indicate the *Successful* status in the RNG-RSP message and allow the MS to continue with network entry. When ranging is successful, the BS normally uses the CDMA allocation IE in the UL-MAP to indicate a dedicated bandwidth allocation for the MS to transmit the RNG-REQ message directly without having to contend for bandwidth. However, if the transmitted ranging code requires further adjustments, the RNG-RSP will include a *Continue* status and include the suggested parameter adjustments allowing the MS to fine-tune its transmission before the next attempt. The adjustment may continue until all parameters are within the approved range of the BS and the MS receives a *Successful* ranging response. If the signal quality of the MS is still unacceptable after a number of corrections, the BS may choose to terminate the ranging process and force the MS to scan for a new channel by transmitting the RNG-RSP message with the *Abort* status. The BS may provide another preamble index or DL frequency overwrite in the *Abort* ranging response to help the MS find an alternate BS without performing the autonomous network scan again.

On a *Successful* ranging attempt, the BS assigns an UL bandwidth allocation for the MS to transmit the ranging request (RNG-REQ) message that indicates its MAC address, requested

DL burst profile and MAC version. The algorithm in selecting the DL burst profile is implementation specific, but ones that maximize the link throughput are commonly used. The BS then responds with the ranging response (RNG-RSP) message including additional adjustments, if any, and the assigned CIDs for management connections in each direction. A CID is a 16-bit value that uniquely identifies the unidirectional mapping of a connection, which is used in transporting management signaling or actual data units from the higher layer belonging to an individual MS to an equivalent peer in the MAC of the BS. The first assigned CID is a basic CID, which can be used to uniquely identify the MS during its active status with the selected BS. Prior to transmitting the RNG-REQ message, the MS remains anonymous to the involved BS during the entire ranging adjustments. On receiving the globally unique 48-bit universal MAC address that was assigned during the manufacturing, the BS is able to identify and map the MS to the basic CID. Within the same network, the basic CID cannot be reassigned as long as the current MS is still active. The basic CID is used to transport short and time-critical MAC management messages such as basic capability negotiation, channel measurement report and handover request and response. The second CID assignment is specifically for primary management connections that are used to transport delay-tolerant management messages such as authentication, registration handshakes and connection setup signaling.

During normal operations, the MS is also responsible for performing periodic ranging to ensure that all UL transmit parameters are within the approved range at all times. Periodic ranging is a process controlled by the MS to periodically check its UL parameters with the current serving BS. Similar to initial ranging, the BS does not know the identity of the MS transmitting the periodic ranging code. However, the BS may send unsolicited ranging response messages to request corrections based on measurements that were made on received UL data. On receiving a ranging response with corrections, the MS is not allowed to transmit anything until the radio parameters have been adjusted in accordance with the ranging response message.

6.2.3.2 Capability Negotiation

All communication between the MS and the BS prior to the capability negotiation is very rudimentary and does not utilize any physical layer enhancements. In other words, only the mandatory schemes are used in order to allow as many devices as possible to be able to discover and learn about the network prior to admission and granting access. However, the lack of advanced features also results in network inefficiency and lower system throughput. On completion of successful ranging and the establishment of management connections, the MS sends the basic capability request (SBC-REQ) to start negotiating the use of basic functionalities with the BS. This enables the use of features that help improve the utilization of radio resources and ensures that the network capabilities fulfill the requirements of the MS and vice versa. The BS can allow or deny the request for the use of any particular feature and indicate additional supported capabilities in the basic capability response (SBC-RSP). If any of the required capabilities are not supported, either side may choose to abort the network entry. Otherwise, the MS may continue with authentication and registration. At this point, both the BS and the MS start using the assigned basic CID as the unique identifier for the MS.

Given the inefficiency of mandatory transmission modes and, more importantly, the lack of verifiable user credentials, the list of negotiated capabilities is limited to those relevant to security authentication and physical layer features that are useful prior to or during the registration process. Furthermore, fragmentation of management messages exchanged at this stage is not possible as the feature has not been negotiated. Therefore, it is highly desirable to limit the size of the capability negotiation messages to reduce the scheduling burden. This is also particularly important to the UL transmission because the MS may need to focus its transmission power over a short message to improve the link budget. Typical basic capabilities that are negotiated at this stage include maximum UL transmit power, FFT size, number of H-ARQ channels, MAP capability, MIMO support, PKM version and authorization policy. A more extensive exchange of capabilities is conducted during the registration procedure.

6.2.3.3 Authentication and Registration

The next step in the network entry procedure is to exchange the digital certificate and to agree on the traffic encryption keys and their associated lifetime. On successful authentication and basic capability exchange, the MS needs to register with the network in order to start receiving services. Chapter 4 provides further details on service provisioning. It also needs to determine capabilities specific to the connection setup, such as IP version, the number of DL/UL connections, classification, handover support, mobility feature, ARQ, RoHC and power saving and receive services. The secondary management connection will also be assigned during registration if the MS can be managed by standard-based management messages such as Mobile IP (MIP), Dynamic Host Configuration Protocol (DHCP), Trivial File Transfer Protocol (TFTP) and Simple Network Management Protocol (SNMP).

6.2.3.4 IP Connectivity

If the Mobile IP is being used, the secondary management connection may be used to secure its address. Otherwise, the MS can obtain an IP address and other relevant parameters via the DHCP mechanisms. On successful establishment of IP connectivity, the MS needs to download a configuration file providing a standard interface for vendor-specific operational information. The time of day is also established in order to synchronize time-stamping logged events for the system management.

6.2.3.5 Summary of Key Profile Features

A summary of the key features is given in Table 6.4.

6.2.4 Connection Management and QoS

Once the initial network entry procedure is finished, the MS is ready to start receiving and transmitting data. In Mobile WiMAX, a logical connection is needed for any type of data (management or convergence sublayer (CS)) exchanged between the two MAC entities and there can be multiple connections being used simultaneously. Generally, the requirements associated with the higher layer protocols define the characteristics of an underlying connection carrying CS data. Once a connection is set up and agreed by both entities, it cannot be used

Table 6.4 Network entry and initialization key profile features [147]

Category	Feature	BS support	MS support
Authorization policy support	802.16 authorization policy support	Y	Y
PKM version support	PKMv1 support	N	N
PKM version support	PKMv2 support	Y	Y
PKMv2 authorization policy support – initial network entry	No authorization	Y	Y
PKMv2 authorization policy support – initial network entry	EAP-based authorization	Y	Y
PKMv2 authorization policy support – initial network entry	EAP-based authorization and authenticated (EIK) EAP-based authorization	N	N
PKMv2 authorization policy support – initial network entry	RSA-based authorization	N	N
PKMv2 authorization policy support – initial network entry	RSA-based authorization and authenticated (EIK) EAP-based authorization	N	N
PKMv2 authorization policy support – initial network entry	RSA-based authorization and EAP-based authorization	N	N
Supported cryptographic suites	No data encryption, no data authentication and 3-DES, 128	Y	Y
Supported cryptographic suites	CBC-mode 56-bit DES, no data authentication and 3-DES, 128	N	N
Supported cryptographic suites	No data encryption, no data authentication and RSA, 1024	N	N
Supported cryptographic suites	CBC-mode 56-bit DES, no data authentication and RSA, 1024	N	N
Supported cryptographic suites	CCM-mode 128-bit AES, CCM-mode, 128-bit, ECB-mode AES with 128-bit key	N	N
Supported cryptographic suites	CCM-mode 128-bit AES, CCM-mode, AES key wrap with 128-bit key	Y	Y

Supported cryptographic suites	CBC-mode 128-bit AES, no data authentication, ECB mode AES with 128-bit key	N	N
Supported cryptographic suites	MBS CTR-mode 128-bit AES, no data authentication, AES ECB-mode with 128-bit key	N	N
Supported cryptographic suites	MBS CTR-mode 128-bit AES, no data authentication, AES key wrap with 128-bit key	N	N
Message authentication code mode	No message authentication	Y	Y
Message authentication code mode	HMAC	N	N
Message authentication code mode	CMAC	Y	Y
Security association	Support of static SA	Y	Y
Security association	Support of dynamic SA	Y	Y
Security association	Support of primary SA	Y	Y
SA service type	Unicast	Y	Y
SA Service Type	Group multicast service	N	N
SA Service Type	MBS services	N	N
Certificate profile	X.509 MS certificate for device authorization	N	N
Certificate profile	X.509 manufacturer certificate	N	N
Certificate profile	X.509 BS certificate profile	N	N
Multicast broadcast re-keying algorithm (MBRA)	MBRA for group multicast service	N	N
Multicast broadcast re-keying algorithm (MBRA)	MBRA for MBS service	N	N

for other purposes. For example, a MAC management message can never be transported over a connection that was set up for normal traffic. Similarly, basic, primary or secondary management connections cannot be used to transport any type of CS data. For bearer or data services, Mobile WiMAX uses the mapping of a service flow onto a distinct transport connection in order to efficiently manage the data transport and guarantee the associated QoS requirements.

The concept of associating a service flow to the transport connection is essential to QoS provisioning in Mobile WiMAX. A connection provides a unidirectional logical link between the MAC interfaces of the BS and the MS while a service flow is initiated and maintained at a higher layer entity regardless of an underlying logical link. The QoS requirements specified for the connection, such as the transmission ordering and scheduling, can be defined according to the parameters of the service flow. For example, during normal operations, a transport connection for a particular service flow can be established with the negotiated QoS parameters that both sides can support. When necessary, the flow parameters for existing connections can also be modified or deleted through the exchange of MAC management messages. Note that the BS is only capable of applying the set of QoS to data traffic and is not in control of the QoS policy for the MS. Section 4.4 provides further details on the network-level QoS policy and management.

6.2.4.1 Connection Setup

Not only is a service flow crucial to the setup and maintenance of connections in Mobile WiMAX, but it also provides the association mechanism for packets traversing the MAC interface with the set of agreed QoS parameters such as bandwidth, latency and jitter. In other words, a service flow is a unidirectional transport of packets, either in DL or UL, which are characterized by a set of specific QoSs. A service flow is assigned a 32-bit service flow identifier (SFID) that is unique within the network domain of the ASN-GW. Therefore, unlike the CID that is unique within the BS, the SFID remains unchanged even when the MS roams through multiple BSs connected through the ASN-GW. It should be noted that there is no relationship between a WiMAX's service flow and the concept of an *IntServ* flow defined by the IETF Integrated Services (IntServ) Working Group [1]. WiMAX uses a two-phase activation model, similar to conventional telephone applications, where the resources for a particular flow are first admitted and then, as necessary, activated to indicate the actual commitment of resources. In addition, a service flow may be activated and deactivated as many times as necessary to conserve resources. For example, a VoIP session being placed on hold may become inactive while still being admitted, so that the resources used for the call can be temporarily allocated to other purposes. Once the conversation has resumed, a flow can be reactivated. Since the admission control was already done when the call was admitted, the request for active QoS parameters within the range of the previous reservation is guaranteed to succeed, allowing the flow to become active again. A service flow can also be provisioned and immediately activated, which means a QoS parameter set is provided by other network entities such as the network management system. Alternatively, a service flow may be dynamically created and immediately activated. In the latter case, the two-step process is skipped and the service flow is immediately available upon authorization.

There are three types of service flows: provisioned, admitted and active. A service flow that is either admitted or active is mapped onto a connection identified by a 16-bit CID. The mapping of the 32-bit SFID and the CID is also unique within the network. In general, the service

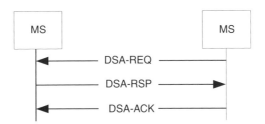

Figure 6.12   Message flow of BS-initiated dynamic service addition

flows are set up and initiated by the BS. However, there are instances, such as a switch virtual connection (SVC) from an ATM network or other MS-initiated events, that require the MS to initiate the creation of a service flow [153]. Additional explanations regarding the policy and usage of service flows in the network architecture are provided in Sections 4.4.3 and 4.4.4.

The establishment and maintenance of service flows in both the UL and DL directions is performed by a three-way handshake involving the request, respond and acknowledgement and must be approved by the authorization and policy module such as the service flow authorization within the ASN-GW (see Section 4.4.3). For example, a service flow creation initiated by the BS uses the dynamic service addition request (DSA-REQ) message containing the SFID, the associated CID and a set of admitted or active QoS parameters. The MS responds with the dynamic service addition response (DSA-RSP) indicating whether it can support this type of service or whether it requires additional adjustments. To finally confirm the negotiated parameters, the BS confirms the transaction by sending the dynamic service addition acknowledgement (DSA-ACK) message and the setup is complete. The MS may initiate the service flow creation in a similar manner. On verifying that the requested service flow and supporting capability are authorized, the BS creates the SFID and CID to be included in the DSA-RSP and the MS concludes the transaction with the DSA-ACK message. Figure 6.12 shows the signaling exchange for the BS-initiated DSA.

For the maintenance of existing service flows, the dynamic service change (DSC) transaction is used to modify the admitted and active QoS parameter sets for both provisioned and dynamically created service flows. Similar to the DSA procedure, the DSC transaction also requires the request–response–acknowledge (REQ–RSP–ACK) sequence. If no longer in use, a service flow may be deleted using the dynamic service deletion (DSD) process, which requires only the request–response sequence. When a service flow is deleted, all resources reserved for this connection are also released and available for other uses.

6.2.4.2 Bandwidth Request and Allocation

An allocation of bandwidth in the DL direction is relatively simple as the BS has complete knowledge regarding scheduling constraints and bandwidth demand through the packet classifier and handling of its own management messages. Since the demand for bandwidth at the MS side may fluctuate, an explicit negotiation is often necessary. Once a connection is established, the MS has a number of ways to notify the BS that it has data to transmit on the UL and thus requires additional transmission bandwidth. The exact methods depend on the connection type and whether the MS has an existing UL grant. A request for bandwidth can be transmitted during any UL allocation, except during the initial ranging period. As the parameters of the service flows may not be available to the scheduler, the CID is used in a bandwidth request sent by the MS to identify the priority as well as other QoS requirements associated with the request. In addition, the requested bandwidth size, for which the UL bandwidth is intended, is also specified in the request. Although a bandwidth request is specific to a particular connection, an UL grant is not tied to any connection and is allocated per MS. Therefore, the BS may aggregate multiple requests and grant a single UL allocation per subscriber. In addition, the granted bandwidth by the BS, unless guaranteed by an UL scheduling service during the service flow provisioning, may actually be smaller than the requested amount where there are scheduling constraints or other unexpected incidents such as lost requests. Nevertheless, the MS is free to use an allocated UL grant for the connections, for which it has requested bandwidth or any other connections according to the demand that could change since the last bandwidth request.

A request for UL bandwidth may be incremental or aggregate. An incremental bandwidth request indicates the additional quantity of bandwidth required for a specific connection, while an aggregate request overwrites the previous perception of required bandwidth for the connection. The MS may send a standalone bandwidth request header with no payload or piggyback a bandwidth request to an existing MAC PDU. These mechanisms require an existing bandwidth grant that the MS may have already received in response to previous bandwidth requests or responses to bandwidth polling. With a bandwidth request header, the requested type (i.e. incremental or aggregate), the CID and the requested size are included. If the piggyback method is used, the incremental bandwidth request needs to be for the same connection in which the request is transmitted. When the MS does not have a sufficient bandwidth allocation to either transmit a bandwidth request or piggyback a request, it needs to fall back to a contention-based bandwidth request that has a similar procedure to the transmission of ranging codes. Specifically, the MS transmits a bandwidth request ranging code, randomly selected from the subset of ranging codes reserved for bandwidth requests, in a ranging slot, and randomly selected from the appropriate ranging region. Since the same bandwidth ranging code and ranging slot may be selected by other MSs, collisions may occur and the contention resolution procedure based on a truncated binary exponential backoff scheme is needed before another bandwidth ranging code can be transmitted. On successfully receiving a bandwidth request ranging code, the BS will provide a sufficient UL grant to transmit a bandwidth request header. Similar to other ranging procedures, the MS remains anonymous during bandwidth request ranging and thus the UL grant can only be identified by the transmitted ranging code and the corresponding slot. It should be noted that the contention-based bandwidth request procedure is not as reliable and, most importantly, not as efficient as other bandwidth request methods. This is due to the two-step process where the MS must first contend for bandwidth that is only sufficient to transmit a bandwidth request.

6.2.4.3 Scheduling Service

The scheduling and grant mechanism allocates bandwidth according to the set of QoS parameters that quantify aspects of its behavior associated with a connection. This is particularly important for the UL service where the BS scheduler anticipates the different needs of bandwidth for subordinate MSs. In the DL, scheduling is relatively simple as the BS is

Table 6.5 Connection management and QoS key profile features [147]

Category	Feature	BS support	MS support
Connection management	Dynamic service flow creation – BS initiated	Y	Y
Connection management	Dynamic service flow creation – MS initiated	Optional	N
Connection management	Dynamic service flow change – BS initiated	Y	Y
Connection management	Dynamic service flow change – MS initiated	Optional	N
Connection management	Dynamic service flow deletion – BS initiated	Y	Y
Connection management	Dynamic service flow deletion – MS initiated	Y	Y
Data delivery services	Unsolicited grant service (UGS)	Y	Y
Data delivery services	Real-time variable rate (RT-VR) service	Y	Y
Data delivery services	Non-real-time variable rate (NRT-VR) service	Y	Y
Data delivery services	Best effort (BE) service	Y	Y
Data delivery services	Extended real-time variable rate (ERT-VR) service	Y	Y
Data delivery services	Multiple ertPS support using CQICH codeword	Y	Y
Data delivery services	ertPS resumption bitmap extended subheader support	Optional	Optional
Bandwidth request	Incremental bandwidth request using BW request header	Y	Y
Bandwidth request	Aggregate bandwidth request using BW request header	Y	Y
Bandwidth request	Bandwidth request using grant management subheader	Y	Y

the only entity that transmits and schedules all DL connections. Once the bandwidth is allocated, the BS communicates the specific assignment to the target MS in the MAP messages that are broadcast in every frame. Four UL scheduling service types are defined in Mobile WiMAX and summarized in Table 6.5 below:

1. **Unsolicited grant service** (UGS) is designed to support real-time service flows for fixed size, periodic data traffic such as constant bit-rate VoIP or T1/E1. The dedicated resources are constantly granted to fulfill strict latency and throughput constraints without the need for additional bandwidth requests or polls. The sizes of these grants are large enough to fit the demand of data associated with the service flow or may be even larger at the discretion of the BS. The required QoS parameters are minimum reserved traffic rate, maximum latency, tolerated jitter, UL grant scheduling type, SDU size for fixed length SDU, request/ transmission policy and unsolicited grant interval. The BS may receive the status updates

of the UGS flow and compensate for any mismatches via the grant management subheader (GMSH) sent by the MS.

2. **Real-time polling service** (rtPS) is designed to support service flows with variable-sized data packets on a periodic basis. This service offers real-time polling via the periodic unicast bandwidth grant opportunities and is suitable for variable bit-rate applications that generate fluctuating traffic payload such as compressed streaming video such as MPEG and VoIP with silence suppression. Although polling introduces additional overheads, this service allows the MS to adjust the requested bandwidth to optimally match the variation in the expected throughput without wasting resources. Since polling is periodically available, the MS is not allowed to use any contention-based bandwidth request on a rtPS connection. The required QoS parameters for this service are minimum reserved traffic rate, maximum sustained traffic rate, maximum latency, UL grant scheduling type, request/transmission policy and unsolicited polling interval.

3. The **extended rtPS** (e-rtPS) service is an extension of rtPS and further allows the combination of UGS and rtPS UL scheduling by providing unsolicited guaranteed unicast grants with additional polling. This allows the MS to change the size of the additional UL allocations based on its current demand. Periodic UL allocations can be used for requesting additional bandwidth or actual data transfer. The MS may also use a contention-based bandwidth request or send a special codeword over the channel quality indication channel to indicate addition demand for bandwidth. The required QoS parameters for e-rtPS are minimum reserved traffic rate, maximum sustained traffic rate, maximum latency, UL grant scheduling type, request/transmission policy and unsolicited grant interval.

4. **Non-real-time polling service** (nrtPS) is designed to support delay-tolerant traffic with variable sizes. It offers unsolicited polls on a regular basis in addition to contention-based and unicast bandwidth requests in order to ensure that requests for bandwidth can be communicated even when the network is congested. However, the polling interval is normally in the order of one second or less, which is simply sufficient to maintain the delay-insensitive connections such as FTP and HTTP but unable to support any types of real-time traffic. The required QoS parameters are minimum reserved traffic rate, maximum sustained traffic rate, traffic priority, request/transmission policy and UL grant scheduling type.

5. **Best effort** (BE) service is suitable for data traffic that has no minimum QoS requirement and is handled on a resource-available basis only. The BS allocates bandwidth, according to the available resources and scheduling constraints, to this service in response to contention-based and unicast bandwidth requests from the MS. There are no guarantees of minimum throughput or delay. The only required QoS parameters for this service are request/transmission policy and UL grant scheduling type.

6.2.4.4 Summary of Key Profile Features

A summary of the key features is given in Table 6.5.

6.3 Mobility Support

The specification [1] for the air interface of Mobile WiMAX is designed to support mobile data services at a speed ranging from stationary, through pedestrian and up to vehicular. In addition, it aims to support the ITU-R mobile QoS classes, mobile terminals, hierarchical cell structure and handover between WiMAX BSs [154]. One important requirement of the mobility support is that the level of QoS should not be compromised due to the movement of the MS within the cell. Furthermore, while moving from the coverage of one cell to another (i.e. during the handover), the MS should be able to obtain continuing connectivity that is sufficient to satisfy the minimal QoS constraints defined for the handover. In short, the objective is to minimize the data loss and delay involved in the handover procedure. Lastly, given the power limitations resulting from the small form factor of mobile terminals and other battery constraints, power saving mechanisms such as sleep and idle mode operations are crucial to support the mobility of Mobile WiMAX.

6.3.1 Handover

Handover is a process during which a MS migrates from the air interface of one BS to another air interface provided by another BS. In the context of this book, handover to another air interface is restricted to the same radio technology. In other words, another air interface refers to another MAC instantiation of Mobile WiMAX and does not include different technologies such as Wi-Fi, 2G (GSM, CDMA) or 3G (WCDMA, HSPA). Although there are ongoing standardization activities related to handover between WiMAX and other radio technologies in multiple places such as the WiMAX Forum and IEEE 802.21, they are beyond the scope of this book. Handover typically occurs when the MS experiences signal fading or higher levels of interference due to movements or the dynamics of the surroundings and needs to seek a new serving BS in order to sustain the QoS requirements. As the MS moves away from the coverage of the current serving BS, there is an increasing risk of service disruption that may result in negative user experiences. Without handover, the MS would have to completely disconnect from the previous serving BS and reinitiate the lengthy network entry procedures. With proper handover procedures, the switching to the new BS can be made seamless.

In general, service disruptions depend on the handover type and whether it allows the MS to reconnect to the new BS before breaking off the connection with the current serving BS. Specifically, break-before-make handover (also commonly known as hard handover) restricts the MS to connect to only one BS at any given time and hence induces a brief period of service discontinuity. Although the MS needs to go through network re-entry at the new BS, the network may share the MS's credentials and prepare the potential target BSs for the expedited entry of the MS to minimize the disruption time. On the other hand, make-before-break handover (or soft handover) enables the MS to register and maintain connections with multiple BSs at the same time while performing handover. Soft handover usually provides a smoother transition to another cell at the cost of consuming considerably more network and radio resources as multiple connections are simultaneously kept active during the handover. Mobile WiMAX provides two soft handover mechanisms: macro diversity handover (MDHO) and fast base station switching (FBSS).

Handover may be initiated by the MS when it expects better services at another BS. The network or the BS in particular can also initiate a handover when it expects that the MS may be better served by a neighboring BS. A BS-initiated handover may be performed as part of radio resource management or load balancing, during which some MSs are asked to move from a heavily loaded serving BS to another BS with a lighter load. A specific algorithm as well as the radio and network parameters used to initiate a handover at both the MS and BS are usually implementation specific but generally include important criteria such as the level of QoS, load factor, cell size and the expected transmit power. Nevertheless, the IEEE specification [1] outlines the necessary handshake signaling and basic conditions to trigger and execute the handover. Even though the procedures for both break-before-make and make-before-break handover operations are defined in the IEEE 802.16 standard, only hard handover is supported in the Mobile System Profile Release 1.0 [146] due to its simplicity. In addition, hard handover requires much less overhead in both network backhaul and over the air signaling. Therefore, the handover procedure described subsequently is focused on hard handover and the corresponding optimization steps that allow Mobile WiMAX based on the system profile Release 1.0 [146] to meet the requirements for handover performance.

To reduce any noticeable distractions to the end users, the Service Provider Working Group (SPWG) requirements limit the maximum service disruption time to 150 ms [142]. With recent advances in error concealment for real-time applications such as VoIP and video streaming, the interruption is barely noticeable by the end users. The requirement on the disruption time is translated into roughly 30 radio frames, in which the MS, all involved BSs and ASN-GWs have to complete the handover process. As handover incurs a considerable amount of signaling overhead and briefly interrupts ongoing services between the MS and the BS, it is desirable that handover occurs as infrequent as possible even when the MS moves through the network. On the other hand, it is often necessary for the network or the MS to make a handover decision in order to sustain the required level of QoS. Since the moving pattern or the future channel conditions cannot be predicted, such decisions have to be based on available information or current statistics. Both sides of the radio interface (i.e. the MS or the BS) have different means to get access to such information, which is required for a handover initiation. Therefore, depending on the circumstances, handovers may be initiated either by the MS or BS, or by another network node.

In general, when the network is initially rolled out, the coverage is often spotty and therefore maintaining connectivity becomes the most critical operating factor. Consequently, the MS usually stays with one serving BS as long as possible and only performs handover when it is about to lose a connection or experiences unsatisfactory QoS. In this scenario, the MS has more complete information regarding the radio conditions than the network and thus is in the best position to determine when to initiate handover. As the network becomes more mature with nearly ubiquitous coverage, radio resource management that helps achieve continuous link quality becomes more essential. For example, the network may prevent the MS from moving to a small-sized BS such as a picocell if the MS is moving at high speed and is likely to require another handover initiation within a short time. In addition, if the load condition at the current serving BS becomes unmanageable, the MS may be asked to hand over to a neighboring BS even though the radio conditions at the current serving BS are more preferable from its perspective. In this scenario, the network should be able to prevent the MS from autonomously initiating a handover back to the previous serving BS and undermine the load balancing attempt by the network. Therefore, as the network becomes fully deployed, operators may prefer to have tighter, more network-centric handover management that restricts the MS's ability to initiate a handover on its own. Another scenario showing the need for network-controlled handover is when a network operator has spectrum availability in both 2.5 and 3.5 GHz. The operator may prefer to distribute the users in both frequency bands. Without network-centric handover management, most MSs may decide to stay with the carrier in the lower frequency because of the more favorable radio propagation, without realizing that they could actually be better served with the carrier in the higher frequency that may have lighter load.

The handover procedure as shown in Figure 6.13 typically involves the follow steps: cell scanning and reselection, handover decision and initialization, network re-entry at the new target BS and termination of MS context at the old serving BS. If the handover cannot be executed successfully, handover cancellation may be used to quickly resume normal operations. Lastly, handover triggers may be used to enforce the uniform behavior of all MSs in the network.

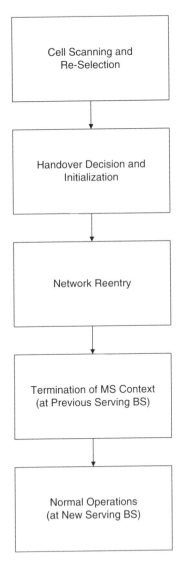

Figure 6.13   Handover procedure

6.3.1.1 Cell Scanning and Reselection

In mobile environments, handover may be triggered by a sudden change in the signal quality, network congestion or new QoS requirements that may require an immediate switch to a new BS. Therefore, a MS needs to be constantly aware of the surroundings in terms of network availability and the associated basic control parameters. By maintaining a list of preferred target BSs obtained during cell scanning and reselection, the MS will be able to quickly initiate handover or negotiate with the current serving BS regarding the target BS for handover. However, due to the number of possible BSs and the associated radio channels, autonomous scanning by the MS can be very time consuming. Contrary to the cell scanning during initial network entry, the MS already has the established service flows and transport connections and cannot afford to go through the lengthy preparation every time it needs to switch to another BS. The re-entry procedure, however, is similar to that of initial network entry. It involves synchronizing the DL reception using the preamble transmitted from the target BS and obtaining basic control information such as the DL-MAP, DCD and UCD messages containing the necessary frame control parameters and channel descriptions. The MS also needs to perform ranging unless its UL transmit parameters are already within the acceptable range of the target BS. Although the MS is able to acquire the preamble and the MAP messages in every frame, the periodicity of DCD and UCD broadcast is usually in the order of 50 ms (i.e. 10 radio frames) or more [146]. Hence, it is time consuming for the MS to obtain all necessary control information, especially from multiple potential target BSs without disrupting the existing communication with the current serving BS.

In order to help the MS with faster learning of the neighbor BS parameters, the serving BS can provide the neighbor advertisement (MOB_NBR-ADV) message that includes information normally contained in the DCD and UCD messages of the neighbors. This message is periodically broadcast in order to increase the reliability with which the MS obtains the latest system parameters from neighboring BSs and thus reduce the amount of disruption time. The serving BS may also grant additional scanning time on the exchange of scanning requests and responses (MOB_SCN-REQ and MOB_SCN-RSP) with the MS. During a scanning period, when the MS scans and monitors the handover suitability of neighboring BSs, the MS is temporarily unavailable to receive DL traffic from the current serving BS. Therefore, it relies on the BS to help maintain the existing connections by buffering incoming data addressed to the MS during these scanning intervals until the next availability period of the MS.

Based on information contained in the neighbor advertisement message along with its own measurements from autonomous scanning, the MS composes and maintains the preferred list of target BSs suitable for handover. The serving BS may provide additional handover parameters that supplement the criteria used at the MS. The network uses the handover parameters to ensure uniform handover behavior within the cell while allowing some variations due to different MS implementations. For instance, if the *hysteresis margin* is specified, the MS needs to add neighbor BSs whose CINRs are larger than the sum of the CINRs of the current serving BS and the hysteresis margin to the list of target BSs in the handover request (MOB_MSHO-REQ). The hysteresis margin does not preclude the MS from using its own criteria in adding more BSs to the list.

The MS normally measures and collects signal quality indicators such as RSSI, CINR and round-trip delay during the scanning intervals. Sometimes, the scanning reports are sent back to the serving BS for the purposes of cell assignment, load balancing and adaptive burst profile adjustments. The cell scanning and reselection process may occur at any time without conjunction to any specific handover decision. In addition to cell scanning, the IEEE 802.16 standard [1] also defines the *association* procedure during scanning that may be used to shorten the re-entry procedure. Association is an initial ranging procedure performed with selected target BSs that allows the MS to acquire and receive feedback on ranging parameters during the scanning intervals. However, the Mobile WiMAX System Profile Release 1.0 [146] does not yet support the association procedure.

6.3.1.2 Handover Decision and Initialization

The handover execution phase begins with the decision to switch the MS from the current serving BS to another BS. Handover can be initiated by the MS or the serving BS. Handover initiation starts with a handover request sent from the originator and is followed by a handover response from the serving BS and/or a handover indication sent from the MS to denote the responses, recommendations and the final handover decision. The handover triggering conditions and the associated actions may be imposed by the BS to ensure the predictable and consistent behavior of all MSs in the network. For example, the occurrences when the absolute CINR of a neighbor BS is greater than a certain threshold may trigger the MS to initiate a handover request. The possible associated actions are not limited to sending the handover requests, they also include adding the target BS to the preferred list or initiating autonomous scanning. As another example of a different triggered action, the MS adds a neighbor BS to the list of potential target BSs if the CINR of a target BS is larger than the sum of the CINRs of the current BS and the hysteresis margin for a predefined scanning time.

For MS-initiated handover, when a need arises for handover at the MS, it initiates the handover sequence by transmitting the handover request (MOB_MSHO-REQ) to the serving BS. The request includes one or more preferred target BSs, to which the MS is interested in handing over. Upon reception of the handover request, the BS, based on its own knowledge and information exchanged with other BSs over the backbone, returns the handover response (MOB_BSHO-RSP) message indicating whether the request is supported or denied. This message may include a list of recommended target BSs together with the 'service level prediction' for each recommended BS. The service level prediction indicates whether the same QoS level for each authorized service flow can be established with a target BS. The criteria that the serving BS uses in recommending a target BS may include the expected MS performance at the target BS, the availability at the target BS and the QoS requirements of the MS. The MS then incorporates information received in the handover response from the BS with its preferences and decides whether to proceed with handover and the final target BS. The handover indication (MOB_HO-IND) message is then sent to the current serving BS to signal whether the MS is about to perform a handover and to which target BS, or to cancel the handover request and return to normal operations.

Similarly, when the BS expects the MS to receive better services from another BS, it can initiate the handover procedure by sending a BS-initiated handover request (MOB_BSHO-REQ). The request may include the list of recommended target BSs to the MS. If the MS agrees with the handover request and is able to perform handover to one of the recommended BSs, it continues with the handover procedure by responding with the handover indication (MOB_HO-IND) message, indicating that it is about to perform handover and the final target BS. Alternatively, if the MS is unable to perform a handover to any of the recommended target BSs listed in the BS-initiated handover request, it can signal the handover rejection using the handover indication message. In the event that the MS would like to hand over to a BS that was not included in the suggested list, it might include the preferred target BS in the handover indication message. Based on the response, the BS can reconfigure the recommended BSs and transmit the handover response (MOB_BSHO-RSP) message with the new list. The MS may again reject or accept the recommendation from the serving BS. If the MS cannot find any suitable BSs for handover, it may reject the request and discontinue the handover procedure.

In order to enforce uniform and predictable handover behavior within a network, one or more BS-specified handover trigger conditions may be used to force the MS to initiate handover. If a trigger condition is met by any MS, it needs to perform the action specified in the trigger condition. These actions may also be performed autonomously by the MS. As mentioned previously, a handover trigger may be set to initiate a handover when the absolute CINR of the current serving BS falls below a specified value. The use of handover triggers, hysteresis margin and BS-initiated handover allows the BS to provide guidance and handover recommendations to MSs within its network. However, with the current Mobile System Profile Release 1.0, the MS has the final decision and may override the recommended handover or the handover triggers.

Figure 6.14 and Figure 6.15 provide examples of handover initiation by the BS and MS respectively, as well as possible decisions by the MS. In Figure 6.14, although the BS requests the MS to hand over to the target BS2, the MS also detects the trigger events for the target BS3. Therefore, it can choose to hand over to the targets BS2, BS3 or any other BS. Figure 6.15 shows the MS-initiated handover procedure when the channel conditions have changed prior to the completion of handover.

The time between the start of handover signaling and the moment when the MS leaves the serving BS may be called the handover preparation phase (see also Section 5.4.1.1). During the

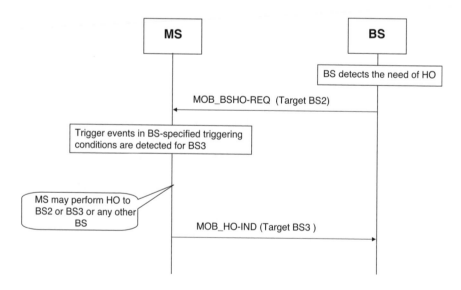

Figure 6.14   Example of handover initiation by BS

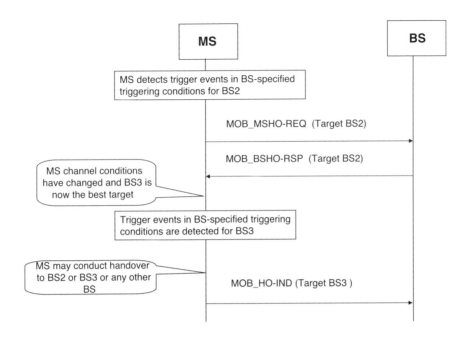

Figure 6.15   Example of handover initiation by MS

preparation phase, the serving BS may use the backbone signaling to prepare one or more potential target BSs about the MS's intent to hand over and supply the MS context over the backbone. It may also initiate the tentative data path setup in the backhaul to one or more target BSs. This will speed up significantly the subsequent network re-entry during the handover action phase (see Section 5.4.1.2) and contributes to handover optimization. It should be noted that the handover decision and initialization handshake does not actually interrupt ongoing connections between the MS and the serving BS. However, the subsequent network re-entry process, which is procedurally comparable to that of initial network entry, is most likely to cause a disruption while the MS breaks the existing connection with the current BS and starts a network re-entry at a new BS.

6.3.1.3 Network Re-entry

In general, the network re-entry to the finally selected target BS process is similar to initial network entry and consists of synchronization and transmission parameter adjustments, capability negotiation, authentication and registration, IP connectivity and connection setup. However, a number of optimization steps are defined to shorten the network re-entry by bypassing certain negotiations between the MS and BS and substituting them with the transfer of existing contexts and negotiations over the backbone between the current serving BS and the selected target BS.

Similar to the initial network entry procedure, the MS needs to ensure that its UL transmit parameters are within the acceptable range of the target BSs. Since the association procedure during scanning is not supported in the Mobile WiMAX System Profile Release 1.0 [146], the MS may need to perform handover ranging upon re-entering the new network.

Contention-based handover ranging and subsequent adjustments are similar to the initial ranging procedure except that a handover ranging code, which is used to indicate to the BS that the MS is attempting a handover, is transmitted instead of an initial ranging code. With successful ranging, the BS will grant a dedicated UL allocation for the MS to transmit the ranging request (RNG-REQ) message and thereby indicate the handover attempt and the previous serving BS-ID. The target BS may already be prepared for the handover during the preceding handover preparation signaling over the backhaul. Otherwise, the target BS can contact the serving BS via the backhaul to request the transfer of MS contexts. The new basic and primary management CIDs are included in the RNG-RSP message from the BS and the MS can continue with the capability negotiation and the rest of network re-entry. Within the same frequency assignment, the serving BS and neighboring BSs are usually time and frequency synchronized. In this case, only minor adjustments of UL parameters of the MS are expected and the parameter adjustment procedure can be significantly shortened even without explicit handover ranging. For target BSs with a different frequency assignment, the WiMAX System Profile Release 1.0 recommends that the MS be able to perform timing, frequency and power adjustments based on both DL reception quality and information contained in the DCD and UCD messages of the target BS without relying on feedback-based ranging [146]. In other words, the MS should be able to perform an open-loop association where the MS exploits the reciprocity of the TDD channel in adjusting the transmit parameters with respect to the neighboring BS.

In order to further expedite the handover process, the BS may use the backbone signaling to prepare one or more potential target BSs for the MS's intent to hand over and supply the MS context over the backbone. One of the most significant handover optimization steps is the coordination and negotiation with each of the recommended target BSs for a non-contention UL grant by the current serving BS over the backbone. The negotiation includes the specific period in which the UL grant is allocated by each recommended target BS. This process, known as 'fast ranging', allows the MS to choose any BS from the recommended list and use the dedicated no-contention UL grant to transmit the ranging request (RNG-REQ) message indicating the handover attempt and the current serving BS-ID. In contrast to contention-based CDMA ranging, fast ranging significantly improves the reliability and speed of network entry because it avoids collisions between multiple MSs trying to perform ranging at the same time. In addition, fast ranging allows the target BS to recognize the MS by the specific ranging code and slot, while the MS would remain anonymous during contention-based ranging. Furthermore, the target BS may verify the identity of the MS by providing the handover ID (HO-ID) for inclusion in the ranging request. The fast ranging procedure also assumes that the channel parameters that the MS acquired prior to the handover through autonomous scanning or the neighbor advertisement message for that BS stay valid and can still be used during the network re-entry. The *action time* field in the handover request/response (MOB_BSHO-REQ/ MOB_BSHO-RSP) from the BS indicates the time after which all recommended target BSs assign the dedicated allocations for the purpose of network re-entry. Although the recommended target BSs may not simultaneously grant the dedicated transmit opportunity, the use of action time along with the relevant BS-specified action time timer defines the lower and upper bounds for which the dedicated grants are expected. If the MS decides to hand over to a target BS not included in the recommended list or the MS is not able to use the fast ranging opportunities provided by the recommended target BSs, it has to resort to contention-based ranging and bandwidth request prior to transmission of the ranging request message.

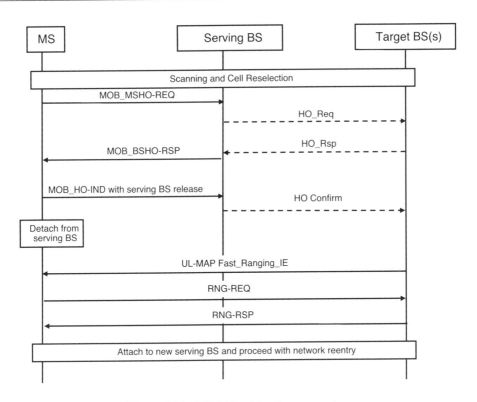

Figure 6.16   MS-initiated handover procedure

On successfully receiving the ranging request (RNG-REQ) message, the BS uses the ranging response (RNG-RSP) message to indicate any transmission adjustments, as well as to assign the new basic and primary management CIDs for the MS to continue with capability negotiation and the rest of network re-entry. Figure 6.16 illustrates the signaling exchanged among the MS, the serving BS and one or more target BSs for MS-initiated handover with fast ranging optimization. Communications to and from the MS are conducted over the reference point R1 (i.e. the radio interface). Backhaul signaling between the serving BS and target BSs is done over the reference points R4, R6 or R8 (see Section 2.2.1).

On successful synchronization and UL parameter adjustments, the network re-entry procedure requires steps similar to the initial network entry procedure. However, the procedure can be significantly shortened if the target BS has been preconfigured during the handover preparation phase as previously described. Specifically, when *handover process optimization* is used, a number of re-entry management messages may be skipped as the target BS expects the transfer of static and dynamic MS contexts from the serving BS. Static context consists of configuration parameters acquired during the initial network entry, security negotiation and service flow information exchanges. Dynamic context contains all counters, timers, data buffer contents and the state machine status. If the MS security credentials are still valid at the new BS, the MS must include the HMAC/CMAC tuples in the ranging request (RNG-REQ) message to verify its authenticity. The target BS, at its discretion, may require the MS to perform the new PKM exchange sequence or full network entry if the identity of the MS cannot be verified.

6.3.1.4 Termination of MS Context

Regardless of the entity that initiates handover, the serving BS still considers the MS to be attached and continues to allocate DL and UL resources to the MS until it receives the handover indication (MOB_HO-IND) message signaling the release of the serving BS. After receiving such indication, the BS starts the *resource retain timer*, which allows the MS to immediately resume normal operations at the current serving BS should the handover not succeed. During this period, the BS still maintains the active connections, the MAC state machine and the associated PDUs for the MS. On expiration of the resource retain timer, all MAC context and PDUs associated with this MS are discarded. In addition, the serving BS may decide on its own after a number of repeated attempts to reach the MS that it is no longer available and may terminate the MS context. In rare cases when the MS abruptly leaves the current serving BS and attaches to a new BS, the reception of a backbone notification of successful attachment of the MS can terminate all MS context with the current BS. The MS cannot expect expedited return at this BS and needs to perform the initial network entry procedure after expiration of the resource retain timer or termination of the MS context.

6.3.1.5 Handover Cancellation

The MS may decide to terminate the ongoing handover at any time for any reason by sending the handover indication (MOB_HO-IND) message with the *Cancel* status to the current serving BS. If the handover is cancelled before the MS indicated release of the serving BS or the handover indication was received by the serving BS prior to expiration of the resource retain timer, the MS can resume normal operations with the current serving BS immediately. However, if the MS cancels the handover after expiration of the timer, it is expected to go through the full network entry procedure.

The MS may use the handover cancellation procedure to recover from a drop during handover, which is defined as the situation when the MS loses communication with the target BS during the handover attempt. Generally, the MS can detect a drop when it loses its DL synchronization or exceeds the ranging retry limit. In addition to performing network re-entry that involves contention-based ranging at a preferred target BS, the MS may choose to return to the current serving BS by transmitting the handover indication (MOB_HO-IND) message with the *Cancel* status. If the cancellation occurs during the resource retain timer, the MS can immediately resume normal operations with the current serving BS.

6.3.1.6 Summary of Key Profile Features

A summary of the key features is given in Table 6.6.

6.3.2 Sleep Mode

In order to prolong the life of battery-operated mobile devices, power management is very important in supporting mobility of WiMAX. Without any power saving, the MS needs to remain active and stay synchronized with the network at all time. Continuing to decode the broadcast control messages as well as monitoring DL traffic in every frame causes a significant decrease in the active/standby time of the devices even when there is no or low traffic. Mobile WiMAX supports two stages of power saving: sleep and idle. The sleep mode allows the MS to dynamically adjust its duty cycle based on the level of activities while maintaining the active

Table 6.6 Handover key profile features [147]

Category	Feature	BS support	MS support
Neighbor advertisement	Neighbor advertisement	Y	Y
Scanning and association	Scanning for cell selection (HO)	Y	Y
Scanning and association	MS requests scanning interval allocations from BS	Y	Y
Scanning and association	Unsolicited scanning interval allocation by BS	Y	Y
Scanning and association	Event-triggered scanning based on serving BS metrics	Y	Y
Scanning and association	MS autonomous neighbor cell scanning	N/A	Y
Scanning and association	Support for association during scanning	N	N
Scanning and association	Mean BS CINR	Y	Y
Scanning and association	Mean BS RSSI	Y	Y
Scanning and association	Relative Rx delay	N	N
Scanning and association	BS round-trip delay	Y	Y
Handover procedure	General HO support	Y	Y
Handover procedure	HO initiated by MS support at MS side	N/A	Y
Handover procedure	HO initiated by MS support at BS side	Y	N/A
Handover procedure	HO initiated by BS support at MS side	N/A	Y
Handover procedure	HO initiated by BS support at BS side	Y	N/A
Handover procedure	HO indication	Y	Y
Handover procedure	Cancellation of HO	Y	Y
Handover procedure	Metric-triggered HO requests	Y	Y
Handover procedure	Resource retention support	Y	Y
Handover procedure	CDMA HO ranging	Y	Y
Handover procedure	Support negotiating of 'HO authorization policy' during HO (i.e. between BSs)	Y	Y
Handover procedure	General FBSS capability	N	N
Handover procedure	General MDHO capability	N	N
Handover optimization	HO optimization support	Y	Y
Handover optimization	Support omission of SBC-REQ management messages	Y	Y
Handover optimization	Support omission of PKM authentication phase except TEK phase	Y	Y
Handover optimization	Support omission of PKM TEK creation phase during re-entry processing	Y	Y
Handover optimization	Support 'full state sharing' – no exchange of network re-entry messages after ranging before resuming normal operations	Y	Y
Handover optimization	Notifying MS of DL data pending	N	N
Handover optimization	OFDMA fast ranging IE	Y	Y

state with the serving BS. On the other hand, the MS can become idle or enter an extended period of power saving while being periodically available for DL broadcast traffic without having to actively register with any particular BS through the use of the idle mode.

When the MS enters the sleep mode and becomes temporarily unavailable for any communication with the BS, part or all of its circuitry may be briefly shut down to conserve energy. In addition, the MS may use the period when it becomes unavailable to perform other tasks such as network scanning or association. Although the sleep mode allows the MS to negotiate the periods of inactivity that can be used for power saving, the MS is still required to register and listen to a broadcast traffic indication message during the availability intervals. Then, the MS can decide whether to continue to sleep, adjust the sleep intervals to suit the current traffic demand, or deactivate the sleep operation and resume normal operations. The capabilities to support the sleep mode and supporting operations are negotiable during the registration process of network entry. The support of the sleep mode is required for both MS and BS that are implemented according to the system profile Release 1.0 [146].

The sleep mode may require different operational parameters such as sleep interval and the activation/deactivation procedure, known as power saving class (PSC). A group of active connections to the same MS sharing similar QoS requirements may be assigned the same PSC in order to reduce the signaling overhead and minimize the scheduling constraints. In addition, a number of PSCs can also be assigned to the same connection in order to provide flexibility in adjusting the durations of sleep and listening intervals according to the traffic demand, which may fluctuate. For example, all BE connections may belong to the same PSC while a few PSCs may be defined for a single real-time voice connection with an adjustable activity level. However, only one of the PSCs, which are uniquely identified by PSC IDs, defined for the same connection can be activated at any time. Since the PSCs are specific to the connections, including management and transport, the actual period of unavailability depends on the PSCs that are active at a particular time.

Figure 6.17 illustrates the period of unavailability for the MS with two different PSCs. An unavailability interval is defined as a time interval that does not overlap with any listening windows, during which the MS remains active, of any active PSCs [1]. Similarly, an availability interval is a time interval that does not overlap with an unavailability interval [1]. As shown in Figure 6.17, if there is one connection that is not associated with any PSCs, the MS is considered to be available at all time (i.e. no unavailability interval for this MS). The assignment (or definition), activation and deactivation of PSCs may be initiated by either

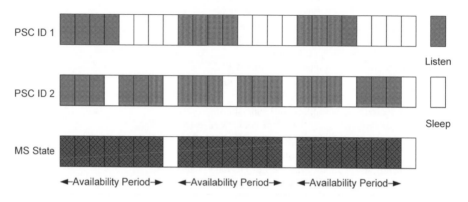

Figure 6.17   MS sleep state over multiple frames with different PSCs

the MS or BS using the sleep request and response messages (MOB_SLP-REQ and MOB_SLP-RSP), power saving encodings in the ranging request and response messages (RNG-REQ and RNG-RSP) or a sleep control header. Depending on the definition of the PSCs, the sleep operations may be deactivated automatically, on being awakened by certain traffic types or explicitly by the management message exchanges. In addition, all ongoing sleep operations with the current serving BS are deactivated, and eventually discarded on the transmission or reception of any handover messages (i.e. MOB_BSHO-REQ, MOB_MSHO-REQ or MOB_HO-IND) and the successful association with the new serving BS.

6.3.2.1 Power Saving Class

There are three power saving classes (PSCs) defined in the standard [1] that specify different parameters and procedures in alternating the sleep and listening windows:

- **PSC of Type I**: This PSC is suitable for delay-tolerant or non-real-time variable bit-rate applications using BE or non-real-time scheduling. It is also the only PSC supported in the system profile Release 1.0 [146]. When PSC Type I is activated, the MS alternates the fixed size listening windows and the variable size sleep windows that start from the minimum value of *Initial-sleep window* and are doubled until the size of sleep windows reaches the maximum value of *Final-sleep window*. During the availability periods, the MS is expected to receive all DL data as in normal operations. The BS can explicitly terminate the active status of this PSC type by sending a positive traffic indication (MOB_TRF-IND) message to indicate pending DL traffic for this class. Similarly the MS can terminate the active sleep status by transmitting a sleep request (MOB_SLP-REQ) message with the deactivation flag or an UL sleep control header. Therefore, the BS can only deactivate the PSC during the listening windows, while the MS has the ability to terminate the PSC at any time. Alternatively, this PSC may be implicitly terminated upon receiving data or a bandwidth request associated with the PSC. Specifically, the sleep mode is deactivated if the traffic wakening flag is set and a MAC PDU or a non-zero bandwidth request containing any CIDs that belong to the PSC is received during the availability periods. Since the BS may not be aware of any immediate demand for bandwidth from the MS, it may allocate an UL grant following the sleep deactivation request to give the MS a transmit opportunity or request additional bandwidth. Since the unsolicited UL grant also serves as confirmation of the BS-initiated sleep deactivation request, the MS needs to respond to the UL grant with a data transmission or a bandwidth request. If it has no UL data to transmit, the bandwidth request may be zero in size.
- **PSC of Type II**: This PSC is suitable for real-time applications with unsolicited grant service (UGS) or other real-time scheduling. The assignment and activation procedures of PSC Type II are similar to those of PSC Type I except that the sleep duration is fixed instead of increasing binary-exponentially. This PSC can only be explicitly terminated by a sleep request/response or other sleep deactivation requests. During the availability periods, the MS is expected to continue its normal operations of all connections contained in the PSC.
- **PSC of Type III**: This PSC is suitable for multicast and management connections that are mostly within control of the BS. Unlike the previous two PSCs, it contains a single sleep window and becomes automatically deactivated at the end of the sleep period. Since the BS is the transmit entity of these connections, it may schedule the MS sleep window to be aligned with the multicast traffic or periodic ranging intervals.

6.3.2.2 Periodic Ranging and Keep-Alive Check in Sleep Mode

Although the sleep mode allows the MS to conserve power and become briefly inactive, the MS still needs to maintain synchronization with the serving BS at all time. Periodic ranging is a process controlled by the MS to ensure all UL transmit parameters are within the acceptable range of the serving BS. The MS can silently extend its availability interval for the purpose of completing CDMA-based periodic ranging until it receives a ranging response with the *Success* status. Since the contention-based ranging procedure does not reveal the identity of the MS (i.e. a random ranging code and slot are used), it is transparent to the BS and does not impact the status of any sleep operations. During periodic ranging, the BS can choose to abort ranging and force the MS to reinitialize its MAC if the physical layer parameters are too poor or the BS needs to verify the MS's credentials. In addition to regularly performing periodic ranging, the MS needs to maintain the current version of the DCD and UCD. This is to allow the MS to be able to promptly resume its normal operations on the deactivation of a PSC. In particular, if the MS does not have the latest version of the DCD and UCD messages, it needs to interrupt the sleep patterns and successfully obtain the current version before returning to the sleep operations.

Since periodic ranging is controlled by the MS and, more importantly, does not allow the BS to keep track of the availability of the MS, the BS may request the MS to perform a keep-alive check for further adjustments and reporting. The keep-alive check mechanism involves the MS transmitting a dedicated ranging request (RNG-REQ) at the specific time indicated by the BS and remaining awake until the adjustment is completed. If no ranging request is received after a number of repeated dedicated UL allocations, the BS may consider the MS inactive and begin removing the MS context from the system. Similarly, if the MS does not receive any ranging response after a number of repeated ranging requests, the MS may consider that it has lost communication with the current BS and begin network re-entry. The BS can also direct the MS to initialize its MAC and go through the initial network entry procedure by aborting the ranging process. In addition, the BS may initiate a keep-alive check by allocating an UL data grant during an availability interval, to which the MS has to respond with an UL transmission. A bandwidth request of size zero may be used to indicate that the MS has no data to transmit and prefer to continue with the current power saving.

6.3.2.3 Summary of Key Profile Features

A summary of the key features is given in Table 6.7.

6.3.3 Idle Mode

During the sleep mode, the MS can conserve energy by suspending communication for selected connections with the serving BS without losing its active status. In those situations where the MS does not expect much traffic and prefers to enter an extended period of unavailability, Mobile WiMAX allows the MS to de-register from the current serving BS and yet be able to promptly re-enter the network on pending data. Specifically, with the idle mode, the MS only periodically listens to DL broadcast control and paging messages without registration to any specific BS. In addition, the MS can transverse through the coverage of multiple BSs with no requirements for handover or maintaining its active status. By restricting the period of availability, the MS can significantly conserve power while being able to resume its normal

Table 6.7 Sleep mode key profile features [147]

Category	Feature	BS support	MS support
Sleep mode	Sleep mode implementation in MS	N/A	Y
Sleep mode	Power saving class (PSC) Type I support	Y	Y
Sleep mode	Support of traffic indication message for PSC Type I	Y	Y
Sleep mode	Traffic-triggered wakening flag	Y	Y
Sleep mode	PSC Type II support	N	N
Sleep mode	PSC Type III support	N	N
Sleep mode	Activation of PSC by unsolicited SLP-RSP message from BS	Y	Y
Sleep mode	DL sleep control extended subheader	Y	Y
Sleep mode	Bandwidth request and UL sleep control header	Y	Y
Sleep mode	Support of periodic ranging in sleep mode	Y	Y
Sleep mode	DL traffic indication by RNG-RSP message	N	N
Sleep mode	Sleep mode multicast CID support at MS	N/A	Y
Sleep mode	Sleep mode multicast CID support at BS	Y	N/A

operations with a reasonable effort. Without the idle mode, the MS would need to go through the full network entry procedure after de-registering from the current serving BS. In addition to ramping down its power consumption, the MS may choose to autonomously scan neighbor BSs, conduct ranging, reselecting its preferred BSs or any other activities during the idle period in which the MS becomes unavailable for any DL traffic. This procedure also benefits the network tremendously by eliminating unnecessary air interface and backhaul signaling that are needed to maintain the connections for an inactive MS.

The concept of idle mode involves a paging group, which consists of a number of BSs having the same paging characteristics that form a continuous geographical area. During the MS paging listening interval, all BSs within the same paging group broadcast similar paging messages indicating the presence of pending DL traffic, or containing requests for location update or full network entry to all MSs belonging to that paging group. Therefore, as long as the MS stays within the geographical area of the paging group it has registered for and is able to decode one of the broadcast paging messages, the MS is considered reachable by the network and should be able to respond promptly to paging. The MS may choose to exit the idle mode and return to its normal operation at any time. Similarly, the idle mode may be interrupted by the BS as specified in the paging message. To provide the different geographical coverage or diverse idle mode parameters such as paging cycles and offsets, the BS may be a member of more than one paging group consisting of different sets of BSs. However, the MS can only belong to one paging group at any time.

6.3.3.1 Idle Mode Initiation

The idle mode can be initiated by the MS or the BS. For the MS-initiated idle mode, the MS sends the de-registration request (DREG-REQ) with a specific action code to de-register from the BS and start the idle mode. The same DREG-REQ with a different action code can also be used to terminate the connection with the BS and then leave the network. The BS uses the de-registration command (DREG-CMD) to indicate its response to the MS, whether the request for idle mode is allowed and whether additional actions are needed. If the idle mode is allowed, paging information such as cycle, offset, paging group ID and paging controller ID will be included in the response message. Similarly, the BS may initiate the idle mode by sending the unsolicited de-registration command (DREG-CMD) to the MS. If the MS cannot enter the idle mode (e.g. it has pending UL data to transmit), it may reject the unsolicited idle mode request. Different action codes in the de-registration request (DREG-REQ) are used by the MS to indicate the rejection or acceptance of the BS-initiated idle mode. Since the execution of the idle mode involves actual de-registering of the MS from the active state, there are timers in place for both the MS and BS to avoid miscommunication. These timers also allow sufficient retransmissions of signaling exchanges during the negotiation of the idle mode before a successful initiation of the idle mode is assumed and management resources, as well as active connections, are removed.

While the MS is in the idle mode, the paging controller and other entities in the ASN (see Section 2.2.1.2) may keep some service and operational information of the MS in order to expedite future network re-entry from the idle mode. However, retained information does not include management CIDs or transport CIDs for unicast connections that have been removed upon de-registering. In order to continue having the idle mode resources retained at the paging controller, the MS needs to perform a periodic location update that is used to identify its continuing presence in the network, or respond to the paging message addressed to itself in a timely manner. If the paging controller cannot determine the presence of the MS, it may decide to remove service and operational information necessary for the MS to execute an expedited network entry.

6.3.3.2 Location Update and Network Re-entry from Idle Mode

On the transmission of the idle mode request command (DREG-CMD) by the BS, both the MS and the paging controller start their own timers, namely the *idle mode timer* and the *idle mode system timer*. These timers are reset after a successful idle mode location update while the MS is still in the idle mode. On expiration of the MS-maintained idle mode timer, the MS expects that all MS service and operational information retained by the paging controller for idle mode management purposes has been discarded and an expedited network entry is no longer possible. Similarly, the paging controller uses the idle mode system timer to keep track of the continuing network presence of the MS and removes all MS service and operational information for idle mode management once this timer has expired or the MS has successfully resumed its normal operations. The MS may choose to perform a location update at any time to avoid expiring these timers. In addition, a location update may be done in order to change the paging cycle for the MS or update its paging group. Since the MS may be moving through the coverage of multiple BSs while in the idle mode, it is important that the MS registers with the correct paging group. This is to continue receiving paging messages and keep its resources at the paging controller for a future re-entry. Similarly, the BS may request the MS to perform a location update when it needs to modify the paging cycle or verify the MS's presence. Furthermore, any of the following five location update conditions can be used to trigger a location update at the MS:

1. **Paging group update:** This condition is met when the MS detects a change in paging group indicated in the DCD or paging advertisement (MOB_PAG-ADV) messages transmitted by the preferred BS.

2. **Timer update condition:** The MS needs to perform a location update prior to expiration of the idle mode timer.

3. **Power-down update:** This location update allows the paging controller to be informed of an orderly power-down of the MS. With a successful power-down location update, all idle mode retained information for the MS is erased from the paging controller. In the absence of a successful power-down update, the paging controller may determine the continuing presence of the MS by repeating the location update polling. Once the paging retry count is exhausted without any response, all idle mode retained information is deleted.

4. **MAC hash skip threshold update:** This condition is met when the MS MAC hash skip counter exceeds the MAC hash skip threshold. The corresponding counters are reset on a successful location update.

5. **Multicast/broadcast service update:** The MS with a multicast/broadcast service (MBS) flow needs to perform a location update when it detects a change in the corresponding MBS zone. MBS zone information is included in the DCD message transmitted by the preferred BS. This update may be skipped if it already has the multicast CID mappings in the selected MBS zone.

The location update procedure involves a network re-entry at the preferred BS via the initial ranging procedure and transmission of the ranging request (RNG-REQ) with the flag indicating a location update request. When the MS includes the valid HMAC/CMAC tuples in the ranging request (RNG-REQ), the BS can authenticate the validity of the MS's security context without having to go through normal authentication usually performed during the initial network entry. On successful validation of the MS context, the BS can respond with a successful location update response including, if necessary, the new paging group ID and MBS service parameters, along with the appropriate HMAC/CMAC tuples in the ranging response (RNG-RSP). Then the location update procedure is finished. Over the backbone, the paging controller is notified of the new location information of the MS as well as any changes in the associated paging group. If the MS chooses not exit the idle mode, it determines the next paging interval and returns to the idle mode. This procedure is known as 'secure location update'. For any reasons including invalid security context, the BS may direct the MS to perform network re-entry from the idle mode via the *unsecure location update* process by indicating a failure of location update response in the ranging response (RNG-RSP).

When the MS does not have the valid HMAC/CMAC tuples, is unable to validate the security context contained in the ranging response (RNG-RSP) from the BS, or there are other indications that the BS requires authentication of the MS's security context, the MS is required to perform an unsecure location update or network re-entry from the idle mode. The network re-entry from idle mode procedure is similar to that of initial network entry except that the MS indicates its re-entry from idle mode intent and the current paging controller ID for the expedited network re-entry. Then the target BS can request MS service and operational information from the paging controller to speed up the re-entry by use of handover process optimization and/or unsolicited basic capability/registration responses (SBC-RSP, REG-RSP). On successful network re-entry from the idle mode, the preferred target BS becomes the new serving BS and the network re-entry procedure concludes with the establishment of normal operations. The paging controller is informed over the backbone of the successful resumption of normal operations and removes the MS from the previously registered paging group.

Table 6.8 Idle mode key profile features [147]

Category	Feature	BS support	MS support
Idle mode	General idle mode functionality	Y	Y
Idle mode	MS-initiated idle mode by DREG-REQ message	Y	Y
Idle mode	BS-initiated idle mode by unsolicited DREG-CMD	Y	Y
Idle mode	Maintain connection information at BS during idle mode initiation process	Y	Y
Idle mode	Request for MS to retain service and operational information by DREG-CMD message	Y	Y
Idle mode	Request from MS to BS to retain service and operational information by DREG-REQ message	Y	Y
Idle mode	Implementation in MS of the reception of periodic transmission of MS MAC address hash in paging message	N/A	N
Idle mode	Implementation in BS of periodic transmission of MS MAC address hash in paging message for an idle MS	N	N/A
Idle mode	BS capability of providing dedicated ranging region and ranging code allocation for location update or network entry of MS in idle mode	N	N/A
Idle mode	MS capability of using dedicated ranging region and ranging code allocation for location update or network entry of MS in idle mode	N/A	N
Idle mode	Paging group update at MS	Y	Y
Idle mode	Timer location update at MS	Y	Y
Idle mode	Power-down location update at MS	Y	Y
Idle mode	MAC hash skip threshold location update at MS	N/A	N
Idle mode	Secure location update	Y	Y
Idle mode	Unsecure location update	Y	Y
Idle mode	Idle mode multicast CID support at MS	N/A	Y
Idle mode	Idle mode multicast CID support at BS	Y	N/A

6.3.3.3 Cell Reselection and Paging

Similar to cell reselection for handover, cell reselection during the idle mode allows the MS to reselect a preferred BS, synchronize, obtain management messages such as DCD, DL-MAP and broadcast paging (MOB_PAG-ADV), and determine the next paging interval. A preferred BS is defined as a BS that the MS selects as the one with the best air interface. A broadcast paging message advertises the paging groups associated with the BS and, if any, notifications of location update polling or network entry request for the specified MSs within each paging group. The MS is identified by the 24-bit MAC address hash. The BS may include the dedicated ranging codes as well as the ranging opportunities for the paged MSs to expedite the ranging procedure.

To ensure that the MS is given sufficient opportunities to receive and respond to the paging indication, the broadcast paging message is retransmitted at the next MS paging listening interval until the MS has successfully performed the action identified in the paging message (i. e. location update or network re-entry) or the number of retries reaches the paging retry count. On expiration of the paging retry count, the paging controller considers that the MS is no longer available and deletes MS service and operational information. Consequently, the MS is required to perform the initial network entry upon the next communication attempt with a BS.

6.3.3.4 Summary of Key Profile Features

A summary of the key features is given in Table 6.8.

7

Radio Evolution beyond System Profile Release 1.0

7.1 Mobile WiMAX System Profile Release 1.5

The Mobile WiMAX System Profile Release 1.0 [146] has been available since 2006 and the first WiMAX-certified radio equipment became available in 2007. With the WiMAX Forum Network Specification Release 1.0, network operators are able to deploy multi-vendor WiMAX networks with interoperable infrastructure and devices. Based on the advances in radio technologies and lessons learned from real-world deployments, the WiMAX Forum has defined a new set of requirements for evolving Mobile WiMAX [142]. Consequently, the Mobile System Profile Release 1.5 is being developed by the Technical Working Group (TWG). The expected improvements include the new profile for frequency division duplexing (FDD), more efficient radio resource management and higher MAC-level efficiency. All necessary fixes required to support these new features were incorporated in the 2009 revision of the IEEE 802.16 standard, namely IEEE 802.16-2009 [1].

7.1.1 Support of Frequency Division Duplexing

Although the WiMAX Forum Fixed Certification Profiles available since 2006 do support both time division duplexing (TDD) and frequency division duplexing (FDD) modes, the later Mobile Certification Profiles only support TDD [145]. Due to the growing demand for Mobile WiMAX worldwide, there has been interest in deploying the technology in the geographical areas where the paired spectrum (i.e. separated downlink and uplink frequencies) is allowed. Therefore, a number of enhancements including the new FDD frame structure and improved MAC efficiency have been introduced in the Mobile System Profile Release 1.5 to satisfy the needs of operators owning the FDD spectrum and further strengthen the ecosystem of Mobile WiMAX. In addition, a number of changes were incorporated in the development of the IEEE 802.16-2009 standard [1], which is the latest revision of the IEEE 802.16 family of standards and serves as the underlying reference for the Mobile System Profile Release 1.5.

Deploying Mobile WiMAX   Max Riegel, Dirk Kroeselberg, Aik Chindapol and Domagoj Premec
© 2009 John Wiley & Sons, Ltd

The Service Provider Working Group (SPWG) [142] also mandates that the Release 1.5 BSs support both half-duplex FDD and full-duplex FDD terminals that are implemented according to the Mobile System Profile Release 1.5. Consequently, the FDD BSs can only operate in the full-duplex FDD mode. A half-duplex FDD (H-FDD) terminal cannot transmit and receive simultaneously and therefore has to alternate its transmitting and receiving cycles. Not only can the alternating pattern be efficiently implemented with low complexity, but it also resembles TDD operations, therefore allowing faster development of the H-FDD hardware. On the other hand, a full-duplex FDD (F-FDD) device is capable of simultaneous transmission and reception and thus is able to achieve the faster transfer rate as well as better end to end delay. However, it normally requires additional processing including complex filtering to separate interference between the uplink (UL) and downlink (DL) communication. There is generally a tradeoff between the simplicity of a half-duplex implementation and the performance of full-duplex terminals.

Figure 7.1 illustrates the FDD frame structure that is specifically designed to efficiently support H-FDD terminals. In particular, H-FDD terminals are placed into two groups, namely group 1 (G1) and group 2 (G2), operating in two different temporal portion of the frame. The common preamble is located at the beginning of each frame and used by all terminals regardless of their group assignments for synchronization. Therefore, there is no UL allocation during the preamble period. It can be seen in Figure 7.1 that the resource allocations for G1 and G2 are mutually exclusive so that the H-FDD MS can operate in only one group. This arrangement significantly increases the frame utilization by alternating the transmission and reception between the two groups. Specifically, G1 H-FDD terminals can only receive data during the first portion of the DL subframe and consequently transmit during the second portion of the UL subframe. Similarly, H-FDD terminals in G2 transmit and receive during the first portion and the second portion of the UL and DL subframes, respectively. In addition, each group operates independently and has its own control signaling including the frame control header, MAP and DL/UL channel description messages and permutation zones. The BS can also dynamically adjust the fraction of the frame allocated to each group by specifying the number of allocated OFDM symbols in the MAP messages. The flexibility in adjusting the

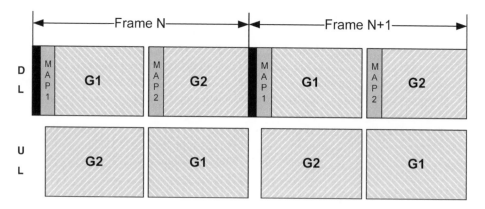

Figure 7.1 FDD frame structure supporting two groups (G1 and G2) with group-specific control signaling (MAP1 and MAP2) and a common preamble at the beginning of each frame

group boundary enables the BS to perform load balancing between the two groups that may serve MSs with the different signal-to-noise ratios (SNRs).

Due to the half-duplex capability, each H-FDD MS can only participate in one group at any given time. Similarly, the BS expects the H-FDD MS to use the resources allocated in the assigned group. Hence, it is important that the BS and MS remain synchronized regarding the group status. For example, when the MS first enters the network (e.g. via the initial network entry, network re-entry or handover), it is required to use G1 as the default. On successful entry or re-entry, the MS may be instructed to remain in G1 or switch to G2 at the discretion of the BS in order to maintain optimal performance as well as network stability. Group switching may occur as often as necessary and is fully under the control of the BS. Failing to switch group as instructed by the BS may cause the MS to lose communication with the BS and eventually have to re-enter the network. As a precaution, the BS may utilize additional mechanisms for detecting the success or failure of the group switch, by request. The BS may assign an acknowledgement channel or *group switch timer* for the specific MS to indicate a successful presence in the new group. The MS may return to the original group after a number of repeatedly failed attempts to join the new group. In the event that the MS fails to decode a number of DL MAP messages and thus is unable to derive the correct group boundary, it has to cease operations in G2 and return to G1 to update boundary information prior to returning to its normal operations in G2.

In addition to group switching, which enables the BS to dynamically adjust the number of MSs associated with each group, the BS can also choose different system parameters to further optimize the overall performance of each group. For instance, the BS may assign a number of terminals having moderate to high SNRs to one group and use the more efficient modulation and coding (less robust) schemes only for this group in order to increase the data rate as well as the bandwidth efficiency. Since grouping can be performed according to the SNR level, the MS may be required to switch to the other group having less efficient modulation and coding (more robust) when it experiences the lower SNR. In general, G1 should provide more robust modulation and coding as it serves as the default group and is expected to be reachable for all MSs in the system. Since the two groups operate independently, the MS should be able to operate in the assigned group as long as it has updated group boundary information as well as channel descriptions such as DCD and UCD. Although the BS may choose the different DCD and UCD parameters for each group, it is more convenient to use the same set of parameters because both groups belong to the same system and the broadcast interval for the DCD and UCD messages is normally in the order of 10 to 100 frames. In addition to the SNR, the BS may choose to assign groups according to other criteria such as the capabilities supported at the MSs.

Although the frame structure shown in Figure 7.1 is optimally designed for H-FDD terminals, F-FDD terminals can easily be supported by mimicking the multi-carrier operations. Specifically, the F-FDD terminal operates with a single MAC instantiation sharing two physical layers (two H-FDD groups) and the BS may assign resources to either or both groups. For example, the full-duplex MS may receive a DL allocation for the same connection in both groups as indicated by the group-specific DL-MAP. Similarly, control signaling such as MAC management messages and bandwidth requests may be used in either group. In turn, a full-duplex MS can have a much shorter turnaround time because there are no alternating transmitting and receiving requirements as in the case of H-FDD terminals. Another advantage of F-FDD operations is that the UL transmissions may be spread over time by utilizing both UL groups to increase the link budget. One possible enabling technique is to fragment the PDU

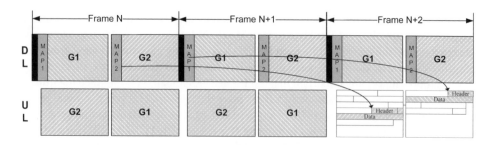

Figure 7.2   Resource allocation in both groups using PDU fragmentation. The arrows indicate the time relevance between the allocations and signalling

and transmit the fragments over the UL bursts as shown in Figure 7.2. The F-FDD capability is negotiated during the network entry procedure; therefore, a full-duplex capable FDD MS has the option to enter the network as a H-FDD or F-FDD one. However, it cannot dynamically switch between the full-duplex and the half-duplex capabilities unless a network entry or re-entry is performed.

7.1.1.1 Summary of Key Profile Features

A summary of the key features in given in Table 7.1.

7.1.2 Persistent Allocation for VoIP

In the Mobile System Profile Release 1.0, a single allocated resource unit uses an information element in the MAP messages (e.g. normal MAP, compressed MAP, H-ARQ MAP) to point to a logical location in the corresponding subframe. With consistently small and periodic data such as VoIP, the amount of overhead required to signal the allocation may be burdensome and reduces the bandwidth efficiency. For example, the adaptive multi-rate (AMR) codec operating at 12.2 kbit/s produces a 44-byte packet every 20 ms during the activity period and an 18-byte packet every 160 ms during the period of inactivity [150]. However, the MAP overhead required to carry one VoIP packet in each direction could easily reach 6 bytes [149], [154] excluding the fixed part of the MAP. Therefore, the signaling inefficiency may become the limiting factor, especially for networks supporting a large number of VoIP users. Persistent

Table 7.1   FDD key profile features [155]

Category	Feature	BS support	MS support
FDD	A 5 ms frame with two standalone half-duplex groups	Y	Y
FDD	Support of full-duplex operations	Y	Optional
FDD	Support of MAP-based signaling for dynamic group partitioning	Y	Y
FDD	Alignment of DCD and UCD parameters in both groups	Y	Y
FDD	Full-duplex UL allocation using PDU fragmentation	Y	Optional

allocation for VoIP, also known as persistent scheduling, takes advantage of the predictable nature of these small data packets and allows a MAP information element to persistently point to the same logical allocation. In other words, a single persistent allocation can assign resources for a specific connection in multiple frames in advance and thus eliminate the need for having multiple information elements in many frames. In addition, the new persistent information element allows an allocation for both persistent and non-persistent traffic in the same message. It is shown in [149] that persistent allocation for VoIP can reduce the MAP overhead by 30–45% and thus increase the overall VoIP capacity by 15%.

Since the level of modulation and coding that is appropriate for carrying the data bursts may change very quickly during the interval in which the resource is pre-allocated, resources assigned too far in the future may become wasteful. As the channel conditions fluctuate, overly conservative scheduling (i.e. allocating more resources than necessary) reduces the packing efficiency. On the other hand, overly aggressive scheduling (i.e. insufficient resource allocation) may result in non-usable allocation as the supplied resource is below the current demand. Furthermore, re-arranging a persistent allocation too often defeats the purpose of reducing the signaling overhead that is associated with each information element. Therefore, the BS needs to find a balance in maximizing the packing efficiency while minimizing the signaling overhead. Such scheduling algorithms that are specifically designed for certain applications or deployment models are beyond the scope of the standard as well as the WiMAX Forum certification programs.

Not only does persistent scheduling require sophisticated design and optimization at the BS, the MS also needs to synchronize its persistent allocations with the BS. Unlike regular non-persistent allocations that only schedule transmissions on a frame-by-frame basis, losing any allocation or deallocation requests may result in resources becoming useless until the next round of allocation and deallocation requests. Therefore, a number of error handling mechanisms are incorporated in the IEEE 802.16 standard and the Mobile System Profile Release 1.5 to ensure robust allocation and deallocation signaling. In particular, acknowledgement channels are used to provide a two-way signaling handshake. Some examples of the error handling procedures are described below and may optionally be used in conjunction with the persistent allocation feature:

- **MAP acknowledgement (MAP ACK)**: MAP ACK is a single-bit indication that indicates successful reception of a persistent allocation or deallocation request. It is transmitted by the MS in the UL fast feedback channel as specified in the request sent by the BS. Similar to dynamic resource allocation (i.e. non-persistent allocation used in Release 1.0), the BS implicitly detects a successful allocation in the absence of MAP ACK when the MS transmits the UL burst as scheduled. Likewise, a successful deallocation is implied when the persistent allocation is no longer used as indicated by the deallocation request.
- **MAP negative acknowledgement (MAP NACK)**: Contrary to MAP ACK, which uses a positive acknowledgement in response to a successfully received persistent control message, MAP NACK allows the MS to feed back a negative acknowledgement when it does not receive an expected allocation or deallocation request. For example, the MS transmits MAP NACK when the current persistent allocation has expired and it has not received a new allocation for the persistently scheduled connection. MAP NACK is also transmitted using the UL fast feedback channel as specified by the BS and may be shared by multiple MSs.

Table 7.2   Persistent allocation key profile features [155]

Category	Feature	BS support	MS support
Persistent allocation	Persistent allocation support	Optional	Y
Persistent allocation	Error handling procedures: allocation/ usage of MAP ACK channel for persistent allocation	Optional	Y
Persistent allocation	Error handling procedures: allocation/ usage of MAP ACK channel for persistent de-allocation	Optional	Y
Persistent allocation	Error handling procedures: MAP NACK channel allocation/Usage	N	N
Persistent allocation	Flexible persistent H-ARQ region	Optional	Y

7.1.2.1 Summary of Key Profile Features

A summary of the key features in given in Table 7.2.

7.1.3 Load Balancing and Seamless Handover

For WiMAX systems based on Release 1.0 [147], the MS has nearly full control over handover decisions. Specifically, the MS may choose to initiate or execute a handover with little intervention from the network or, in some cases, even against the recommendations from the network. This flexibility enables simple radio resource management and network interactions during the initial deployments where the insufficient coverage remains a primary concern. As the MS roams through those networks, the serving BS may not have as complete a knowledge as the MS regarding the radio conditions, especially at cell edges where handovers normally occur. Therefore, the MS in this case should be able to decide whether to stay with the current serving BS or switch to another preferred BS according to its own criteria. In addition, since the MS may have to execute the handover at the last minute in order to maintain its active connection, the current serving BS, in some cases, may not be aware that the MS has left the network. As a result, the BS has to retain the system resources that are no longer needed for that MS until it has received a notification over the backbone from the new serving BS or it has determined that the MS is no longer available.

7.1.3.1 Load Balancing

One major issue with MS-centric handover in Mobile WiMAX Release 1.0 is scalability. As the network continues to grow, a hierarchical or overlapping cell structure may be deployed in order to provide ubiquitous coverage and ensure that all users receive the best possible services. For example, one network operator may utilize multiple picocells within the same geographical area of a large macro cell. Therefore, the network may prefer to have a fast moving MS stayed with the larger cell so that it does not need to perform multiple handovers while moving through the coverage area while letting static and nomadic users connect to a smaller cell with lighter load. Without network-controlled handover, many MSs may perceive a high-power cell as more preferable, resulting in unnecessary network congestion. The use of multiple carriers with

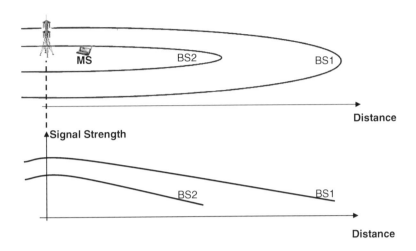

Figure 7.3   Overlapping cell structure with collocation of BSs with different sizes and transmit power

different frequency assignments is another example of an overlapping cell structure that could benefit from load balancing. Figure 7.3 illustrates an overlapping cell structure with collocated BSs having different transmit power and coverage. Similarly, without network-centric handover the majority of MSs may end up using the overly crowded BS2 based on the higher signal strength.

The serving BS, in conjunction with neighboring BSs, may attempt to balance the load among themselves in order to maintain the network stability as well as the optimal level of services. However, without any tight control of the handover behavior of the MS, MS-initiated handovers may inadvertently interfere with those decisions initiated by the network for the purpose of load balancing. For instance, the network may initiate a handover for a few MSs along the cell edge of a heavily loaded cell to a nearby BS. According to Release 1.0, although the BS may initiate a handover request, the MS is not required to execute the request to hand over to one of the suggested BSs. Even if the MS were to follow the request and perform the handover to one of the suggested BSs, it could immediately attempt to hand over to the previous serving BS or another heavily loaded BS. The back-and-forth handover attempts, also known as the ping-pong effects, create unnecessary signaling and instability in network control and management.

In the WiMAX Mobile System Profile Release 1.5, a few load balancing techniques are incorporated to provide more control to the serving BS regarding the handover decisions and initiations, especially in a loaded network. First, the *BS-controlled MS-initiated handover or BS-controlled handover*, restricts the ability of the MS to hand over autonomously. Specifically, when the BS initiates a handover request, the MS is required to comply with the request and move to one of the recommended target BSs as specified in the request. Only in those situations where the MS is unable to perform the handover (e.g. insufficient signal strengths from all recommended target BSs) may the MS reject the handover request and stay with the current serving BS. However, it cannot choose to hand over to another target BS outside the recommended list or not meet the criteria set by the BS. Additionally, when the MS initiates a handover, the serving BS needs to agree with the choices of target BSs proposed by the MS as indicated in the handover response. The above BS-controlled restrictions effectively eliminate the ping-pong effects and give the network much tighter control of the handover behavior of all MSs within its control. In severe cases, in which the MS experiences a sudden change in the radio condition or is about to lose the connection to the current serving BS, the MS is allowed to initiate a handover only to the target BSs meeting the trigger conditions.

Unlike handover mechanisms used in Release 1.0 as shown in Figure 6.14 and Figure 6.15, BS-controlled handover restricts the handover options to the recommended BSs or those meeting the trigger conditions. Figure 7.4 and Figure 7.5 illustrate BS-controlled handover when the BS and MS initiate the handover procedure, respectively. Since the trigger conditions are set and broadcast to all terminals by the serving BS, the network can also expect a more uniform behavior even in the cases of MS-initiated handover. Although BS-controlled handover is a mandatory feature in Release 1.5, legacy terminals (i.e. Release 1.0 implementation) may or may not be able to comply with BS-controlled handover unless this capability has been explicitly indicated during the capability negotiation. As with any other features, the BS may choose to abort or continue with the network entry attempt by the MS according to the present capabilities.

For the terminals that are already registered and active in the network, BS-controlled MS-initiated handover provides a predictable mechanism for the network to balance the load among neighboring BSs. Similar to other handover procedures, it cannot be used with the terminals that are trying to enter the network. Therefore, Release 1.5 includes additional load balancing features that indicate to the MS during the network entry that the BS is currently busy. For example, the heavily loaded BS can redirect the new MS to another BS during network entry or prevent the MS from re-entering the same BS for a specified time.

The second load balancing feature uses the preamble index and/or downlink frequency override that redirects the MS to a neighboring BS during the initial network entry, handover

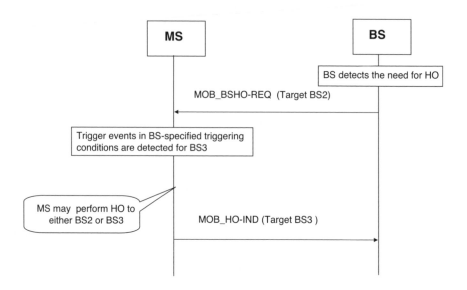

Figure 7.4   Example of BS-controlled, BS-initiated handover

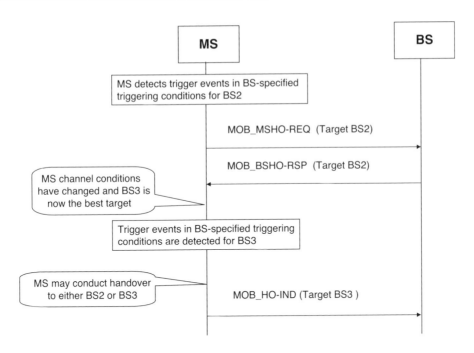

Figure 7.5   Example of BS-controlled, MS-initiated handover

attempt or network re-entry. The override serves as an indication that the BS is currently busy and the MS should receive better service at the recommended BSs. It also helps eliminate the need to a lengthy BS-controlled handover later on. On receiving the preamble index override or the frequency override in the ranging response (RNG-RSP) message from the BS, the MS should perform ranging and network entry with one of the BSs indicated in the override message. Although the MS may choose any other BS for network entry, it is likely that the loaded BS has coordinated with neighboring BSs and only recommended the BSs that are willing to accept new terminals. Lastly, the *ranging abort timer* can also be used to balance the load. The BS may include the ranging abort timer in the ranging response (RNG-RSP) with the *Abort* status to prevent the MS from re-entering, or more specifically repeated ranging at this BS, until the ranging abort timer has expired. The last mechanism avoids unnecessary network entry attempts while the BS is still overloaded, yet allows the MS to return at a later time. A combination of these load balancing techniques may be used to increase the efficiency of radio resource management.

7.1.3.2 Seamless Handover

In Mobile WiMAX Release 1.0, only hard handover and optimized hard handover are supported (see 6.3.1). Since the handover procedure involves a number of management messages exchanged among the MS, the serving BS and one or more target BSs, it is desirable to shorten the number of handover messages and thus reduce the handover interruption time. Additionally, the reduction in the number of management messages can actually improve the robustness of handover execution, which is normally performed when the MS is at its weakest points such as the cell edge or blind spots. Most importantly, these enhancements can positively contribute to the quality of delay-sensitive applications such as VoIP or H-ARQ data. Therefore, *seamless handover* is incorporated in Release 1.5.

The key elements of seamless handover [144], which is based on *optimized handover* in Release 1.0, are the preassignment of the CID prior to the actual re-entry at the target BS and the ability to start data transmissions before completion of security authentication. Specifically, the current serving BS may negotiate with the potential target BSs over the backbone and preassign a block of CIDs that the MS can immediately use upon successful re-entry at the target BS. The CID preassignment, which includes basic CID, primary management CID and transport CID, eliminates the sequence of the ranging request and response (RNG-REQ/RSP) normally used to assign the CID and authenticate the MS's credentials upon successful re-entry. In addition, the *seamless handover action time* can optionally be used to indicate the time after which the preallocated CIDs are valid at the target BS. In addition, the target BS may immediately schedule the transmissions of data packets in both the UL and DL directions after a successful re-entry. However, the authentication steps normally performed during the ranging requests and responses need to be finished before the receiver can pass the data packets to the higher layer for processing.

7.1.3.3 Summary of Key Profile Features

A summary of the key features in given in Table 7.3.

7.1.4 Location-Based Services

In Mobile WiMAX Release 1.0, there is no assistance from the network regarding the location-based services (LBS). Not only does the LBS enable exciting location-based applications such as navigation systems and mobile e-commerce, but the network may also use the location in providing information related to network entry or regulatory restrictions. For example, GSM phones may use the country code that is broadcast by the BSs to determine the appropriate spectrum masks based on the country-specific regulations. Although many modern terminals

Table 7.3 Load balancing and seamless handover key profile features [155]

Category	Feature	BS support	MS support
Load balancing	Load balancing using BS-initiated HO (includes BS-controlled HO indication)	Y	Y
Load balancing	Load balancing using preamble index and/ or DL frequency override	Optional	Y
Load balancing	Load balancing using ranging abort timer for initial network entry, handover and re-entry from idle mode.	Y	Y
Seamless handover	Seamless HO support	Optional	Optional
Seamless handover	Data exchange before RNG-REQ/RSP transaction	Optional	Optional

Table 7.4 LBS key profile features [155]

Category	Feature	BS support	MS support
LBS	Location information	Optional	Optional
LBS	Timing information	Optional	Optional
LBS	BS frequency accuracy information	Optional	Optional

are already equipped with GPS, there are operational limitations, especially when there is no direct line of sight to the GPS satellites. In addition, Mobile WiMAX devices with a very small form factor, such as a USB dongle, may not have a built-in positioning system. Therefore, the BS according to Release 1.5 may optionally broadcast location information about itself as well as about neighboring BSs in order to assist the MS with LBS. Such information may include geographical data (e.g. latitude, longitude, altitude), GPS time and BS transmit frequency accuracy. The MS can use broadcast data along with any autonomously obtained information such as GPS data to speed up the acquisition of the current location or to approximate its location when GPS is not available. Additionally, the broadcast location from the BSs may help the MS to expedite the channel scanning during network entry or re-entry. For example, on powering up, the MS may use the broadcast location or the last known location instead of an exhaustive scan to determine the preferred channels.

7.1.4.1 Summary of Key Profile Features

A summary of the key features is given in Table 7.4.

7.1.5 Coexistence with Wi-Fi and Bluetooth

Given the multi-radio connectivity seen in mobile devices nowadays it is increasingly common for WiMAX devices to be used in conjunction with other radio access technologies such as Wi-Fi or Bluetooth. The combination of WiMAX and Wi-Fi, for example, may be used in customer premises equipment (CPE) that uses a WiMAX link as a backbone while providing indoor coverage over Wi-Fi. Similarly, a Bluetooth headset may be used to provide a handfree connectivity to a WiMAX terminal. Although both Wi-Fi and Bluetooth equipment operate in the unlicensed spectrum of 2.4 GHz, they may actually interfere with a WiMAX terminal operating in 2.3 and 2.5 GHz, especially when these radio modules are physically collocated within the same device. Due to different medium access protocols as well as the hidden node problem, the resulting interference may significantly degrade the performance of all parties involved. Therefore, a number of coexistence mechanisms are included in Release 1.5. Further details on these mechanisms including on that modifies the sleep patterns of the MS to provide transmit and receive opportunities for Wi-Fi and Bluetooth systems [143]. For instance, the BS is not allowed to allocate any DL or UL data during the sleep windows of the PSC that has been explicitly defined for coexistence.

8

Outlook

In the previous chapters, we explain the functional blocks and protocol features that can be used in deploying a Mobile WiMAX network based on the WiMAX Forum Release 1.0 and 1.5 specifications as well as the published IEEE 802.16 standard. These specifications cover the basic radio and network architecture including a set of add-on standalone features. In summary, the following fundamental features are discussed:

- Splitting access (ASN) and core (CSN) entities to allow them to be deployed by different business entities, mapping to NAP and NSP.
- Strong security and policy control as well as support for numerous network services like IP mobility based on an AAA framework with roaming support.
- Subscription management including dynamic over-the-air (OTA) activation and provisioning of WiMAX-enabled retail devices.
- Network support for quality of service (QoS), location-based services, or emergency calls.
- Integrated hooks for enabling the 3GPP-defined IMS service over WiMAX networks.
- ASN-internal mobility as well as network-wide IP mobility with different options for the IP mobility protocol.
- Radio architecture covering the physical (PHY) and the medium access control (MAC) layer of Mobile WiMAX.

There is work in progress beyond Release 1.5 on improving and optimizing the overall radio and network architecture based on findings from the currently deployed networks based on the previous release. Further efforts are dedicated to extend the list of standardized features that can enrich existing as well as new services in a meaningful and interoperable way.

For the remainder of this chapter, we focus on WiMAX Forum release planning and other ongoing work on a few selected topics in order to share initial insights.

8.1 WiMAX Forum Release Planning

The contents of this book mainly follow the Mobile WiMAX Release 1.0 and 1.5 specifications. There will be further developments after the completion of Release 1.5, which address

Deploying Mobile WiMAX   Max Riegel, Dirk Kroeselberg, Aik Chindapol and Domagoj Premec
© 2009 John Wiley & Sons, Ltd

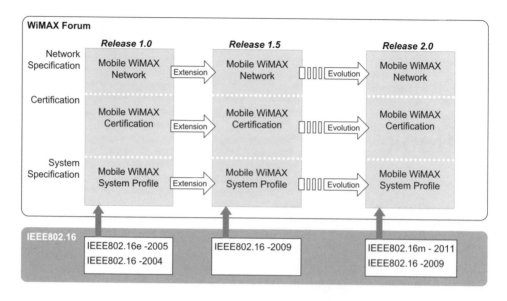

Figure 8.1   Mobile WiMAX technology roadmap

enhancements to the radio interface, the network and roaming architecture, and the services provided on top of the plain IP communication layer of WiMAX.

Release 1.0 of Mobile WiMAX established the basic functionality of a mobile access network for IP-based services. It utilizes the IEEE 802.16e-2005 OFDMA radio interface together with an IP-based mobile access network. Also, the first release of the WiMAX roaming architecture as well as certification and network interoperability testing are incorporated in Release 1.0.

After the basics of the Mobile WiMAX technology have been laid down with the completion of the Release 1.0 specification, Release 1.5 extended the initial specification framework to provide a complete set of specifications addressing all the required functions of fixed, nomadic, portable and mobile deployments for all kinds of business applications worldwide.

Having reached a functionally complete, or at least nearly complete, specification framework with Release 1.5, Mobile WiMAX will continue to evolve with upcoming Release 2.0 as shown in Figure 8.1. The IEEE 802.16m amendment to the radio standard will provide a remarkable increase in the performance, and essentially more throughput and lower latency, of the mobile networking technology. Release 2 based on IEEE 802.16 m is aimed at fulfilling all the requirements for membership in the IMT-Advanced family.

8.2 Network Architecture Evolution

Essentially, as an intermediate step toward Release 2.0, a number of improvements to the network architecture are planned. Let us take a brief look at some of the approaches on how to evolve the fundamental network architecture where the focus lies mostly on improving the ASN architecture.

With the WiMAX network Release 1.5, a fully standardized R4 interface between ASN-GWs is available. In more detail, in the ASN architecture, as we showed in several of the previous chapters, individual instances of ASN-GW functions that serve the same MS session are relocated individually to a different ASN-GW when the MS moves. Such a functional split, commonly referred to as the 'distributed model', avoids relocation of all the control functions serving the MS. It mainly extends the data path at the time of handover to a BS that is controlled by a different ASN-GW. Relocation can be done later on. This takes place for example by moving the authenticator function to the new ASN-GW using a re-authentication procedure. It will restore the optimal data and signaling path for the MS, as explained in Section 5.3.

Such architectural flexibility allows for a wider range of implementation options and possible deployments, but on the other hand it introduces overheads and an additional degree of complexity in the specifications that is certainly not necessary for all deployments. For example, based on the Release 1.5 network standards, it is not possible to do a combined relocation of all the ASN-GW functions at the same time, without the overhead involved in setting up a temporary R4 data path. Also, complexity is typically ranged against interoperability between different equipment vendors, because each vendor may choose to implement a different profile of the overall set of options and because of increasing complexity in the configuration of network equipment.

One effort to address this complexity and allow simpler ASN-internal mobility procedures is to introduce optimizations for the relocation of ASN-GW functions across the R4 interface. This covers both combined relocation of several functions to a new ASN-GW with optimized signaling procedures over R4 and optimizations for the relocation process of individual functions to minimize handover delays or service breaks. One example of the latter is a relocation of the authenticator function without performing re-authentication that would run end to end between the MS and backend AAA server.

When looking at the WiMAX network reference architecture, it becomes clear that the R4 reference point is of major importance for standardized inter-ASN communication, as shown in Figure 8.2. Hence, a NAP that logically partitions the network into several ASNs can benefit

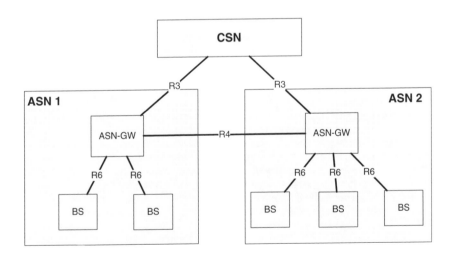

Figure 8.2 NAP deployment with multiple ASNs

from the well-defined R4 communication between the different ASNs and from the flexibility to relocate only parts of the functions serving the MS at the actual time of handover to minimize possible service breaks. At one extreme, the network deployment may consist of a number of integrated base stations (IBSs) that include BS and ASN-GW functionality and comprise a full ASN. In such flat architecture, any handover between BSs requires interaction across R4, because each of the BSs belongs to a different ASN.

Within NAP deployments that define only a single logical ASN domain, the role of the R4 reference point can be less important. In general, the ASN consists of BSs and a set of ASN-GWs controlling the BSs. Cases where each BS is assigned to one single ASN-GW are similar to the above example of multiple ASNs, as shown in Figure 8.2. Any handover of a MS to a BS controlled by another ASN-GW requires an R4 interaction in the ASN. On the contrary, it is also possible that a single BS can communicate with several or all of the ASN-GWs in the whole (single-ASN) NAP. Such a 'macro-ASN' or 'R6-flex' configuration is shown in Figure 8.3, where it is clear that for a handover the target BS can just directly communicate with the ASN-GW of the 'old' BS across the R6 reference point instead of using R4 functionality or relocating ASN-GW functions. Consequently, R4 functionality becomes less relevant, although it can still be useful for things like load balancing.

Connecting a BS to several ASN-GWs in parallel has always been allowed as a valid NAP deployment option by the WiMAX network specifications. However, when looking at the overall functionality of Release 1.5, important mechanisms for such a deployment are still missing. These mechanisms will form part of future versions of the network specifications. One good example is dynamic ASN-GW discovery for the BS where ASN-GW selection might be based on the actual load of the ASN. During handover, the target BS could simply be informed about the current ASN-GW for the MS session and could dynamically select to use this ASN-GW, avoiding any relocation or data transfer across R4. Furthermore, standardized procedures for redirecting a BS to another ASN-GW would be beneficial. For example, a BS selecting its default ASN-GW during the initial network entry of a MS could be redirected by this ASN-GW to another one during an overload situation.

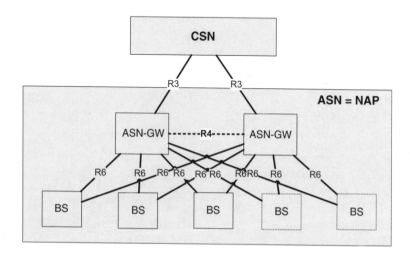

Figure 8.3 NAP deployment with single ASN

Regarding the ASN architecture and IP mobility management, not many enhancements are expected in future WiMAX releases. CSN-anchored mobility was amended in Release 1.5 with support for MIP-less deployments in the form of simple IP features. Support for fixed and nomadic deployments means WiMAX has to comply with markets where mobility is not allowed, or the extent of the mobility was limited by the operator. Release 1.5 also brings in support for PMIPv6 to align with current industry trends and as a vehicle to enable interworking with other access technologies, like those defined by 3GPP, which is aligned with the evolved packet core architecture of 3GPP and also supports PMIPv6 for 'non-3GPP' access technologies connecting to a 3GPP core. Also, PMIPv6-based tunneling between the ASN and CSN adds a valuable building block to the important discussion on the transition of IPv4 to IPv6-based transport. Here, numerous solutions and scenarios are under discussion in all the relevant organizations, with the IETF providing the central building blocks to choose from. For WiMAX, the main requirements that will be addressed by the next release of network specifications, to alleviate the IPv4 exhaustion concerns of large operators, will cover supporting appropriate tunneling methods through the WiMAX network where PMIPv6 comes in handy, and enabling dual-stack operation.

The above enhancements to Release 1.5 mostly fulfill the current requirements for IP-based mobility management in WiMAX networks. So any changes in this area are unlikely to include any new options. However, there could be some enhancements in the area of IP mobility options that are widely deployed, like PMIP, to provide better guidance for vendors and operators.

8.3 Support for Femtocell Deployments

One feature that is not yet covered by the Release 1.5 radio and network architecture, but is likely to receive standardized support within the WiMAX network as part of the subsequent release, is support for femtocell deployments to leverage an interoperable ecosystem for such a technology. The underlying use case for an operator deploying femto support is to hand out specific CPE-like devices called femto access points (FAPs) to customers. These devices act as very small WiMAX BSs in the customers' home premises, small offices, or even public locations, serving for example up to four connected devices at the same time. The FAPs connect over standard broadband connections like a DSL or cable line to the WiMAX operator's infrastructure. Many realistic use cases are envisioned for such femto devices, though it is not yet clear which of those will appear on the market, especially since Wi-Fi technology is dominantly available today and can be used to cover an overlapping set of cases. For the sake of providing a technical view, we will consider the basic case where FAPs can significantly improve indoor coverage, or enable local coverage in remote locations that are not reachable by any of the operator's macro BSs.

Femto is a hot topic nowadays, and several standards bodies like 3GPP and 3GPP2 have already developed initial frameworks on how to integrate FAPs (called *Home Node B* in the 3GPP standards) into their existing network infrastructures.

Femto-related efforts in WiMAX are based on a detailed set of requirements developed by the WiMAX Forum service provider community [109]. Besides an anticipated substantial overlap with commonly accepted requirements, such as those compiled by the Femto Forum [111] and architectural building blocks for femtocell systems for 3GPP systems [110], the femtocell architecture has to take a number of WiMAX-specific aspects into account.

These include the split into access and core networks that, if run by different operators, leads to specific aspects such as the design of security functions to protect the femto deployments. In contrast to the protection of all data exchanged between the FAP and WiMAX network that can be handled by the ASN, there is a common understanding that the 'subscription' for the FAP is owned by the CSN operator and the profile data is stored on a CSN AAA server. Such a femto subscription profile may be combined with a standard WiMAX subscription that just adds femto services to the regular WiMAX service offered to this subscriber. On the contrary, it may also be a femto-only subscription.

Practically speaking, femto support is an optional feature that may not be suitable for all operators and most likely will not be implemented by all equipment vendors. In theory, there are various deployment options, including:

- Adding femto support to an existing regular WiMAX deployment. The scenarios include network equipment serving both macro BSs and FAPs at the same time, with just a logical separation between them, or physically separated network equipment that keeps the FAP and macro BS control separate.
- A standalone network (femto-only operator).
- A NAP supporting both macro BS and FAP operation in a NAP-sharing deployment. The NAP can simultaneously serve standard NSPs and NSPs that support only femto operation.

Hence, it makes sense to realize such femto support in the WiMAX network architecture by keeping the domains for standard WiMAX network operation and for femto-specific support logically separate. This is depicted in Figure 8.4, where the network reference architecture for femto is extended by a femto ASN and CSN. The ASN-GW is shown as a new network element and is kept logically separate from the standard ASN-GW, which in fact keeps the macro BS

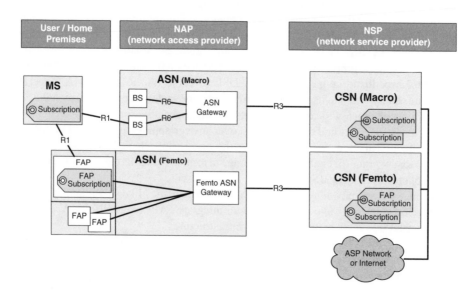

Figure 8.4   Femto support in the network architecture

and the FAPs in separate domains of the NAP, or even accommodates a femto-only NAP. Actually, this is an excellent example of how the concept of ASNs as logical domains within the same NAP can be used to flexibly support different kinds of femto deployments. To keep standard WiMAX ASN functionality and femto-specific operation separate, the NAP chooses a network design that splits the two groups of functionalities into two separate ASNs, so all BSs belong to one ASN and all FAPs belong to the other. However, it is still up to the operator to choose (as well as the vendor's implementation choice) whether to deploy a physically separate ASN-GW for the femto-only part or whether to realize the network with a single ASN-GW that can handle both logical domains in a single physical entity. The situation is similar for the femto CSN. Whether this is in fact realized by a dedicated AAA server, or whether an operator handles regular WiMAX subscribers and femto within the same AAA server, is a deployment decision.

As we can also see in Figure 8.4, one of the main architectural challenges is that the FAP is part of the ASN, so the NAP operator, when allowing FAPs into the network, in fact extends its network infrastructure into the home premises of the customers. The main technical impact is that this results in numerous security requirements to ensure that a FAP, not being under the physical control of the operator, cannot be modified to bypass charges for service usage, negatively impact network operation and availability by injecting malicious or mal-configured control signaling, or perform other inappropriate action. These requirements are mainly addressed by strong authentication of the FAP by the ASN, by an additional authentication of the FAP subscription with the CSN, and by strong cryptographic protection of all data and control information exchanged between the FAP and ASN. In practice, the latter is solved by IPsec-based protection that is also used by femto CPE units following the 3GPP specifications. In addition, it is important for the FAP vendor and the operator to ensure that the integrated security functions of the FAP cannot easily be modified or bypassed by persons with physical access to the FAP. This is typically ensured by implementing protection methods based on a secure and trusted environment within the FAP.

8.4 IEEE 802.16m and Relay Support

Multi-hop relaying and multi-node cooperation in wireless networks have emerged as important research topics in recent years due to the increasing demand for the higher data rate and even greater ubiquitous coverage. Traditional infrastructure-based systems suffer from the weaker signal strength and thus lower signal-to-noise ratio (SNR) as terminals move away from the BS. In addition, the allocation of spectrum for the third (3G) and fourth generations (4G) of cellular networks in the higher frequency bands also worsens the propagation loss, resulting in faster decaying signals. Reducing the cell size is a common practice to offset the higher path loss or poor coverage at cell edges; however, it essentially increases the number of BSs required to cover the same geographical area. This, in turn, negatively affects the investment in new BSs as well as the operating expenditure such as backhaul connections. Therefore, network scalability, infrastructure and operational costs as well as demand for high data rate with ubiquitous coverage are driving the requirements for the next generation of mobile broadband wireless. A multi-hop relay network, which divides the link between the source and the destination into a number of shorter connections with better link quality, has been extensively researched and tested for the last decade. It was first standardized as part of a cellular broadband system as the IEEE 802.16j standard [158] in May 2009. An example of a simple multi-hop relay is shown in

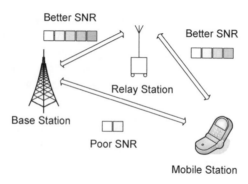

Figure 8.5 Placement of relay station (RS) to overcome the SNR limitations

Figure 8.5. The direct link with a poor SNR between the source (BS) and destination (MS) is replaced with a two-hop connection with a much better SNR.

IEEE 802.16j specifies the extended air interface supporting the multi-hop relaying capability. In order to maintain backward compatibility, the multi-hop relaying functionality is transparent to the MS. In other words, the MS is not aware of whether it is connected to a legacy BS, a relay-enabled BS or an actual relay station. In addition, to maintain the point-to-multipoint topology of Mobile WiMAX, all data communication has to be originated or terminated at the BS and thus direct communications between two relay stations or two terminals are not allowed. It is expected that the WiMAX Forum will later create the system profile and associated interoperability testing to support the multi-hop relay extension for Mobile WiMAX.

In addition to the Release 1.5 enhancements and multi-hop relaying, the next major evolution of Mobile WiMAX will be based on IEEE 802.16m [157], which is being standardized and expected to be ratified before 2011. In order to meet the stringent IMT-Advanced requirements, the followings are examples of the enhancements that are being discussed:

- **Superframe structure**: A number of smaller frames are aggregated to support new features such as primary and secondary preamble and superframe header. This multi-level control signaling carries short-term and long-term system configuration information.
- **Single-user MIMO, multi-user MIMO and multi-BS MIMO**: The minimum antenna configuration in the downlink is 2×2 and 1×2 in the uplink.
- **Lower latency with faster and more reliable MAC management messages**: H-ARQ can be applied to management messages and the MAC header is shortened to make it more efficient for small payload.
- **Better MS power management**: The sleep cycle can be dynamically adjusted, even within a sleep cycle, to match the traffic patterns and H-ARQ operations. A two-step paging procedure is introduced to increase the efficiency of the idle mode.

References

[1] IEEE 802.16-2009, 'IEEE Standard for Local and Metropolitan Area Networks, Part 16: Air Interface for Fixed and Mobile Broadband Wireless Access Systems, Amendment for Physical and Medium Access Control Layers for Combined Fixed and Mobile Operation in Licensed Bands', June 2009.

[2] WiMAX Forum: http://www.wimaxforum.org/.

[3] B. Aboba, L. Blunk, J. Vollbrecht, J. Carlson, and H. Levkowetz, 'Extensible Authentication Protocol (EAP)', RFC 3748, June 2004. http://www.ietf.org/rfc/rfc3748.txt.

[4] IEEE 802.16-2004, 'IEEE Standard for Local and Metropolitan Area Networks, Part 16: Air Interface for Fixed and Mobile Broadband Wireless Access Systems'.

[5] B. Adoba, M. Beadless, J. Arkko, and P. Eronen, 'The Network Access Identifier', RFC 4282, December 2005. http://www.ietf.org/rfc/rfc4282.txt.

[6] C. Rigney, S. Livingston, A. Rubens, and W. Simpson, 'Remote Authentication Dial In User Service (RADIUS)', June 2000. http://www.ietf.org/rfc/rfc2865.txt.

[7] B. Aboba and P. Calhoun, 'RADIUS (Remote Authentication Dial In User Service) Support for Extensible Authentication Protocol (EAP)', RFC 3579, September 2003. http://www.ietf.org/rfc/rfc3579.txt.

[8] M. Chiba *et al.*, 'Dynamic Authorization Extensions to Remote Authentication Dial In User Service (RADIUS)', RFC 3576, July 2003. http://www.ietf.org/rfc/rfc3576.txt.

[9] P. Eronen, T. Hiller, G. Zorn, 'Diameter Extensible Authentication Protocol (EAP) Application', RFC 4072, August 2005. http://www.ietf.org/rfc/rfc4072.txt.

[10] C. Perkins (ed.), 'IP Mobility Support for IPv4', RFC 3344, August 2002. http://www.ietf.org/rfc/rfc3344.txt.

[11] P. Calhoun, J. Loughney, E. Guttman, G. Zorn, and J. Arkko, 'Diameter Base Protocol', RFC 3588, September 2003. http://www.ietf.org/rfc/rfc3588.txt.

[12] P. Calhoun, G. Zorn, D. Spence, and D. Mitton, 'Diameter Network Access Server Application', RFC 4005, August 2005. http://www.ietf.org/rfc/rfc4005.txt.

[13] WiMAX Forum WMF-T33-001-R010v04, 'WiMAX Forum Network Architecture (Stage-3: Detailed Protocols and Procedures)', Release 1.0, Version 4, February 2009.

[14] WiMAX Forum, 'WiMAX Forum Overall Deployment, Applications and OAM; WMF Credentials (X.509); Device Certificate Profile', DRAFT-T12-002-R010v02-B Draft Specification, June 2009.

[15] WiMAX Forum, 'WiMAX Forum Overall Deployment, Applications and OAM; WMF Credentials (X.509); Server Certificate Profile', DRAFT-T12-001-R010v02-B Draft Specification, June 2009.

[16] WiMAX Forum, 'WiMAX Forum Device PKI Certificate Policy Approved Specification', Version 1.0.3, May 2008.

[17] WiMAX Forum, 'WiMAX Forum Server PKI Certificate Policy Approved Specification', Version 1.0.2, May 2008.

[18] WiMAX Forum, 'WiMAX Forum Overall Deployment, Applications and OAM; WMF Credentials (X.509); Certificate and Certificate Revocation List (CRL) Profile', DRAFT-T12-003-R010v02-B Draft Specification, June 2009.

[19] WiMAX Forum, 'WiMAX Forum Overall Deployment, Applications and OAM; WMF Credentials (X.509); Online Certificate Status Protocol (OCSP) Profile', DRAFT-T12-004-R010v02-B Draft Specification, June 2009.

[20] WiMAX Forum, 'WiMAX Forum Network Architecture; Architecture, detailed Protocols and Procedures; WiMAX Over-The-Air General Provisioning System Specifications', WMF-T33-103-R015v02, November 2009.

[21] WiMAX Forum, 'WiMAX Forum Network Architecture; Architecture, detailed Protocols and Procedures; WiMAX Over-The-Air Provisioning & Activation Protocol based on OMA DM Specifications', WMF-T33-104-R015v02, November 2009.

[22] WiMAX Forum, 'WiMAX Forum Network Architecture; Architecture, detailed Protocols and Procedures; WiMAX Over-The-Air Provisioning & Activation Protocol based on TR-69', WMF-T33-105-R015v01, November 2009.

[23] W. Simpson, 'PPP Challenge Handshake Authentication Protocol (CHAP)', RFC 1994, August 1996.

[24] B. Aboba, D. Simon, 'PPP EAP TLS Authentication Protocol (EAP-TLS)', RFC 2716, October 1999.

[25] B. Aboba, D. Simon, R. Hurst, 'The EAP-TLS Authentication Protocol', RFC 5216, March 2008.

[26] P. Funk and S. Blake-Wilson, 'EAP Tunneled TLS Authenticated Protocol version 0 (EAP-TTLSv0)', IETF RFC 5281, August 2008.

[27] G. Zorn, 'Microsoft PPP CHAP Extensions, Version 2', RFC 2759, January 2000.

[28] J. Arkko and H. Haverinen, 'Extensible Authentication Protocol Method for 3rd Generation Authentication and Key Agreement (EAP-AKA)', RFC 4187, January 2006.

[29] EAP method types: http://www.iana.org/assignments/eap-numbers.

[30] T. Clancy and H. Tschofenig, 'EAP Generalized Pre-Shared Key (GPSK)', IETF RFC 5433, February 2009.

[31] ITU-T Recommendation X.509 (2005), ISO/IEC 9594-8:2005, 'Information technology – Open Systems Interconnection – The Directory: Public-key and attribute certificate frameworks'.

[32] D. Cooper et al., 'Internet X.509 Public Key Infrastructure Certificate and Certificate Revocation List (CRL) Profile', RFC 5280, May 2008.

[33] J. Jonsson, B. Kaliski, 'Public-Key Cryptography Standards (PKCS) #1: RSA Cryptography Specifications Version 2.1', RFC 3447, February 2003.

[34] M. Myers et al., 'X.509 Internet Public Key Infrastructure Online Certificate Status Protocol - OCSP', IETF RFC 2560, June 1999.

[35] A. Deacon and R. Hurst, 'The Lightweight Online Certificate Status Protocol (OCSP) Profile for High-Volume Environments', IETF RFC 5019, September 2007.

[36] S. Blake-Wilson et al., 'Transport Layer Security (TLS) Extensions', IETF RFC 4366, April 2006.

[37] National Institute of Standards and Technology (NIST), 'Recommendation for Key Management – Part 1: General (Revised)', NIST Special Publication 800-57, March 2007.

[38] WiMAX Forum X.509 ordering process: http://www.wimaxforum.org/resources/pki.

[39] 3GPP TS 33.102, 'Technical Specification Group Services and System Aspects; 3G Security; Security Architecture (Release 8)', June 2008.

[40] 3GPP TS 33.221, 'Technical Specification Group Services and System Aspects; Generic Authentication Architecture (GAA); Support for subscriber certificates (Release 7)', December 2007.

[40] Open Mobile Alliance (OMA), 'DRM Architecture', OMA-AD-DRM-V2_1-20070724-C Candidate Version 2.1, July 2007. http://www.openmobilealliance.org.

[42] Open Mobile Alliance (OMA), 'OMA DM Protocol, Version 1.2', OMA-TS-DM_Protocol-V1_2, January 2005. http://www.openmobilealliance.org.

[43] Broadband Forum (formerly DSL Forum), 'CPE WAN Management Protocol v1.1', TR-069 amendment 2, December 2007. http://www.broadband-forum.org.

[44] Broadband Forum, 'Internet Gateway Device Data Model for TR-069', TR-98 amendment 1, December 2006. http://www.broadband-forum.org.

[45] IANA assignments for AAA parameters: http://www.iana.org/assignments/aaa-parameters/aaa-parameters.xhtml.

[46] https://standards.ieee.org/regauth/BOPID/IEEERegistrationAuthority_BOPID.html.

[47] WiMAX Forum, 'WiMAX Forum Network Architecture; Protocols and Procedures for Location Based Services', WMF-T33-110-R015v01, November 2009.

[48] Open Mobile Alliance, 'Secure User Plane Location V 2.0 Enabler Release Package', July 2008. http://member.openmobilealliance.org/ftp/Public_documents/LOC/Permanent_documents/OMA-ERP-SUPL-V2_0-20080627-C.zip.

[49] Open Mobile Alliance, 'UserPlane Location Protocol Version 2.0', June 2006. http://member.openmobileal-liance.org/ftp/Public_documents/LOC/Permanent_documents/OMA-TS-ULP-V2_0-20080627.

[50] P. Eronen and H. Tschofenig, 'Pre-Shared Key Ciphersuites for Transport Layer Security (TLS)', IETF RFC 4279, December 2005.

[51] M. Barnes (ed.), 'HTTP Enabled Location Delivery (HELD)', IETF draft, Version 16, August 2009.

[52] M. Thomson and J. Winterbottom, 'Revised Civic Location Format for Presence Information Data Format Location Object (PIDF-LO)', IETF RFC 5139, February 2008.

[53] S. Blake et al., 'Architecture for Differentiated Services', IETF RFC 2475, December 1998.

[54] S. Hanks et al., 'Generic Routing Encapsulation (GRE)', IETF RFC 1701, October 1994.

[55] WiMAX Forum, 'WiMAX Forum Network Architecture; Architecture, detailed Protocols and Procedures; Policy and Charging Control', WMF-T33-109-R015v01, November 2009.

[56] 3GPP TS 23.203, 'Technical Specification Group Services and System Aspects; Policy and Charging Control Architecture (Release 7)', Version 7.8.0, September 2008.

[57] 3GPP TS 23.203, 'Technical Specification Group Services and System Aspects; Policy and Charging Control Architecture (Release 8)', Version 8.3.1, September 2008.

[58] J. Rosenberg et al., 'SIP: Session Initiaton Protocol', IETF RFC 3261, June 2002.

[59] 3GPP TS 23.228, 'Technical Specification Group Services and System Aspects; IP Multimedia Subsystem (IMS); Stage 2 (Release 7)', Version 7.13.0, September 2008.

[60] H. Sinnreich and A. Johnston, Internet Communications Using SIP: Delivering VoIP and Multimedia Services with Session Initiation Protocol, John Wiley & Sons Inc., August 2006.

[61] D. Sisalem, J. Kuthan, U. Abend and H. Schulzrinne, SIP Security, John Wiley & Sons Ltd., December 2008.

[62] J. Rosenberg and C. Jennings, 'The Session Initiation Protocol (SIP) and Spam', IETF RFC 5039 (informa-tional), January 2008.

[63] H. Schulzrinne, 'The tel URI for Telephone Numbers', IETF RFC 3966, December 2004.

[64] Meteoalarm – altering Europe for extreme weather: www.meteoalarm.eu.

[65] IETF ECRIT work group homepage: http://www.ietf.org/html.charters/ecrit-charter.html.

[66] H. Schulzrinne et al., 'Extensions to the Emergency Services Architecture for dealing with Unauthenticated and Unauthorized Devices', IETF draft, November 2008.

[67] 3GPP TS 32.240, 'Technical Specification Group Services and System Aspects; Telecommunication manage-ment; Charging management; Charging architecture and principles (Release 8)', Version 8.4.0, September 2008.

[68] C. Perkins and P. Calhoun, 'Authentication, Authorization, and Accounting (AAA) Registration Keys for Mobile IPv4', IETF RFC 3957, March 2005.

[69] M. Thomson and J. Winterbottom, 'Location Measurements for IEEE 802.16e Devices', IETF draft, June 2009.

[70] 3GPP TS 33.280, 'Technical Specification Group Services and System Aspects; Telecommunication manage-ment; Charging management; Advice of Charge (AoC) service (Release 8)', Version 1.0.0, November 2008.

[71] C. Rigney, 'RADIUS accounting', IETF RFC 2866, June 2000.

[72] H. Hakala, L. Mattila, J.-P. Koskinen, M. Stura and L. Loughney, 'Diameter Credit-Control Application', IETF RFC 4006, January 2005.

[73] A. Lior, P. Yegani, K. Chowdhury, H. Tschofenig, and A. Pashalidis, 'Prepaid Extensions to Remote Authentication Dial-In User Service', IETF draft, November 2008.

[74] J. Mailley, R. Garcia, S. Whitehead, G. Farrell, 'Phone Theft Index', Palgrave Macmillan Security Journal, 21, 212–227, 2008. http://www.palgrave-journals.com/sj/journal/v21/n3/pdf/8350055a.pdf.

[75] GSMA's GSM World Web Page on handset theft. http://www.gsmworld.com/our-work/public-policy/protect-ing-consumers/handset_theft.htm.

[76] J. Menezes, P. van Oorschot, and S. Vanstone, Handbook of Applied Cryptrography, CRC Press, 1997. http://www.cacr.math.uwaterloo.ca/hac/.

[77] D. Johnson, C. Perkins, and J. Arkko, 'Mobility Support in IPv6', RFC 3775, June 2004. http://www.ietf.org/rfc/rfc3775.txt.

[78] R. Stewart, 'Stream Control Transmission Protocol', RFC 4960, September 2007. http://www.ietf.org/rfc/rfc4960.txt.

[79] R. Koodli and C. Perkins, 'Mobile IPv4 Fast Handovers', RFC 4988, October 2007. http://www.ietf.org/rfc/rfc4988.txt.

[80] E. Fogelstroem, A. Jonsson, and C. Perkins, 'Mobile IPv4 Regional Registration', RFC 4857, June 2007. http://www.ietf.org/rfc/rfc4857.txt.

[81] K. El Malki, 'Low-Latency Handoffs in Mobile IPv4', RFC 4881, June 2007. http://www.ietf.org/rfc/rfc4881.txt.

[82] K. Leung, G. Dommety, P. Yegani, and K. Chowdhury, 'WiMAX Forum/3GPP2 Proxy Mobile IPv4', IETF draft,, November 2008. http://tools.ietf.org/html/draft-leung-mip4-proxy-mode-10.

[83] S. Gundavelli, K. Leung, V. Devarapalli, K. Chowdhury, and B. Patil, 'Proxy Mobile IPv6', RFC 5213, August 2008. http://www.ietf.org/rfc/rfc5213.txt.

[84] S. Deering, 'ICMP Router Discovery Messages', RFC 1256, September 1991. http://www.ietf.org/rfc/rfc1256.txt.

[85] R. Droms, 'Dynamic Host Configuration Protocol', RFC 2131, March 1997. http://www.ietf.org/rfc/rfc2131.txt.

[86] C. Perkins, 'IP Encapsulation within IP', RFC 2003, October 1996. http://www.ietf.org/rfc/rfc2003.txt.

[87] D. Farinacci, T. Li, S. Hanks, D. Meyer, and P. Traina, 'Generic Routing Encapsulation (GRE)', RFC 2784, March 2000. http://www.ietf.org/rfc/rfc2784.txt.

[88] G. Dommety, 'Key and Sequence Number Extensions to GRE', RFC 2890, September 2000. http://www.ietf.org/rfc/rfc2890.txt.

[89] P. Yegani et al., 'GRE Key Extension for Mobile IPv4', IETF draft, September 2007. http://tools.ietf.org/id/draft-yegani-gre-key-extension.

[90] S. Thomson, T. Narten, and T. Jinmei, 'IPv6 Stateless Address Autoconfiguration', RFC 4862, September 2007. http://www.ietf.org/rfc/rfc4862.txt.

[91] R. Droms (ed.), J. Bound, B. Volz, T. Lemon, C. Perkins, and M. Carney, 'Dynamic Host Configuration Protocol for IPv6 (DHCPv6)', RFC 3315, July 2003. http://www.ietf.org/rfc/rfc3315.txt.

[92] A. Hee Jang et al., 'DHCP Option for Home Agent Discovery in MIPv6.', IETF draft, May 2008.

[93] T. Narten, E. Nordmark, W. Simpson, and H. Soliman, 'Neighbor Discovery for IP version 6 (IPv6)', RFC 4861, September 2007. http://tools.ietf.org/html/draft-ietf-mip6-hiopt.

[94] A. Patel, K. Leung, M. Khalil, H. Akhtar, and K. Chowdhury, 'Mobile Node Identifier Option for Mobile IPv6', RFC 4283, November 2005. http://www.ietf.org/rfc/rfc4283.txt.

[95] R. Wakikawa and S. Gundavelli, 'IPv4 Support for Proxy Mobile IPv6', IETF draft, January 2009. http://tools.ietf.org/html/draft-ietf-netlmm-pmip6-ipv4-support-09.

[96] C. Kaufman, 'Internet Key Exchange (IKEv2) Protocol', RFC 4306, December 2005. http://www.ietf.org/rfc/rfc4306.txt.

[97] A. Patel, K. Leung, M. Khalil, H. Akhtar, and K. Chowdhury, 'Authentication Protocol for Mobile IPv6', RFC 4285, November 2005. http://www.ietf.org/rfc/rfc4285.txt.

[98] 3GPP TS 23.402, 'Technical Specification Group Services and System Aspects; Architecture enhancements for non-3GPP accesses (Release 8)', January 2009.

[99] WiMAX Forum NWG, 'WiMAX Forum Network Architecture; Universal Services Interface (USI). An Architecture for Internet + Service Model', WMF-T33-115-R015v01, November 2009.

[100] Open Mobile Alliance (OMA), 'Mobile Location Protocol 3.3', OMA-TS-MLP-V3_3-20080627-C Candidate Version 3.3, June 2008. http://www.openmobilealliance.org.

[101] WiMAX Forum, 'WiMAX Forum Network Architecture; Architecture, Detailed Protocols and Procedures; WiMAX-SIM Application on UICC', WMF-T33-114-R015v01, November 2009.

[102] ETSI TS 102.220, 'Smart Cards; UICC-Terminal Interface; Physical and Logical Characteristics (Release 7)', version 7.11.0, July 2008.

[103] ETSI TS 102.310, 'Smart Cards; Extensible Authentication Protocol support in the UICC (Release 8)', version 8.0.0, September 2008.

[104] IEEE Std 802.1X-2004, 'IEEE Standard for Local and Metropolitan Area Networks: Port-based Network Access Control', IEEE Computer Society, December 2004.

[105] WiMAX Forum WMF-T48-001-v01, 'WiMAX Forum Roaming Models Whitepaper', April 2009.

[106] USA National Emergency Number Association (NENA), 'Functional and Interface Standards for Next-Generation 9-1-1 (i3)', Version 2.0, January 2009.

[107] M. Thomson and J. Winterbottom, 'Revised Civic Location Format for Presence Information Data Format Location Object (PIDF-LO)', IETF RFC 5139, February 2008.

[108] M. Thomson, J. Winterbottom, and H. Tschofenig, 'GEOPRIV Presence Information Data Format Location Object (PIDF-LO) Usage Clarifications, Considerations, and Recommendations', IETF RFC 5491, March 2009.

[109] WiMAX Forum, 'Requirements for WiMAX Femtocell Systems', WMF-T31-010-R016v01 Femtocell Requirements, WiMAX Forum SPWG, April 2009.

[110] 3GPP TR 23.830, 'Technical Specification Group Services and System Aspects; Architecture aspects of Home NodeB and Home eNodeB (Release 9)', version 0.5.0, June 2009.

[111] The Femto Forum: http://www.femtoforum.org/femto/.

[112] WMF-A11-001-v01, 'WiMAX Forum Documentation Structure and Identification', May 2009.

[113] WMF-A11-002-v01, 'WiMAX Forum Release Structure and Process', May 2009.

[114] 3GPP TS 33.210, '3G security; Network Domain Security (NDS); IP network layer security', June 2009. ftp:// ftp.3gpp.org/Specs/html-info/33210.htm.

[115] WMF-T32-001-R015v01, 'WiMAX Forum Network Architecture, Stage 2, Base Specification', November 2009.

[116] WMF-T33-001-R015v01, 'WiMAX Forum Network Architecture, Stage 3, Base Specification', November 2009.

[117] IEEE Registration Authority: Details on the allocation of IEEE 802.16 Operator ID: http://standards.ieee.org/ regauth.

[118] WMF-T40-001-R010v01, 'WiMAX Forum Roaming Guidelines', May 2009.

[119] WMF-T40-002-R010v01, 'WiMAX Forum WiMAX Roaming Interface Code', May 2009.

[120] WMF-T42-001-R010v02, 'WiMAX Forum WRI Stage-2 Overview', May 2009.

[121] WMF-T43-001-R010v01, 'WiMAX Forum WRI Stage-3 Overview', May 2009.

[122] WMF-T42-002-R010v02, 'WiMAX Forum WRI-Stage-2 AAA-Proxy', May 2009.

[123] WMF-T43-002 R010v01, 'WiMAX Forum WRI Stage-3 AAA-Proxy', May 2009.

[124] WMF-T42-003-R010v02, 'WiMAX Forum WRI Stage-2 Wholesale-Rating', May 2009.

[125] WMF-T43-003-R010v01, 'WiMAX Forum WRI Stage-3 Wholesale-Rating', May 2009.

[126] WMF-T42-004-R010v02, 'WiMAX Forum WRI Stage-2 Clearing', May 2009.

[127] WMF-T43-004-R010v01, 'WiMAX Forum WRI Stage-3 Clearing', May 2009.

[128] WMF-T42-005-R010v02, 'WiMAX Forum WRI Stage-2 Financial-Settlement', May 2009.

[129] WMF-T43-005-R010v01, 'WiMAX Forum WRI Stage-3 Financial-Settlement', May 2009.

[130] WMF-T42-006-R010v02, 'WiMAX Forum WRI Stage-2 Interconnection', May 2009.

[131] WMF-T48-001-R010v01, 'WiMAX Forum Roaming Agreement Template', September 2008.

[132] WMF-T48-002-R010v01, 'WiMAX Forum Roaming Agreement Annex', September 2008.

[133] Metro Ethernet Forum, 'MEF 6.1 – Ethernet Services Definition – Phase 2', April 2008.

[134] IEEE STD 802.1Q-2005, 'Virtual Bridged Local Area Networks'.

[135] IEEE STD 802.1ad-2005, 'Amendment 4: Provider Bridges'.

[136] IEEE STD 802.1ah-2008, 'Amendment 7: Provider Backbone Bridges'.

[137] Broadband Forum TR-101, 'Migration to Ethernet-Based DSL Aggregation', April 2006.

[138] H. Jeon, M. Riegel, and S. Jeong, 'Transmission of IP over Ethernet over IEEE802.16 Networks', IETF draft (work in progress). http://www.ietf.org/id/draft-ietf-16ng-ip-over-ethernet-over-802-dot-16-10.txt.

[139] 3GPP TS 23.401: 'LTE; GPRS enhancements for E-UTRAN access', March 2009.

[140] 3GPP TS 23.402: 'UMTS; LTE; Architecture enhancements for non-3GPP accesses', March 2009.

[141] 3GPP TS 23.203: 'Policy and charging control architecture', March 2009.

[142] WiMAX Forum, 'SPWG Requirements for Release 1.5', July 2008. http://wimaxforum.org/sites/wimaxforum. org/files/documentation/2009/080717_SPWG_Req_Release_1.5CLEAN.pdf.

[143] Co-located coexistence support: http://ieee802.org/16/maint/contrib/C80216maint-08_092r2.doc.

[144] HO Latency Reduction: http://ieee802.org/16/maint/contrib/C80216maint-08_016r8.doc.

[145] WiMAX Forum White Paper, 'The WiMAX Forum Certified^{TM} Program', September 2008, http://www. wimaxforum.org/sites/wimaxforum.org/files/document_library/wimax_forum_2008_certification_white_ paper_final.pdf.

[146] WiMAX Forum, 'WiMAX Forum Network Architecture; Architecture, Detailed Protocols and Procedures; Emergency Services Support', WMF-T33-102-R015V02, November 2009.

[147] WiMAX Forum, 'Mobile System Profile Specification', Release 1-IMT-2000 Edition, August 2009. http:// www.wimaxforum.org/sites/wimaxforum.org/files/technical_document/2009/07/WMF-T23-007-R010v02_ MSP-IMT-2000.pdf.

[148] H. Yaghoobi, 'Scalable OFDMA Physical Layer in IEEE 802.16 WirelessMAN' *Intel Technology Journal* **8**, 201–212 2004.

[149] F. Wang *et al.*, 'Mobile WiMAX Systems: Performance and Evolution', *IEEE Communications Magazine*, pp. 41–49, October 2008.

[150] M.-H. Fong *et al.*, 'Improved VoIP Capacity in Mobile WiMAX Systems Using Persistent Resource Allocation', *IEEE Communications Magazine*, pp. 50–57, October 2008.

[151] S.M. Alamouti, 'A Simple Transmit Diversity Technique for Wireless Communications' *IEEE Journal on Selected Areas in Communications,* vol. 16(8), pp. 1451–1458, 1998.

[152] A. Maleki-Tehrani, B. Hassibi, and J.M. Cioffi, 'Adaptive Equalization of Multiple-Input Multiple-Output (MIMO) Channels' *Proceedings of the IEEE International Conference on Communications*, vol. 3, pp. 1670–1674, 2000.

[153] C. Eklund *et al.*, 'IEEE Standard 802.16: A Technical Overview of The WirelessMan Air Interface for Broadband Wireless Access', *IEEE Communications Magazine*, pp. 98–107, June 2002.

[154] IEEE 802.16, 'Mobile System and Proposal Evaluation Requirements', January 2003. http://ieee802.org/16/tge/docs/80216e-03_01.pdf.

[155] WiMAX Forum, 'Mobile System Profile Specification', Release 1.5, August 2009. http://www.wimaxforum.org/sites/wimaxforum.org/files/technical_document/2009/07/WMF-T23-001-R015v01_MSP-Common-Part.pdf.

[156] T. Hardie, A. Newton, H. Schulzrinne, and H. Tschofenig, 'LOST: A location-to-service translation protocol', IETF RFC 5222, August 2008.

[157] IEEE 802.16, 'P802.16m: Project Authorization', December 2006. http://standards.ieee.org/board/nes/projects/802-16m.pdf.

[158] IEEE 802.16j-2009, 'IEEE Standard for Local and metropolitan area networks - Part 16: Air Interface for Broadband Wireless Access Systems, Amendment 1: Multihop Relay Specification', June 2009.

Index